The Handbook of Environmental Chemistry

Founding Editor: Otto Hutzinger

Editors-in-Chief: Damià Barceló • Andrey G. Kostianoy

Volume 86

More information about this series at http://www.springer.com/series/698

Polish River Basins and Lakes – Part I

Hydrology and Hydrochemistry

Volume Editors: Ewa Korzeniewska · Monika Harnisz

With contributions by

R. Augustyniak · R. Babko · E. Bajkiewicz-Grabowska ·
E. Bogacka-Kapusta · D. Borowiak · S. Chmiel · A. Choiński ·
R. J. Chróst · B. Czarnecki · R. Dondajewska · S. Głowacki ·
R. Gołdyn · W. Golus · A. Górniak · K. Grabowska · J. Grochowska ·
T. Kaliński · A. Kapusta · B. Kiersztyn · P. Klimaszyk ·
K. Kowalczewska-Madura · M. Kwidzińska · M. Łopata ·
M. Markowski · Z. Michalczyk · K. Mięsiak-Wójcik ·
A. Napiórkowska-Krzebietke · K. Nowiński · K. Parszuto ·
R. Piotrowicz · M. Ptak · W. Siuda · J. Sposób · K. Stawecki ·
A. Świątecki · R. Tandyrak · B. Zdanowski

Springer

Editors
Ewa Korzeniewska
The Faculty of Environmental Sciences
University of Warmia and Mazury
Olsztyn, Poland

Monika Harnisz
The Faculty of Environmental Sciences
University of Warmia and Mazury
Olsztyn, Poland

ISSN 1867-979X ISSN 1616-864X (electronic)
The Handbook of Environmental Chemistry
ISBN 978-3-030-12122-8 ISBN 978-3-030-12123-5 (eBook)
https://doi.org/10.1007/978-3-030-12123-5

This Springer imprint is published by the registered company Springer Nature Switzerland AG
The registered company address is: Gewerbestrasse 11, 6330 Cham, Switzerland

The Handbook of Environmental Chemistry Also Available Electronically

The Handbook of Environmental Chemistry is included in Springer's eBook package *Earth and Environmental Science*. If a library does not opt for the whole package, the book series may be bought on a subscription basis.

For all customers who have a standing order to the print version of *The Handbook of Environmental Chemistry*, we offer free access to the electronic volumes of the Series published in the current year via SpringerLink. If you do not have access, you can still view the table of contents of each volume and the abstract of each article on SpringerLink (www.springerlink.com/content/110354/).

You will find information about the

- Editorial Board
- Aims and Scope
- Instructions for Authors
- Sample Contribution

at springer.com (www.springer.com/series/698).

All figures submitted in color are published in full color in the electronic version on SpringerLink.

Aims and Scope

Since 1980, *The Handbook of Environmental Chemistry* has provided sound and solid knowledge about environmental topics from a chemical perspective. Presenting a wide spectrum of viewpoints and approaches, the series now covers topics such as local and global changes of natural environment and climate; anthropogenic impact on the environment; water, air and soil pollution; remediation and waste characterization; environmental contaminants; biogeochemistry; geoecology; chemical reactions and processes; chemical and biological transformations as well as physical transport of chemicals in the environment; or environmental modeling. A particular focus of the series lies on methodological advances in environmental analytical chemistry.

Series Preface

With remarkable vision, Prof. Otto Hutzinger initiated *The Handbook of Environmental Chemistry* in 1980 and became the founding Editor-in-Chief. At that time, environmental chemistry was an emerging field, aiming at a complete description of the Earth's environment, encompassing the physical, chemical, biological, and geological transformations of chemical substances occurring on a local as well as a global scale. Environmental chemistry was intended to provide an account of the impact of man's activities on the natural environment by describing observed changes.

While a considerable amount of knowledge has been accumulated over the last four decades, as reflected in the more than 150 volumes of *The Handbook of Environmental Chemistry*, there are still many scientific and policy challenges ahead due to the complexity and interdisciplinary nature of the field. The series will therefore continue to provide compilations of current knowledge. Contributions are written by leading experts with practical experience in their fields. *The Handbook of Environmental Chemistry* grows with the increases in our scientific understanding, and provides a valuable source not only for scientists but also for environmental managers and decision-makers. Today, the series covers a broad range of environmental topics from a chemical perspective, including methodological advances in environmental analytical chemistry.

In recent years, there has been a growing tendency to include subject matter of societal relevance in the broad view of environmental chemistry. Topics include life cycle analysis, environmental management, sustainable development, and socio-economic, legal and even political problems, among others. While these topics are of great importance for the development and acceptance of *The Handbook of Environmental Chemistry*, the publisher and Editors-in-Chief have decided to keep the handbook essentially a source of information on "hard sciences" with a particular emphasis on chemistry, but also covering biology, geology, hydrology and engineering as applied to environmental sciences.

The volumes of the series are written at an advanced level, addressing the needs of both researchers and graduate students, as well as of people outside the field of "pure" chemistry, including those in industry, business, government, research establishments, and public interest groups. It would be very satisfying to see these volumes used as a basis for graduate courses in environmental chemistry. With its high standards of scientific quality and clarity, *The Handbook of Environmental Chemistry* provides a solid basis from which scientists can share their knowledge on the different aspects of environmental problems, presenting a wide spectrum of viewpoints and approaches.

The Handbook of Environmental Chemistry is available both in print and online via www.springerlink.com/content/110354/. Articles are published online as soon as they have been approved for publication. Authors, Volume Editors and Editors-in-Chief are rewarded by the broad acceptance of *The Handbook of Environmental Chemistry* by the scientific community, from whom suggestions for new topics to the Editors-in-Chief are always very welcome.

Damià Barceló
Andrey G. Kostianoy
Editors-in-Chief

Preface

The book *Polish River Basins and Lakes* is based on the scientific developments and results obtained by Polish scientists within many years of research related to the management of catchment areas of lakes and river basins in the context of global change. It consists of two volumes: Part I: *Hydrology and Hydrochemistry* and Part II: *Biological Status and Water Management*. Complementing each other, the volumes constitute the first such comprehensive study on changes in the chemical as well as the biological status of Polish surface waters. Presented by almost 100 Polish researchers, the environmental topics cover a wide range of disciplines and several main study areas, e.g. chemistry, hydrology, hydrochemistry, biology, ecology, microbiology, ichthyology and water management.

The two parts of the book contain 35 chapters. The first volume refers to Polish river basins and Polish lakes' catchments, anthropogenic pollution sources and chemical pollution of water and sediments, the evaluation of chemical dynamics and the impact on climate change projections. The second volume deals with the assessment of biological status of numerous Polish rivers and lakes. All of the quality elements associated with aquatic ecosystems including the macrophytes, phytoplankton, zooplankton, macroinvertebrates, fish and, as a matter of growing concern, invasive alien aquatic species are evaluated in this volume. Also presented are the general state of biodiversity of Polish surface waters, a set of conservation and restoration practices as well as the review of protected sites within the basins and catchments areas.

The authors hope that the book content will be of interest to environmental chemists, geologists, hydrologist, biologists, students and surface water managers as well as the general public.

We would like to thank the authors of this book for their valuable contribution and efforts to create its chapters. We would also like to underline the importance of the suggestions and the recommendations given by Prof. Bogusław Zdanowski during the process of the realization of the book.

Finally we would like to extend our sincere thanks to Prof. Damià Barceló who invited us and inspired to create this book.

Olsztyn, Poland Ewa Korzeniewska
Olsztyn, Poland Monika Harnisz
November 2018

Contents

Current Climatic Conditions of Lake Regions in Poland and Impacts on Their Functioning

Andrzej Górniak

Contents

Abstract The condensed characteristics of the climate of Poland are presented with particular emphasis on the areas of lake districts. More detailed studies were provided on 19 lakes with daily hydrological and thermic observations in the calendar years 2000–2016. Climate warming is clearly observed in observational temperature data, with increases of 0.2–0.4°C per decade. Slightly lower air temperatures and a higher annual sum of precipitation in relationship to neighboring regions are specific for the Polish Lakelands. Mean annual air temperature in the eastern part is 1.5°C lower than in the western regions. Much greater variation of precipitation and climatic water balance was noted, where the Wielkopolskie and Lubuskie Lake Districts have a permanent water deficit that is not observed in the rest of the Lakelands. Polish postglacial lakes are characterized by generally small fluctuations in water level, whereas variations in artificially regulated lakes are higher than in lakes with a natural regime, as well as a higher frequency of decreasing water level trends. Only a few lakes have a statistically significant increase of lake water temperature, mostly in northeastern Poland. A simulation indicated that in year 2050 the annual lake water temperature will be 1.1–1.9°C

A. Górniak (✉)
Department of Hydrobiology, Institute of Biology, University of Białystok, Białystok, Poland
e-mail: hydra@uwb.edu.pl

E. Korzeniewska, M. Harnisz (eds.), *Polish River Basins and Lakes – Part I*,
The Handbook of Environmental Chemistry 86,
https://doi.org/10.1007/978-3-030-12123-5_1

higher than at present. We can expect more evident changes in lake functioning in spring and summer, shifting lakes toward higher trophy, but dystrophic lakes will remain more humic when their hydrology is stable.

Keywords Climate · Global change · Lakes · Tendency · Water temperature

1 Introduction

The thin gas cover of the Earth, the atmosphere, has been constantly evolving, as has that for other spheres, during its existence over several billion years. On the geological scale, changes in the atmosphere reflect the spatial changes of the land, as well as its biotic structure. After the postglacial restructuring of the environment, a zonal climate system was created on Earth. The climatic conditions varying in different locales are a vector of six fundamental climatic factors: latitude, Earth–Sun relationships, position on the continent, atmospheric and oceanic circulation, topography, and local features [1].

Until recently, it was recognized that the weather system in a given place for at least 30 years can be considered as characteristic for the geographic regions or smaller parts of the continents. Global climatic classification by Köppen (or its modifications), Thornthwaite, Holdrige, and Budyko are used to show climatic differences of the Earth [1]. In Poland, the W. Okołowicz climatic classification has a tradition of long use [2].

Global weather observations in the twentieth century, constantly improved, expanded, and automated, indicate climate change and the need for constant monitoring. The World Meteorological Organization (IPCC groups) recently suggested climate evaluation in 10-year periods to better recognize the direction and rate of climate change [3].

Constantly increasing human activity and its various effects are considered the most important factors in the recorded changes of the Earth's climate in time and space. In addition to natural causes, anthropogenic causes of climate change on a global scale are primarily air pollution (emission of heat, gases, dusts), deforestation, and alterations in the water cycle. Direct effects are the mechanisms of global warming, such as local urban heat islands, melting of the ice cover, smog, and desertification, which are contributing to difficulties in economic management and reducing the scale and value of environmental services. Natural disasters, which are now occurring with greater frequency, will increase social unrest and provoke political tensions on a national or international scale [3].

Climatic changes do not remain unrelated to natural environmental processes, both chemical and biological, which directly or indirectly affect human beings to different degrees. The hydrosphere is considered the most sensitive part of the geosystem during climate changes, because it is an energy storage system, a carrier of elements or chemical compounds during circulation and between different organisms, and thus regulates the ecosystem energy budget.

The aims of this present study are the minimized characteristics of the Polish climate and its changes, with particular emphasis on lake districts, where freshwaters are more frequent. Two of the many functional aspects of the lakes, that is, the dynamics of water resources and the thermic regime, were selected to show the emerging effects of climate change in Poland. Considerations for the prediction of the effects of climate change on the functioning of lakes in the near future are summarized.

2 Material and Methods

Daily weather observations and hydrological data came from the database of the Institute of Meteorology and Water Management in Warsaw and data published by the Governmental Statistical Office of Poland [4]. The authors' calculations of monthly and annual means of climatic parameters for seven meteorological stations were made from data of the years 2000–2016, when evident climate changes are seen. Chosen stations represent longitudinal differentiation of climatic conditions in postglacial Lakeland regions in Poland. I have added a Włodawa station from the Polesie Lubelskie region (Southeast Poland), where the small lakes groups occur (Fig. 1, Table 1). Evapotranspiration for each month and station in the calendar years 2000–2016 was calculated using the Ivanov formula, which is frequently used in hydroclimatic studies.

For detailed analysis, I selected 19 lakes, including 6 lakes where the water level is modified by man for waterway communication or energy production. In this group the dimictic lakes dominate; only the Łebsko Lake is polymictic, where Baltic

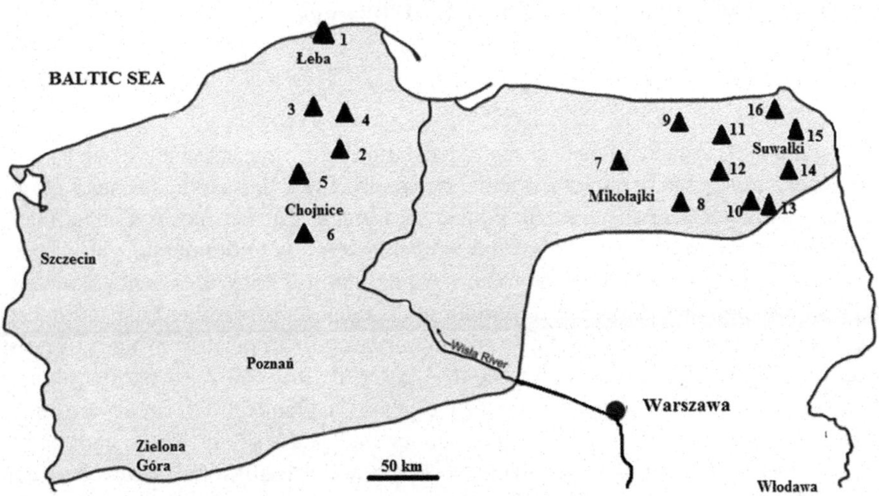

Fig. 1 Location of meteorological stations and lakes under study

Table 1 Description of locations from which meteorological data originated in relationship to Polish lake districts

Meteorological stations	Station elevation [m a.s.l.]	Lakeland names
Suwałki	184	Suwalskie
Mikołajki	128	Mazurskie
Chojnice	164	Pomorskie-East
Łeba	2	Shoreland
Poznań	87	Wielkopolskie
Zielona Góra	192	Lubuskie
Szczecin	1	Pomorskie-West
Włodawa	177	Łęczyńsko-Włodawskie

seawater can flow into the lake during storms. Selected lakes include the largest in Poland, Mamry Lake, and also the deepest, Hańcza Lake. For these lakes, almost complete series of water level up to the year 2017 were available, with daily water temperature measurements also available for 16 lakes. For 2 lakes only, Wigry and Hańcza, longer series of water temperature are presented as examples of much longer trends. The mean monthly lake water level and water temperature (measured at depth 0.4 m of the inshore zone) were calculated in calendar years. Data of the Hurrell North Atlantic Oscillation (NAO) Index for the winter period (December–March) originated from the USA National Center for Atmospheric Research [5].

Data computation utilized the statistical program Past [6]. A Mann–Kendall test was used for trend detection in time data series, in which Z (statistic test) and p (level of significance) were computed.

3 The Climate of Poland and Its Changes

3.1 Climatic Conditions of Poland

The terrain of Poland allows unrestricted latitudinal air exchange, whereas free meridian movement of air is difficult, especially from the south, because of the mountain range. The location of Poland in Central Europe means that climatic conditions have features typical of the temperate zone with domination of western air mass circulation as well as the main wind direction. The Azores anticyclone and Iceland low pressure center are important in the creation of weather types, although the continental air masses (arctic from the northeast or tropical from the southeast direction) also periodically reach Poland. A polar maritime air mass has the highest frequency (more than 60% of days) in the year; the frequency of arctic masses is three times smaller, nearly a 12% polar continental mass of air. These geographic features cause the climate of Poland to experience a transitional nature between oceanic and continental varieties. This nature is manifested by increasing maritime features toward the west and increasing continentalism toward the east. As a consequence, Okołowicz [2], in his climate classification, distinguished in Poland a temperate transitional climate and also a local mountain climate for the Carpathian and Sudety Mountains in the southern part of the country. The transition of the climate of Poland from the continental to the oceanic variety is also manifested in the existence of two additional seasons, early winter and early spring.

An important feature of the climate of Poland is the significant variability of weather from day to day, as well as large differences in weather at the same time of the year but in different years.

The latitudinal layout of the geographic regions is also reflected in the thermal and humidity diversity of the country (Table 2). Current spatial variation of mean annual air temperature (in the past decade, 2001–2010) is in the range of 7.9°C in mountain regions, 8.5°C in uplands, 8.8°C in the lowlands, and 8.7°C in shorelands [7]. Latitudinal thermal differentiation of the country is more prominent than longitudinal. Decrease of air temperature amplitude from east to west is also characteristic. In the western part of the country, there are no frosts for 4 to 5 months a year, while in the eastern part of the country only in July and August do frosts not occur. Hot days (T_{max} >25°C) occur from 15 to 20 times a year on the coast and from 25 to more than 30 times a year in the south in the center of the Poland. Heat waves are noted to be very rare, about 3 to 8 occurrences in the years 2000–2010. Air temperature variability in space and time is connected to the duration of the thermal growing season, which varies from 200 days in northeast Poland to 240 days in the southwest regions [8].

Mean annual precipitation in Poland increases from 450 mm in the Wielkopolskie Lakelands to 900 mm in submontane and Pomeranian Lakelands regions, and up to 1,800 mm in the highest parts of the Tatry and Sudety Mountains. Summer precipitation is the highest (maximum, July); in the lowlands it is about 20% of the annual

Table 2 Latitudinal (A) and longitudinal (B) variability of mean air temperature and precipitation in Poland

				Temperature		Precipitation	
Group	Stations	Latitude N	Longitude E	1971–2000	2001–2016	1971–2000	2001–2016
A	Łeba	54.75	17.53	7.7	8.5	632	699
	Suwałki	54.13	22.95	6.3	7.2	591	634
	Chojnice	53.70	17.55	7.3	8.1	547	633
	Toruń	53.03	18.58	8.1	8.9	528	568
	Łódź	51.73	19.40	8.0	8.8	571	590
	Częstochowa	50.82	19.10	8.0	8.9	617	645
	Kraków	50.08	19.78	8.1	8.9	662	684
	Zakopane[a]	49.30	19.95	5.4	6.3	1,107	1,163
B	Terespol	52.07	23.62	7.5	8.4	512	551
	Włodawa	51.55	23.53	7.5	8.4	515	577
	Warszawa	52.17	20.97	8.1	9.0	519	566
	Kalisz	51.73	18.08	8.4	9.3	507	492
	Poznań	52.42	16.83	8.5	9.4	507	545

Data from Governmental Statistical Office [4]
[a]Mountain station (elevation 840 m a.s.l.)

total and in the Lakelands and uplands more than 30%. Precipitation occurs mainly in the form of rain; sleet and snow are only about 12% of the annual total precipitation. The total duration of rainfall in Poland is 10% of the year, and precipitation occurs on 150 to 180 days in a year [9]. An increase in the ratio of precipitation in the cold period to precipitation in the warm period is observed. The precipitation yield is the highest in the summer, and periodically extreme precipitation causes local floods (flash floods) or regional floods, as, for example, in the summers of 1997, 2005, and 2010.

Snow cover in Poland (except in mountainous areas) occurs from October to April; the longest duration, nearly 100 days, occurs in the eastern lake districts, and the shortest, about 40 days, in the west [10]. The winter snow season is the sum of several periods with a snow cover, and the height of the snow cover differs greatly between consecutive winters. Snow cover duration indicates an 8-year periodicity in Poland [11].

The sum of evapotranspiration in the growing season from April to September exceeds 500 mm in most areas of Poland, with a mean of 520 mm. It is commonly assumed also that periods with drought occur in Poland once every 4 to 5 years [12], especially in the southern part of the Polish Lakelands (Wielkopolskie, Lubuskie). Drought usually begins in western Poland, moves through the central part, and eventually reaches the eastern side of the country. A tendency toward an increasing number of dry days is observed.

The spatial diversity of cloudiness in the country is not great, around 5.0–5.6 oktas on the octant scale; only in the mountains is this higher. Minimal rates of cloud cover are in May and August; the maximum values fall in November and December. Full and nearly full overcast skies (7–8 oktas[1]) are the most frequent condition in Poland, near 35–40% of the days in a year [13]. The largest number of cloudless skies was recorded in March, April, and August. Mean annual sunshine duration (in hours) is more variable than cloudiness, with the range of 1,400–1,650 h/year showing decreasing values from east to west.

Three first-order climates exist in Poland in the light of the global climate map with the scheme by Köppen [14]: these are C-mesothermal and D-microthermal, both in the wet suborder (constantly moist), with warmest months <22°C, and the H-mountain type of climate. Climate Cfb (mesothermal with significant precipitation in all seasons and warmest month averaging below 22°C) is dominant over Dbf (microthermal with warm summer, fully humid with snow cover in winter): the latter occurs only in the northeastern part of Poland.

3.2 *Climate of the Polish Lakelands*

Polish lake districts are slightly elevated in relationship to the surrounding regions, and thus are characterized by slightly lower air temperatures and an increased annual sum of precipitation. Currently, the mean annual air temperature is between 8.1 and 9.6°C; only the most eastern outskirts of the Suwałki Lake District are definitely cooler (Table 3). The highest thermal variation was observed in the coolest and

[1]Scale of cloud cover is measured in oktas.

Table 3 Rates of selected climatological parameters in the Polish Lakelands in the years 2000–2016

Parameters	Suwałki	Mikołajki	Chojnice	Poznań	Włodawa	Szczecin	Zielona Góra
Annual air temperature	7.2	8.1	8.2	9.5	8.4	9.6	9.4
Annual precipitations	618	615	628	547	584	571	599
Annual evapotranspiration	500	536	540	704	599	586	722
Annual sun duration [h]	1,715	1,797	1,877	1,868	1,827	1,719	1,733
Sun duration (May–Aug)	997	1,025	1,040	1,038	1,034	955	929
Cloudiness [octants]	5.2	5.2	5.5	5.1	5.3	5.1	5.2
Days with snow cover	80	65	54	39	66	28	46
Days with rain	118	121	133	122	109	142	126
Days with snow	52	49	44	33	50	29	41
Days with storm	22	27	25	22	30	19	28
Days with wind >10 m/s	26	8	28	29	30	30	5
Wind [m/s]	3.6	3.1	5.5	3.6	3.7	3.8	3.0

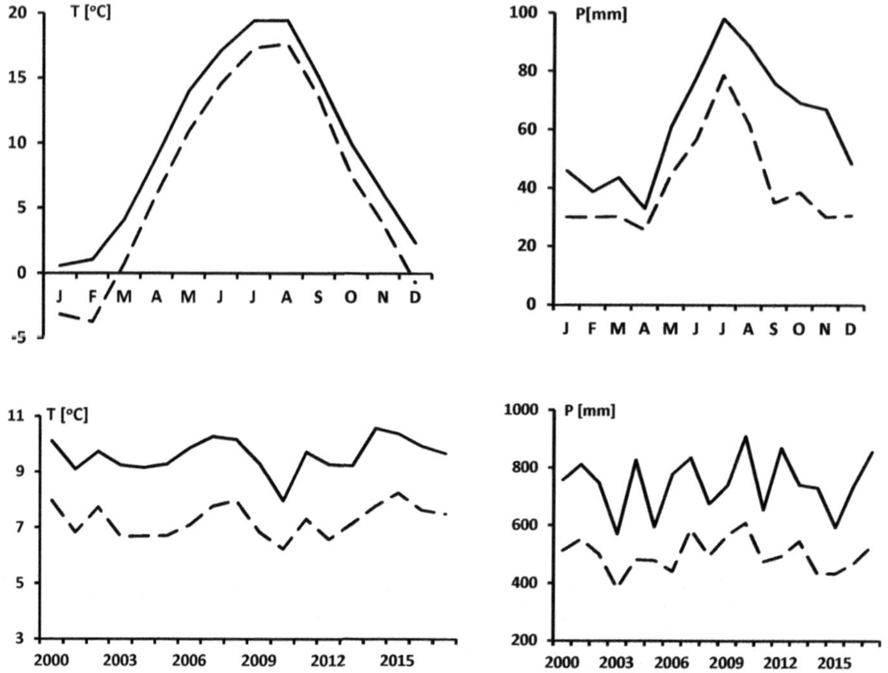

Fig. 2 Seasonal (*upper panel*) and multiannual (*lower panel*) variability of air temperature and precipitation in Lakelands regions in years 2000–2016 (minimum and maximum data for stations in Table 3)

hottest periods (Fig. 2). Pluvial differences of the Polish Lakelands with the highest intensity are observed in the second part of the year, with the maximal variation in July and the least in April.

The current climatic water balance (precipitation – evaporation) of the Polish Lakelands is very different in different locations: the eastern and more northern parts have a positive balance (Suwałki, 118 mm; Mikołajki, 79 mm), but the Lakelands located to the west and south have a negative water balance (Poznań, 257 mm) (Table 3).

This difference in balance is the result of the seasonal differences of the relationship of precipitation (P) to evaporation (E): in the Suwałki Lake District the predominance of evaporation over rainfall lasts on average 3 months (April–June), whereas in the western and southern parts of the Polish lake districts this pattern continues for 6 months, from April to September (Fig. 3).

Another climatic difference between the eastern and southwestern lake districts concerns the seasonality of cloudiness and sun duration, wherein the eastern parts have better solar conditions then the western part (Fig. 3, Table 3). Also, sunshine duration has decreased in the past 17 years in the eastern lake districts, although in the remaining areas this tendency was not observed (Figs. 4 and 5).

Also, anemometric conditions varied in the Polish Lakelands, where the East Pomeranian Lakelands and the shorelands area (not presented in Table 3) have a

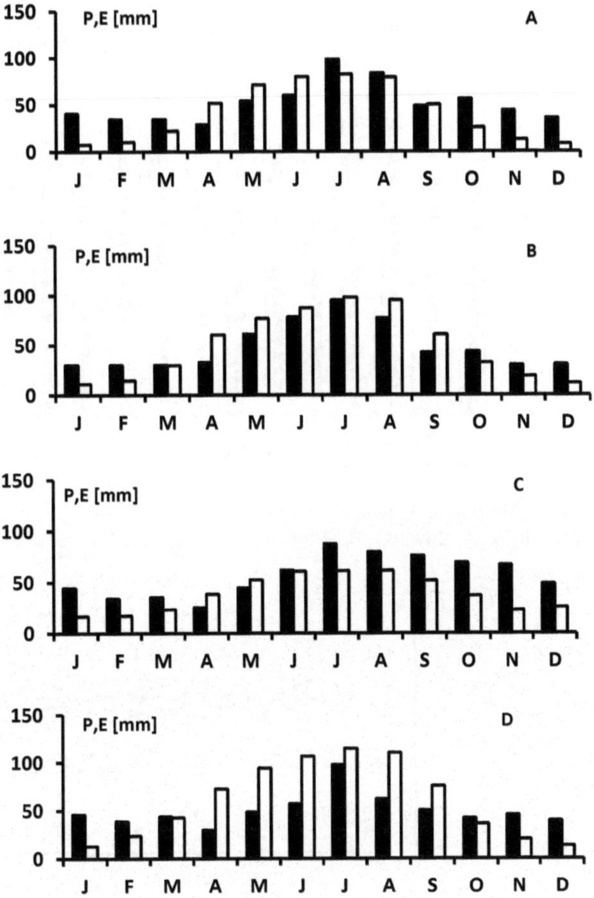

Fig. 3 Mean monthly precipitation (*black bars*) and evapotranspiration (*white bars*) in the years 2000–2016 in selected meteorological stations: (**a**) Suwałki, (**b**) Włodawa, (**c**) Łeba, (**d**) Zielona Góra

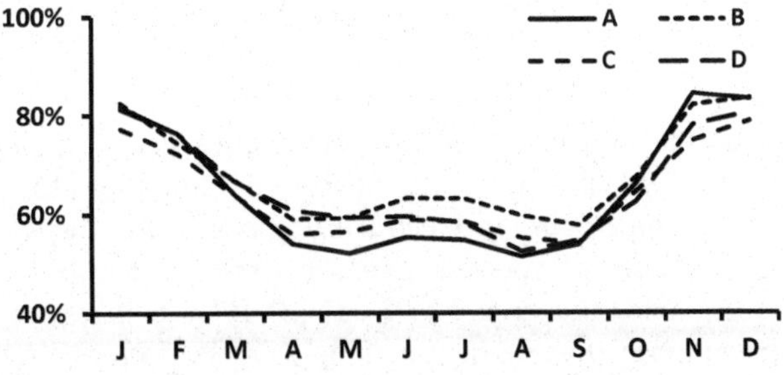

Fig. 4 Mean monthly cloudiness data (recalculated from octant scale to % of sky) in selected stations in the period 2000–2016: A, Suwałki; B, Chojnice; C, Poznań; D, Włodawa

significantly higher average wind speed in a year than occurs in other locations. In the period from October to March, wind speed was about 20% higher than in the period from April to September (Fig. 6). Moreover, in Mikołajki and Łeba stations in the analyzed period 2000–2014, a significant increase of mean annual wind velocity was noted.

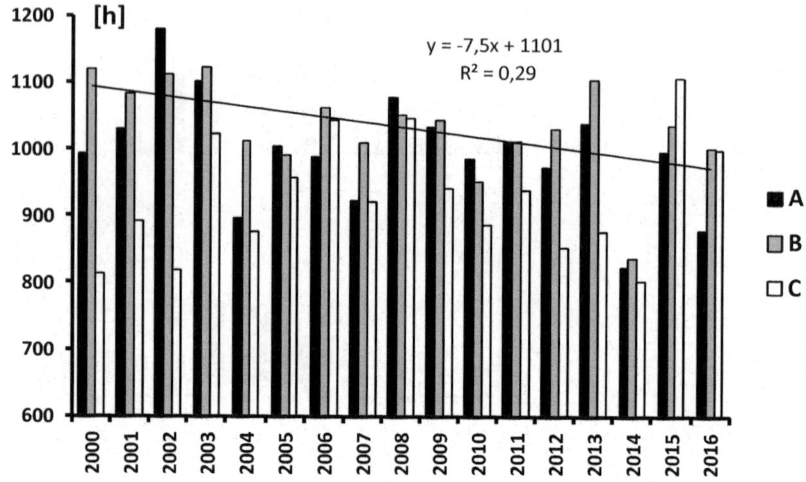

Fig. 5 Multiannual changes of sun duration in May–August periods in selected stations in years 2000–2017: A, Suwałki; B, Włodawa; C, Zielona Góra; trend line for Włodawa station

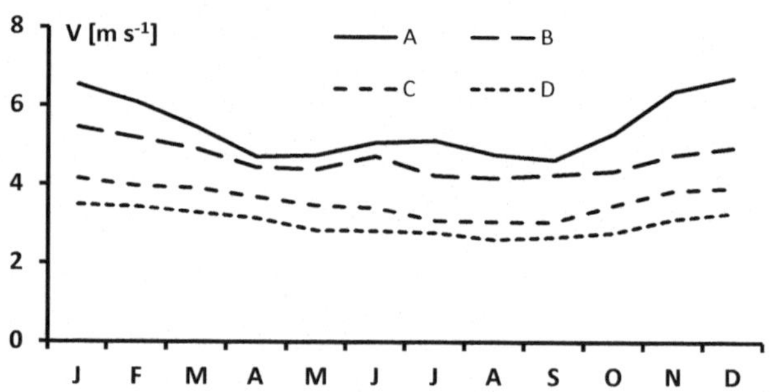

Fig. 6 Mean monthly wind speed [m s^{-1}] in the different lake districts in Poland in the years 2000–2016: A, Chojnice; B, Łeba; C, Suwałki; D, Zielona Góra

As the latest Intergovernmental Panel on Climate Change (IPCC) report emphasizes, the climate of the earth continues to warm, and the rising temperature trend is superimposed on the inherent natural variability [15]. The recent trend of annual mean temperature in Central Europe exceeds the global mean land trend. The temperature is expected to rise by about 1.2°C by 2035 (probability, 50%) and precipitation to increase by 5–7% [3]. IPCC authors point to a very likely increase in the number of warm days and nights and a decrease in the number of cold days and nights. Also indicated are general increases in the intensity and frequency of extreme precipitation, especially in winter [16].

Wójcik and Miętus [7] noted that changes of annual air temperature in Poland show a statistically significant increase with an average rate exceeding 0.2°C per decade. In the decade 2001–2010, the rate in February was more than 0.5°C/10 years. The highest rate of temperature increase was detected in northern Poland, including the Lakeland regions, whereas the lowest rate of increase is observed for the highlands and the Sudety Mountains.

In the short analyzed period of 17 years (2000–2017), by climate analyses no statistically significant changes were noted in the majority of the present climate parameters of Polish lake districts. Such changes can be detected only over a longer period of time. It is widely recognized that global warming began to increase at the end of the 1970s, and from that time on, increasing environmental effects could be expected during climate change [3].

4 Polish Lakes Under Climatic Changes

4.1 Lake Water Storage Changes

One of the effects of climate change is the transformation of the water cycle on the lands, including lake water resources. As Bajkiewicz-Grabowska emphasized [17], the reaction of lakes to changes in the nature of the water cycle can vary depending on location in the water system, conditioned by the geological structure, and especially the relationship between lakes and groundwater. Therefore, in the 19 Polish lakes analyzed, both a statistically significant decreasing tendency of the level of Lakes Jasień, Raduńskie, and Dzierzgoń and an increase in the amount of water in Lakes Sępoleńskie, Rospuda, and Studzieniczne were noted (Table 4, Fig. 7). However, for the majority of the lakes analyzed there are no significant changes in water level attributable to climate change, because in many of these the water level is artificially regulated within waterways or for the operation of hydro-electric plants [18]. The noted changes in the water abundance of these lakes are usually caused by human activity and result from climate change to only a small extent.

It is interesting to note that for lakes with larger areas and with artificially regulated water levels (lakes italicized in Table 4), a decreasing tendency is detected, although this is not significant statistically.

The mean annual amplitudes of water level in these regulated lakes were also higher than in unregulated lakes, in which this amplitude is rather lower than 30 cm and the standard deviation of water level variation is less than 20 cm (Table 4). The monthly amplitude of lake water level was small, in the range 5–20 cm, except in Łebskie Lake, which is located in the shoreline and is hydrologically both a catchment and one impacted by the Baltic Sea. Variability of lake water level results from the different shape of the catchment and the types of geological structures in postglacial areas; in consequence, the intensity and amount of flowing water from the basin to the lakes varies [19]. Mean annual amplitude of lake water level has no regular fluctuations as in the Jasień Lake (Fig. 8C), where these are the smallest from all lakes analyzed. Only in two lakes, Sępoleńskie and Charzykowy, have I detected a significant decrease of amplitude rates (Fig. 8A, B). This condition has probably resulted in a negative water balance in past years, when groundwater accumulated

Table 4 Basic morphological data of lakes studied and parameters of lake water levels in the years 2000–2016

		Area	Volume	Mean depth	Elevation	Mean monthly amplitude	SD	Max amplitude	Water level tendency
No.	Lakes	ha	mm^3	m	m a.s.l.	cm	cm	cm	
1	Łebsko	7,142	118	1.6	0.9	38.8	14.1	180	
2	Wdzydze	1,417	221	15.2	133.8	8.7	15.1	73	(–)
3	Jasień	577.2	283	5.0	112.7	5.0	5.5	36	–
4	*Raduńskie*	363	60	15.5	161.6	7.0	5.7	43	–
5	Charzykowskie	1,364	135	9.8	120.2	8.6	11.2	71	(–)
6	Sępoleńskie	158	8	4.8	112.8	13.0	10.9	76	+
7	Dadaj	978	121	12.0	122.5	12.8	24.0	136	
8	*Mikołajskie*	424	56	11.2	116.7	7.2	13.0	70	(–)
9	*Mamry*	9,851	1,003	9.8	115.8	7.8	13.4	74	(–)
10	Selmęt Wielki	1,208	99	7.8	120.7	9.1	23.4	76	
11	Litygajno	155	10	6.0	132.8	14.1	10.1	122	
12	Ełckie	382	57	15.0	118.9	15.5	14.5	101	
13	*Rajgrodzkie*	1,503	143	9.4	118.4	19.2	30.5	147	(–)
14	*Studzienicze*	244	22	8.7	123.4	7.3	5.9	39	+
15	Wigry	2,118	337	15.8	131.9	5.9	10.3	51	(+)
16	Hańcza	292	120	38.7	227.3	10.7	11.9	77	
17	Dejguny	765	93	12.0	116.8	6.6	14.1	73	
18	Dzierzgoń	834	296	6.4	81.5	11.4	29.7	151	–
19	Rospuda F	323	50	15.0	170.3	7.3	10.8	64	+

Italicized lake names indicate regulation of water level
SD, standard deviation of all monthly water levels
Brackets indicate tendency not significant by Mann–Kendall test

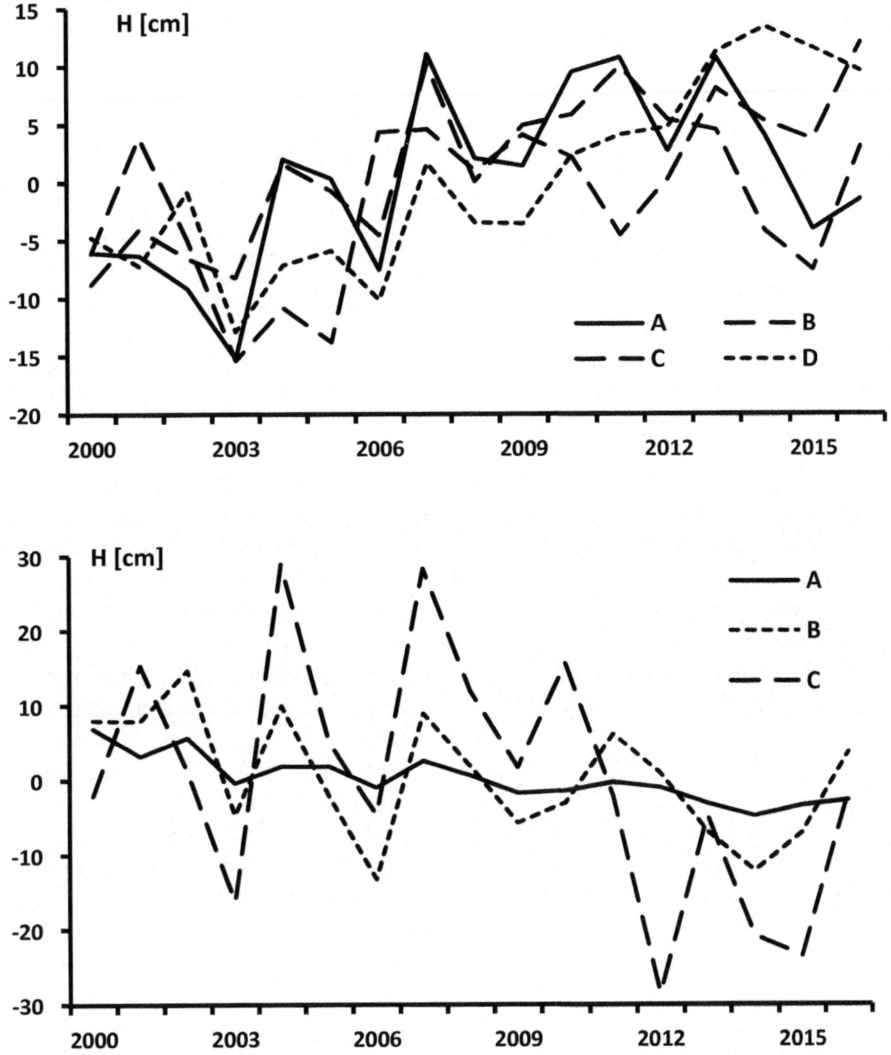

Fig. 7 Changes of annual lake water level in the years 2000–2016: *upper*, lakes A, Wigry; B, Sępoleńskie; C, Mikołajskie; D, Rospuda; *lower*, lakes A, Raduńskie; B, Rajgrodzkie; C, Wdzydze

new rain portions, filling its own resources and limiting the load in lakes. A similar water level regime with no regular fluctuations was observed in lakes in northeastern Poland [20], located in the Łęczyńsko–Włodawskie lake district [21, 22].

The varied response of lakes to atmospheric water input and/or surface runoff is even better seen in the analyses of monthly water level runs (Fig. 9). The earliest lake hydrological reaction was observed in the East Pomeranian Lakes (group C) with a maximum water level in February or March, except Charzykowy Lake, and in March or April in most Polish lakes. Lakes from the Suwałki group have a similar

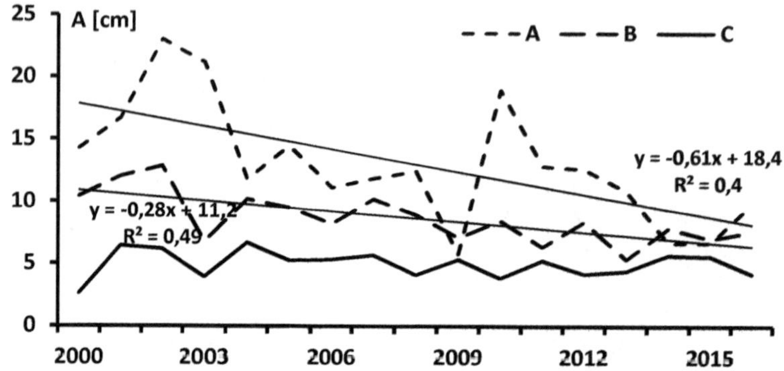

Fig. 8 Multiannual changes of mean annual amplitudes of water level in selected lakes: A, Sępoleńskie; B, Charzykowskie; C, Jasień; in the years 2000–2016

hydrological system. In the other groups presented in Fig. 9, there is a distinct type of lake with a maximum of water seen later, such as Ełckie Lake, Rajgrodzkie Lake, or Charzykowskie Lake. Łebskie Lake has a completely different hydrological seasonality with a maximum in December and January. An equally exceptional lake is Jasień Lake (in the E group in Fig. 9), with a very uniform level of water throughout the year that very rarely reacts to the inflow of atmospheric waters.

Summing up, it can be stated that Polish postglacial lakes are characterized by generally small fluctuations in their water level. Only in lakes with artificially regulated water level can there be significant amplitudes of variations, for example, Lake Rajgrodzkie, or a reduced amplitude of level fluctuations, for example, Lakes Mamry and Mikołajskie. As most of these are through-flow lakes, they are included in water systems in which local hydrogeological conditions determine the rate of water circulation and react differently to both atmospheric supply and anthropogenic changes in water discharge [17, 18, 23]. Weyhenmeyer [24] showed that a response to climate change seen in one lake may be very different in another. In the majority of analyzed lakes there are no long-term tendencies of changes in water level, which indicates a generally limited reaction of lakes to progressive climate changes because local river–lake systems exhibit high self-regulation abilities. Therefore, these systems are relatively resistant to climate-driven changes, which confirms that lakes can act as sentinels of climate change [25].

During long-term water deficits in lake basins (drought), the network of streams connecting the lakes frequently disappears. For lakes with a smaller surface area droughts are a significant threat, because the small underground sub-basin is not able to compensate for the loss of evaporating water. In large lakes, the share of horizontal groundwater supply is significant, because their basins are characterized by substantial underground retention and unit outflows are two or more times higher than observed in lowland basins [17, 18]. Lakes can show various hydrological functions even in the same system, whether drainage, retention, or convertible, depending on precipitation rate in a given period [18]. It is thus explained why

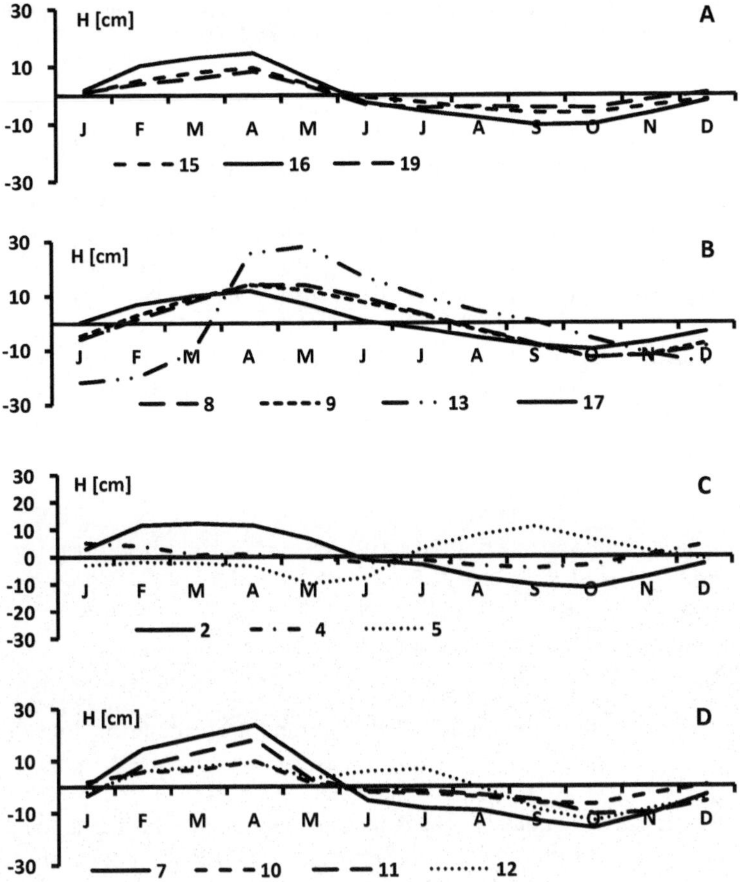

Fig. 9 Mean monthly lake water level (in relationship to mean level) in the years 2000–2016. Lake numbers as in Table 4: A, Suwałki group; B, Mazurian group; C, Pomeranian–East group; D, Ełk group; E, specific lakes group

lakes in the same climatic location have varying tendencies, as shown by the data presented here.

4.2 *Lake Water Temperature Regime and Recent Tendency*

Surface and epilimnetic water temperatures, which can be highly correlated with regional-scale air temperatures, exhibit a rapid and direct response to climatic forcing, making epilimnetic temperature a useful indicator of climate change [25, 26]. In the Polish lakes the recent mean water temperature is in the range

Table 5 Mean of selected monthly, annual, and maximum temperatures and results of Mann–Kendall time series test of lake water temperature in the years 2000–2016

		Mean water temperature [°C]						Mann–Kendall test		
No.	Lakes	Feb	May	Jul	Nov	Year	Max.	*Z*	*p*	Significance
1	Łebsko	1.5	14.0	19.7	5.4	9.8	25.2	2.16	0.03	+
2	Wdzydze	1.9	13.5	20.4	7.5	10.4	25.9	1.86	0.06	–
3	Jasień	1.7	14.5	20.6	8.0	10.7	25.4	1.20	0.23	–
4	*Raduńskie*	1.8	12.7	19.3	6.8	9.9	24.8	1.40	0.16	–
5	Charzykowskie	1.6	13.7	20.2	7.2	10.3	26.1	3.31	0.00	+
6	Sępoleńskie	1.5	15.3	21.3	6.9	10.9	26.8	1.90	0.06	–
7	Dadaj	1.3	14.3	21.3	7.0	10.5	27.5	2.98	0.00	+
8	*Mikołajskie*	1.1	13.2	20.8	7.5	10.3	27.2	2.11	0.03	+
9	*Mamry*	1.0	13.4	20.5	6.1	9.8	26.9	2.54	0.01	+
10	Selmęt Wielki	1.2	14.2	21.1	6.6	10.3	26.6	1.92	0.05	–
11	Litygajno	1.1	14.2	21.3	5.9	10.2	26.7	1.87	0.06	–
12	Ełckie	1.3	14.8	21.4	6.8	10.5	27.1	0.37	0.71	–
13	*Rajgrodzkie*	1.3	14.4	21.7	6.8	10.5	28.1	0.96	0.33	–
14	*Studzienicze*	1.3	14.6	21.4	6.0	10.3	27.4	2.52	0.01	+
15	Wigry	1.3	14.1	21.5	6.6	11.5	27.6	2.34	0.02	+
16	Hańcza	0.9	10.9	20.0	4.9	9.0	26.2	1.04	0.30	–

Italics indicated the lakes with an artificial water level regulation

9–12°C with a monthly minimum in February and a maximum in July (Table 5, Fig. 10).

In the period 2000–2016, the maximal water temperature of 28.1°C was noted in Rajgrodzkie Lake. It should be noted that the discussed lake water temperature results come from one-time daily measurements at 6.00 (UT - universal time) and do not take into account the variation of temperature that occurs commonly throughout the day. During the day in spring and summer, the amplitude of water temperature change in the epilimnion of large lakes can reach several degrees Celsius (Górniak, unpublished data).

The highest monthly surface water temperature variation among Polish lakes is noted in April and May: in the rest of the year two changes less than 2°C were seen. The Mann–Kendall test for the period 2000–2017 indicated that only 40% of the lakes have a statistically significant increase of lake water temperature, located mostly in the eastern lake districts (Mazury and Suwałki Lakelands) (Table 5, Fig. 11), where a larger climatic gradient exists. Computed trends of lake water temperature were in the range 0.28–0.38°C per decade for the years 2000–2016. This trend results mainly from higher rates in the summer and spring, but in Jasień Lake from Pomeranian Lakeland, specific higher autumn rates were larger by effects of the groundwater supply (Fig. 10).

Only in Charzykowy Lake have I detected a strong increase of water temperature, with the trend of 0.081°C/year.

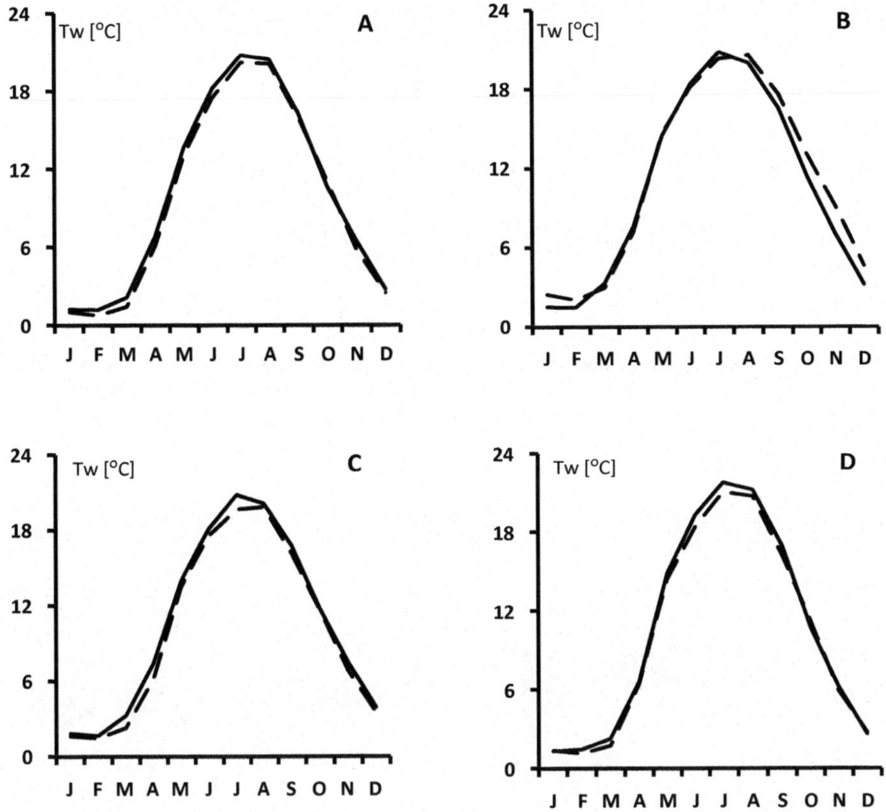

Fig. 10 Seasonality of monthly water temperature in lakes (**a**) Mamry, (**b**) Jasień, (**c**) Charzykowskie, and (**d**) Studzieniczne in two periods: 2000–2007 (*dashed line*) and 2008–2017 (*solid line*)

Skowron [27] documented that the Polish lake water temperature in spring increased fastest among seasons, about 0.2–0.5°C per decade. My results for Polish lakes from the period 2000–2016 are in agreement with global analysis, where national lake water temperature trend has been estimated as 0.34°C per decade in the years 1989–2009 [28] and for air temperature 0.25°C per decade [3].

Across the state of Wisconsin (USA), whole lake water temperatures increased with an average trend of 0.042°C/year (for the years 1990–2012). In large (>50 ha) lakes, the positive temperature trend was similar across all depths. In small lakes (<50 ha) the warming trend was restricted to shallow waters, so they respond differently than large lakes [29].

The recorded increase of air temperature in the spring season and early termination of ice phenomena is in agreement with Choiński et al. [30, 31], who showed that in long-time data series (1951–2010) average ice phenomena duration and ice cover decreased with a rate of 0.5 day/year. In winters with a positive NAO phase, the ice

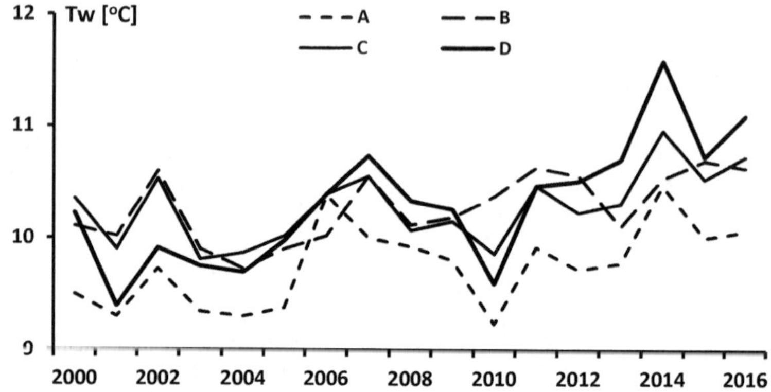

Fig. 11 Changes of annual lake water temperature (depth 0.4 m) in Łebsko Lake (A), Studzieniczne Lake (B), Mikołajskie Lake (C), and Charzykowskie Lake (D) in calendar years 2000–2017

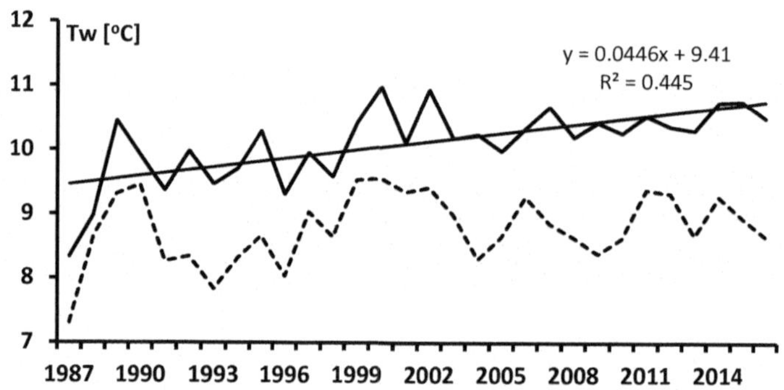

Fig. 12 Long-term variation of annual lake water temperature (depth 0.4 m) in Wigry Lake (*solid line*) and Hańcza Lake (*dashed line*) in calendar years 1987–2016; all data from Institute of Meteorology and Water Management

phenomena duration is significantly shorter than during negative NAO phases [32, 33].

Similar to the tendency for air changes, the analyzed period 2000–2016 is too short to find significant changes in the temperature of lake waters. For the aforementioned period there were no statistically significant trends in changes of Wigry Lake water temperature, whereas for a longer period (1987–2016) the trends proved to be statistically highly significant. However, for Hańcza Lake, the deepest in Poland, analysis for the 17-year and 30-year periods (Fig. 12) did not indicate any significant changes in water thermic regime under the influence of climate changes.

I have noted that there are significant statistical relationships between yearly water temperature in lakes and mean annual air temperature measured at the nearest meteorological station (Table 6). As presented in Table 6, the formulas show a high rate of R^2 and can be used for the prediction of lake water temperature during climate

Table 6 Relationships between annual air temperature and rate of increase of lake water temperature in selected lakes computed for the period 2000–2016

Station	Lakes	R^2	Formula	T_w 2050 [°C]
Suwałki	Studzieniczne	0.51	$0.41x + 7.21$	1.2
	Wigry	0.72	$0.43x + 7.40$	1.3
Mikołajki	Mikołajskie	0.51	$0.38x + 7.21$	1.1
Łeba	Łebskie	0.52	$0.45x + 5.92$	1.4
Chojnice	Sępopolskie	0.60	$0.54x + 6.48$	1.6
	Wdzydze	0.56	$0.48x + 6.50$	1.4
	Charzykowskie	0.50	$0.62x + 5.32$	1.9

R^2, determination index for $p < 0.001$; in formula x, mean air temperature [°C] is used

change. Assuming an increase in average annual air temperature in Poland as presented by Wójcik and Miętus [7], in 2050 the average temperature of lake waters should be expected to increase by 1.1–1.9°C. Predicted changes of water temperature in Wigry Lake using the formula from the multiannual lake water trend line (Fig. 12) and from the relationship between air temperature and lake water temperature were the same, that is, 1.3°C.

Whole lake water temperatures increased across the state of Wisconsin (USA) from 1990 to 2012, with an average trend of 0.042°C/year. In large (>50 ha) lakes, the positive temperature trend was similar across all depths. In small lakes (<50 ha), the warming trend was restricted to shallow waters, so they respond differently than do large lakes [29].

Analysed lake water temperature tendencies of the four seasons in the period 2000–2016 were significant only for a small group of lakes (Table 7), located mainly in the western part of Poland with warmer climatic conditions. Charzykowskie Lake stood out from the others because in all seasons the trends were statistically significant. Also, in Jasień Lake negative lake water temperature tendencies (all statistically significant, $p < 0.001$) clearly manifested a change in the water cycle in the catchments. In this specific lake a higher winter water supply of the lake from groundwater, with a constant temperature (6–8°C), gives a "warming" effect. On the other hand, this may be an effect of the decreased role of surface runoff in the lake supply, which may also result in an increase of heat resources in the epilimnion of the lake. Tendencies of increasing temperature were much higher in the warm season (summer, autumn) than in the colder periods. For most lakes, the winter increase of water temperature was about 0.3–0.4°C per decade, close to the rate of air temperature change indicated earlier by Wójcik and Miętus [7].

Confirmation of significant changes of winter thermal conditions during climate warming in Central Europe is also found an increase in western circulation connected with the positive phase of NAO [32, 33]. I have found a significant correlation between the Hurell NAO Index and water temperature in most of the lakes under study (Table 6). NAO effects in Europe are known; also, most Polish lakes react with increased winter water temperature and reduced duration and intensity of ice phenomena [33–35].

Table 7 Tendency of lake water temperature for seasons in years 2000–2016 and determination coefficient (r^2) in correlation between mean lake water temperature in the period December–March and Hurrel NAO Index for these months

	Lake water tendencies [°C/year]				DJFM (period from December to March)	
Lakes	Spring	Summer	Autumn	Winter	r^2	p
Łebsko	0.04			0.033	**0.43**	**0.003**
Wdzydze		0.027		0.037	**0.30**	**0.031**
Jasień			**−0.180**	**−0.860**	0.01	
Raduńskie		**0.041**			**0.40**	**0.003**
Charzykowskie	**0.08**	**0.085**	**0.078**	**0.073**	**0.39**	**0.005**
Sępoleńskie		0.036		0.029	**0.37**	**0.007**
Dadaj		**0.068**	**0.056**	0.029	0.09	
Mikołajskie	0.03			0.034	0.15	
Mamry	0.04	0.068		0.036	0.16	
Selmęt Wielki					**0.42**	**0.003**
Litygajno	0.03			0.035	**0.29**	**0.021**
Ełckie	0.02				**0.29**	**0.039**
Rajgrodzkie					0.02	
Studzienicze			0.034	0.036	0.12	
Wigry					0.17	
Hańcza					**0.26**	**0.031**

Data from USA National Center for Atmospheric Research
Bold font indicates statistically significant tendencies ($p < 0.05$) in the Kendall–Mann test

4.3 Predicted Climate Change Impact on Lake Functioning

Lakes are commonly affected by multiple interacting stressors [36, 37], which could confound the signals from climate change. However, variations in surface water temperatures were highly synchronous (coherent) and related to fluctuations in air temperature, as also documented earlier for European lakes located more southwest from Poland by Dokulil et al. [38]. The same authors emphasize physical lake properties as a beginning of freshwater response to climate-driven change, which launched cascades down from physical parameters via chemical and nutrient variables to biological entities. The main and possible factors and climate-related drivers changing freshwaters have been identified and classified by Adrian et al. [25].

I have used their proposals of estimation of possible changes in Polish lakes under global warming with respect to seasonality and response features such as increase, decrease, shift, and meaningful fluctuations (Table 8). It should be mentioned that all freshwater ecosystems have a certain memory (physical, biochemical or biocenoitic) of past extreme events, the effects of which may appear with some delay, with an intensity not always predictable.

Moreover, the effects of changes in climatic conditions on hydrobionts are complex, difficult to distinguish from other influences, and not easy to generalize [3].

Table 8 Predictions of future changes in selected variables in Polish lowland lakes during climate changes

	Climate-related responses	Months											
		Jan	Feb	Mar	Apr	May	Jun	Jul	Aug	Sep	Oct	Nov	Dec
1	Water level			◊	◊	◊	◊	◊	◊				
2	Water temperature		↔				↑	↑	↑		↔		
3	Stratification					↔							
4	Ice phenomena duration	↔									↔		
5	Turbidity (littoral)					◊							
6	Oxygen (meta- and hypolimnion)			◊	◊		↓						
7	TOC littoral			↑	↑			◊	◊	◊		◊	
8	Diatom development			↔									
9	Cyanobacteria blooms						↑						
10	Primary productivity			↑			↑						

↔ – shift in time, ↓ – decrease, ↑ – increase, ◊ – large fluctuations
TOC total organic carbon

All types of variables used in this proposal can be classified in four groups: hydrological (1), physical (2–5), chemical (6–7), and biocenoitic (8–10) (limited to producers only). According to the present scenario, summer and spring will be the seasons with significant changes in the functioning of lakes. Climate signals such as changes in water level, temperature, and ice-off time in the spring will determine the variable response of diatom multiplication in high dissolved silica and total organic carbon (TOC) resources [39]. TOC resources will also be an important source of nutrients for planktonic organisms. The rate of spring diatom development, determined by the time of ice melting, is also affected in the spring primary production of lake photic zones. Wagner and Adrian [40] pointed out that the spring regime of abiotic and biotic lake parameters will manifest more clearly, with lesser effects appearing in summer.

I predicted an intensive summer changes in lake functioning affected by climate change. The forecast lake water temperature increase and the accompanying extremely high precipitation, together with the deep mixing of the epilimnion, will result in acceleration of the circulation of matter. Also, increasing warming of the epilimnion will cause changes in stratification, but the magnitude of stratification responses to climate change across lakes will be associated with the lake average temperature and morphometry, but not with warming rates [41].

Catchment origin nutrient load to the littoral zone will slowly become an important driver for the increase of lake water trophy, being especially more frequent in oligo- and mesotrophic lakes. Rigosi et al. [42] highlighted that nutrients had a larger role in oligotrophic lakes whereas temperature was more important in mesotrophic lakes.

In many scenarios more substantial cyanobacterial development in eutrophic and hypertrophic lakes [42] is predicted. As lakes become more eutrophic, the *Cyanobacteria* will become more sensitive to the interaction of nutrients and temperature, but ultimately nutrients are the most important predictor of cyanobacterial biovolume. Similarly, *Microcystis* is more sensitive to temperature than is *Anabaena* [42]. Linking the increased allochthonous organic load to the lakes with trophic increases and the browning process during climate change is becoming more and more common [43–45]. On the North American continent, the greatest effects of these processes are visible in increase of the murky lakes type, according to nutrient-color status [46], which in Poland were previously referred as humo-eutrophic (auxotrophic) [47]. This lake type exhibits high chlorophyll *a* and total phosphorus concentrations and cyanobacterial densities [44].

It is highly probable that climate change in Poland will contribute to an increasing number of humo-eutrophic lakes, especially in the group of small lakes surrounded by eutrophic wetlands. It will be accompanied by an increase in CO_2 emission and the preservation of strong heterotrophic ecosystems. On the other hand, during climate changes in small oligotrophic dystrophic lakes surrounded by bogs, an increase in TOC concentration, change in water color, and the formation of a stronger thermal stratification of these lakes are predicted [48]. Climate change may also lead to a change to a mictic type for some of the large, previously dimictic, lakes, when ice cover becomes permanently absent.

5 Conclusions

As presented here, the effects of global change on Polish freshwaters, narrowed to only the main stressors and the first trophic level in aquatic ecosystems, clearly show that lakes will be exposed to several changes. Previous Polish research on warmed lakes in Poland (presented in this volume also) has indicated a significant change in the trophic structure of the aquatic ecosystem created by increasing lake water temperatures. Hillbricht-Ilkowska [49], more than 25 years ago, pointed out a wide spectrum of future and possible lake changes based on earlier observations from heated lakes in the Konin region in Poland. All her suggestions are still valid and find confirmation in current research [24, 36, 38]. She has predicted that during temperature increases a change will lead to the permanent elimination of cold and stenothermic species, reducing the species diversity of planktonic organisms and other groups such as Insecta, Mollusca, or fish species that prefer cold waters. A new arrangement of thermal and mictic lake conditions may create opportunities for mass colonization of alien species or massive development of species that so far are few in number.

In addition to typical natural changes in lake functions, changes in the extent of lakes and wetlands are expected. This potential must be included in local zoning plans.

References

1. Rohli RV, Vega AJ (2018) Climatology4th edn. Jones & Bartlett Learning, Burlington
2. Okołowicz W (1969) Basic climatology. PWN, Warsaw
3. Hartmann DL et al (2013) IPCC fifth assessment report, climate change 2013: the physical science basis (Stocker TF et al, eds). Cambridge University Press, Cambridge
4. Governmental Statistical Office of Poland (2017) Environment. Warsaw
5. Hurrell JW, National Center for Atmospheric Research Staff (eds) (2013) The climate data guide: Hurrell North Atlantic Oscillation (NAO) Index (PC-based). Accessed 18 Aug 2018
6. Hammer Ø, Harper DAT, Ryan PD (2001) PAST: paleontological statistics software package for education and data analysis. Palaeontol Electron 4(1):9
7. Wójcik R, Miętus M (2014) Some features of long-term variability in air temperature in Poland (1951–2010). Przegl Geogr 86(3):339–364. (in Polish)
8. Krasowicz S, Górski T, Budzyńska K, Kopiński J (2012) Agricultural characteristics in the territory of Poland. In: Pastuszak M, Igras J (eds) Temporal and spatial differences in emission of nitrogen and phosphorous from Polish territory to the Baltic Sea. National Marine Fisheries Research Institute – Institute of Soil Science and Plant Cultivation – State Research Institute – Fertilizer Research Institute, Gdynia, pp 47–107
9. Kożuchowski K (2017) Precipitations. In: Jokiel P et al (eds) Hydrology of Poland. PWN, Warsaw, pp 36–44. (in Polish)
10. Paszyński J, Niedźwiedź T (1991) Climate. In: Starkel L (ed) Geography of Poland: natural environment. PWN, Warsaw, pp 296–355. (in Polish)

11. Falarz M (2004) Variability and trends in the duration and depth of snow cover in Poland in the 20th century. Int J Climatol 24:1713–1727
12. Łabedzki L (2007) Estimation of local drought frequency in central Poland using the standard ized precipitation index SPI. Irrig Drain 56(1):67–77. https://doi.org/10.1002/ird.285
13. Filipiak J, Miętus M (2009) Spatial and temporal variability of cloudiness in Poland 1971–2000. Int J Climatol 29:1294–1311
14. Kottek M, Grieser J, Beck C, Rudolf B, Rubel F (2006) World map of the Köppen-Geiger climate classification updated. Meteorol Z 15:259–263. https://doi.org/10.1127/0941-2948/2006/0130
15. Kundzewicz Z, Juda-Rezler K (2010) Climate change related risk. Nauka 4:69–76. (in Polish)
16. Kundzewicz ZW, Matczak P (2012) Climate change regional review: Poland. WIREs Clim Change 3:297–311. https://doi.org/10.1002/wcc.175
17. Bąjkiewicz-Grabowska E (2002) Circulation of matter in the river-lake systems. Warsaw University Press, Warsaw
18. Choiński A (1995) Physical limnology of Poland. UAM, Poznań
19. Wrzesiński D, Ptak M (2016) Water level changes in Polish lakes during 1976–2010. J Geogr Sci 26(1):83–101. https://doi.org/10.1007/s11442-016-1256-5
20. Górniak A, Piekarski K (2002) Seasonal and multiannual changes of water levels in lakes of northeastern Poland. Pol J Environ Stud 11(4):349–354
21. Michalczyk Z, Chmiel S, Turczyński M (2011) Lake water stage dynamics in the Łęczna-Włodawa Lake District in 1991–2010. Limnol Rev 11(3):113–122
22. Michalczyk Z, Mięsiak-Wójcik K, Sposób J, Turczyński M (2017) The state of and changes in water conditions in the Łęczna-Włodawa Lake District. Przegl Geogr 89(1):9–28. (in Polish)
23. Polderman NJ, Pryor SC (2004) Linking synoptic-scale climate phenomena to lake-level variability in the Lake Michigan-Huron basin. J Great Lakes Res 30(3):419–434
24. Weyhenmeyer GA (2008) Rates of change in physical and chemical lake variables – are they comparable between large and small lakes. Hydrobiologia 599:105–110
25. Adrian R et al (2009) Lakes as sentinels of climate change. Limnol Oceanogr 54(6):2283–2297
26. Livingstone DM (2003) Impact of secular climate change on the thermal structure of a large temperate central European lake. Clim Change 57:205–225
27. Skowron R (2011) The differentiation and variability of chosen elements of the thermal regime of water in lakes on Polish lowland. Nicolaus Copernicus University Press, Toruń. (in Polish)
28. O'Reilly CM et al (2015) Rapid and highly variable warming of lake surface waters around the globe. Geophys Res Lett 42:10773–10781. https://doi.org/10.1002/2015GL066235
29. Winslow LA, Read JS, Hansen GJA, Hanson PC (2015) Small lakes show muted climate change signal in deepwater temperatures. Geophys Res Lett 42:355–361. https://doi.org/10.1002/2014GL062325
30. Choiński A, Ptak M, Skowron R (2014) Trends to changes in ice phenomena in Polish lakes in the years 1951–2010. Przegl Geogr 86(1):23–40. (in Polish)
31. Choiński A, Ptak M, Skowron R, Strzelczak A (2015) Changes in ice phenology on Polish lakes from 1961–2010 related to location and morphometry. Limnologica 53:42–49
32. Górniak A, Pękala M (2001) Ice phenomena in lakes of north-eastern Poland. Przegl Geofiz 46(1–2):91–109. (in Polish)
33. Wrzesiński D, Choiński A, Ptak M (2015) Effect of the North Atlantic Oscillation on the thermal characteristics of lakes in Poland. Acta Geophys 63(3):863–883. https://doi.org/10.1515/acgeo-2015-0001
34. Marszelewski W, Skowron R (2006) Ice cover as an indicator of winter air temperature changes: case study of the Polish Lowland lakes. Hydrol Sci J 51(2):336–349
35. Marszelewski W, Skowron R (2009) Extreme ice phenomena on the lakes of Northern Poland. Limnol Rev 9(2–3):81–89
36. Yan ND et al (2008) Long-term trends in zooplankton of Dorset, Ontario, lakes: the probable interactive effects of changes in pH, TP, DOC and predators. Can J Fish Aquat Sci 65:862–877

37. Christensen MR, Graham MD, Vinebrook RD, Findlay DL, Paterson MJ, Turner MA (2006) Multiple anthropogenic stressors cause ecological surprises in boreal lakes. Glob Change Biol 12:2316–2322
38. Dokulil MT, Teubner K, Jagsch A, Nickus U, Adrian R, Straile D, Jankowski T, Herzig A, Padisák J (2010) The impact of climate change on lakes in Central Europe. In: George G (ed) The impact of climate change on European lakes. Springer, Dordrecht, pp 387–409
39. Blenckner T et al (2007) Large-scale climatic signatures in lakes across Europe: a meta-analysis. Global Change Biol 13:1314–1326
40. Wagner C, Adrian R (2009) Exploring lake ecosystems: hierarchy responses to long-term change? Global Change Biol 15:1104–1115. https://doi.org/10.1111/j/1365-2486.2008.01833.x
41. Kraemer BM et al (2015) Morphometry and average temperature affect lake stratification responses to climate change. Geophys Res Lett 42:4981–4988. https://doi.org/10.1002/2015GL064097
42. Rigosi A, Cayelan C, Bas W, Ibelings J, Brookes D (2014) The interaction between climate warming and eutrophication to promote cyanobacteria is dependent on trophic state and varies among taxa. Limnol Oceanogr 59(1):99–114
43. Creed I, Bergström AK, Trick CG et al (2018) Global change-driven effects on dissolved organic matter composition: implications for food webs of northern lakes. Global Change Biol 24:3692–3714. https://doi.org/10.1111/gcb.14129
44. Leech DM, Pollard AI, Labou SG, Hampton SE (2018) Fewer blue lakes and more murky lakes across the continental U.S.: implications for planktonic food webs. Limnol Oceanogr. https://doi.org/10.1002/lno.10967
45. De Wit HA et al (2016) Current browning of surface waters will be further promoted by wetter climate. Environ Sci Technol 3:430–435. https://doi.org/10.1021/acs.estlett.6b00396
46. Williamson CE, Morris DP, Pace ML, Olson OG (1999) Dissolved organic carbon and nutrients as regulators of lake ecosystems: resurrection of a more integrated paradigm. Limnol Oceanogr 44:795–803
47. Górniak A (1996) Humic substances and their impact on freshwaters ecosystems functioning. Warsaw University Press, Białystok. (in Polish)
48. Houser JN (2006) Water color affects the stratification, surface temperature, heat content, and mean epilimnetic irradiance of small lakes. Can J Fish Aquat Sci 63:2447–2455. https://doi.org/10.1139/f06-131
49. Hillbricht-Ilkowska A (1993) Lake ecosystems and the global climate changes. Kosmos 42(1):107–121

Polish Rivers as Hydrographic Objects

Elżbieta Bajkiewicz-Grabowska, Maciej Markowski, and Włodzimierz Golus

Contents

Abstract This paper characterises rivers, the major hydrographic objects within the water network of Poland. The authors point out to the diverse nature of the river network resulting from the natural topography of Poland and the fact that the contemporary river network uses valleys inherited from the past. Systems of two largest Polish allochthonous (transit) rivers, the Vistula and the Oder, are characterised. The rivers of Przymorze and the Niemen basin are discussed in less detail. The paper shows that the stability of discharge in Polish rivers, resulting from climate conditions (precipitation exceeding field evaporation over the entire water year), is determined by the underground contributor to the discharge, which is a result of underground water drainage by the river channels. The contribution of underground runoff to the total discharge of Polish rivers is generally stable over the year, and the percentage of surface runoff contributing to the total discharge depends on the morphogenetic zone (mountains, highlands, lowlands) where these rivers flow. Finally, the types of regimes characterising Polish rivers are presented and their spatial differentiation is discussed. Polish rivers are characterised by complex regimes. Most rivers are characterised by the moderately developed nival regime with the high water period in the spring. Some of the lowland rivers have well-developed nival regime, while some lakeland rivers are characterised by the poorly developed nival regime. The regimes of mountain rivers vary.

E. Bajkiewicz-Grabowska (✉) · M. Markowski · W. Golus
Department of Limnology, Institute of Geography, University of Gdańsk, Gdańsk, Poland
e-mail: bajka37@wp.pl

E. Korzeniewska, M. Harnisz (eds.), *Polish River Basins and Lakes – Part I*,
The Handbook of Environmental Chemistry 86,
https://doi.org/10.1007/978-3-030-12123-5_2

Keywords Hydrological regimes of Polish rivers · Oder river system · River channel types · River network density · River valley network · Transport of river bed load · Vistula river system

1 Introduction

The network of river valleys in Poland clearly corresponds to the landforms inherited from the past [1]. The principal variation of this network follows the latitudinal arrangement of the landforms (Fig. 1). The natural topography of Poland is

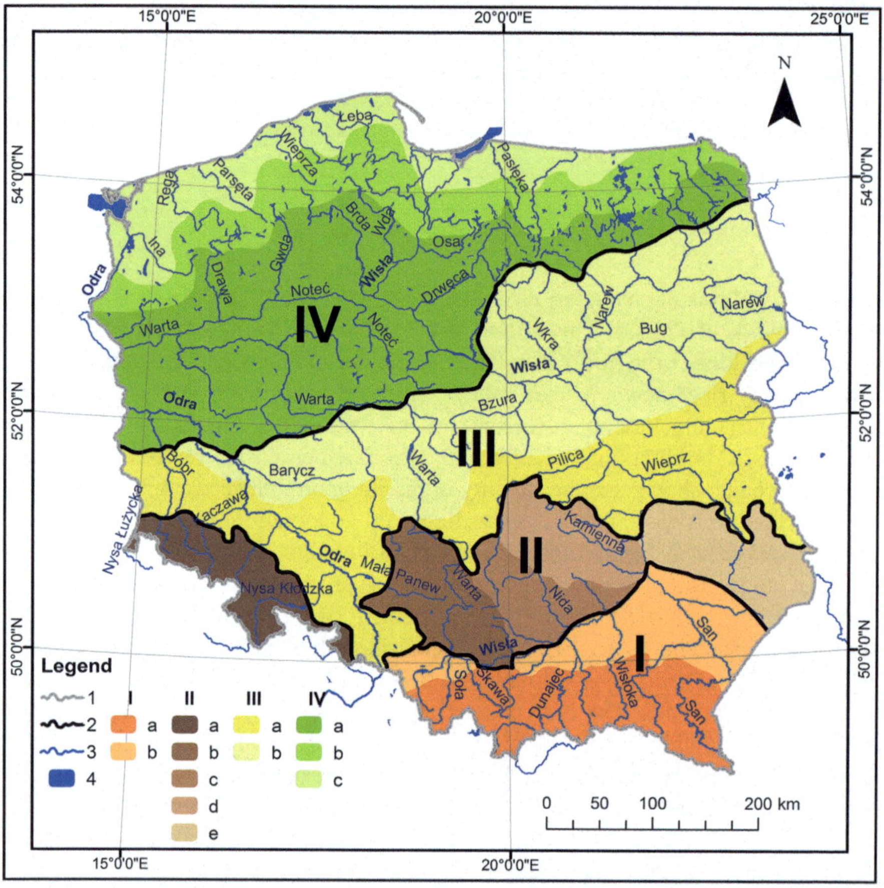

Fig. 1 The main elements of the natural topography of Poland. Explanations: (1) rivers and water reservoirs, (2) state borders, (3) boundaries of geomorphological units, (I) young mountains and piedmont basins (a, Carpathians; b, sub-Carpathian basins), (II) old mountains and uplands (a, Sudeten and Przedgórze; b, Silesian-Cracovian Upland; c, Nida Basin; d, Kielce Upland; e, Lubelsko-Wołyńska Upland), (III) late postglacial areas (a, late postglacial plains; b, lakeless plateaus), (IV) early postglacial areas (a, lake plateaus; b, lakeland hump; c, coastal plains) (adapted from [2])

dominated by lowlands (0–300 m.a.s.l.), which cover 91.3% of the country's area. Moving south, the land rises into hills (300–500 m.a.s.l.), which cover 5.6% of the country's area, and mountains (>500 m.a.s.l.), which cover a mere 3.1% [3]. The latitudinal variation of the configuration of landforms in Poland was determined by the Alpine orogeny, which formed the Carpathian Mountains and created the Carpathian Foredeep and the Meta-Carpathian Swell [4, 5]. In the Pleistocene Epoch, the parallel arrangement of landscape belts was emphasised by the transgressions of decreasing ranges in the Scandinavian Ice Sheet (which commenced approximately 115,000 years ago and lasted for approximately 103,000 years) and by the development of periglacial zones in its forefield [1]. The westbound flow of snowmelt from the melting ice sheets formed the latitudinally arranged belts of marginal paleo-valleys. In the Holocene Epoch, which commenced approximately 10,000 years ago, the contemporary river valleys were considerably remodelled, and the river network was stabilised.

Four morphogenetic regions are distinguished in Poland, namely, young mountains (the Carpathians) and piedmont basins, old mountains (the Sudeten) and highlands, late postglacial regions and early postglacial regions [2], which are only transected by two large river valley systems, that of the Vistula and that of the Oder, which drain water from the mountains in the south to the Baltic Sea in the north (Fig. 1).

The Poland's contemporary network of river valleys, irrespective of the influences resulting from the complex geological past, has to the largest degree been shaped by the changing climate which determines the changes in the hydrological regime of rivers [6]. The river bed and channel characteristics (tortuosity, depth, width) largely depend on the river's hydrological regime (water stage fluctuations, stream flow rate and energy, transport of bed flow), which in turn is determined by climate conditions (precipitation, air temperature). The channels of Polish rivers are largely regulated by hydraulic structures, and the channels of large rivers are additionally secured by levees. Over the past 50–70 years, they have been significantly transformed by man.

The river network of Poland drains 99.7% of the region located within the Baltic Sea catchment basin (Fig. 2). Within this basin, the following two rivers have the largest catchment areas: the Vistula (which rises in the Silesian Beskids and empties into the Gdansk Bay) and the Oder (which rises in the Oder Mountains in the Czech Republic and empties into the Szczecin Lagoon, a part of the Pomeranian Bay). The catchment basins of the Vistula and the Oder drain, respectively, 54% and 33.9% of Poland's surface. The remainder (18.1%) is made up of the catchment basins of the Przymorze rivers (Polish rivers directly draining into the Baltic Sea, namely, the rivers Rega, Parsęta, Wieprza, Słupia, Łupawa, Łeba and Reda) with some of the Vistula delta (Dead Vistula with Motława and its lakeland tributary Radunia and Wisła Śmiała), the basins of the Vistula Lagoon (catchment basins of the following rivers: Szkarpawa z Wisłą Królewiecką, Nogat, Elbląg, Bałda, Pasłęka, Łyna, Węgorapa) and of the Szczecin Lagoon (without the catchment basin of the Oder) and the catchments of the Czarna Hańcza and the Szeszupa (tributaries of the river Niemen). Approximately 0.2% of Poland's surface lies within the catchment basin of

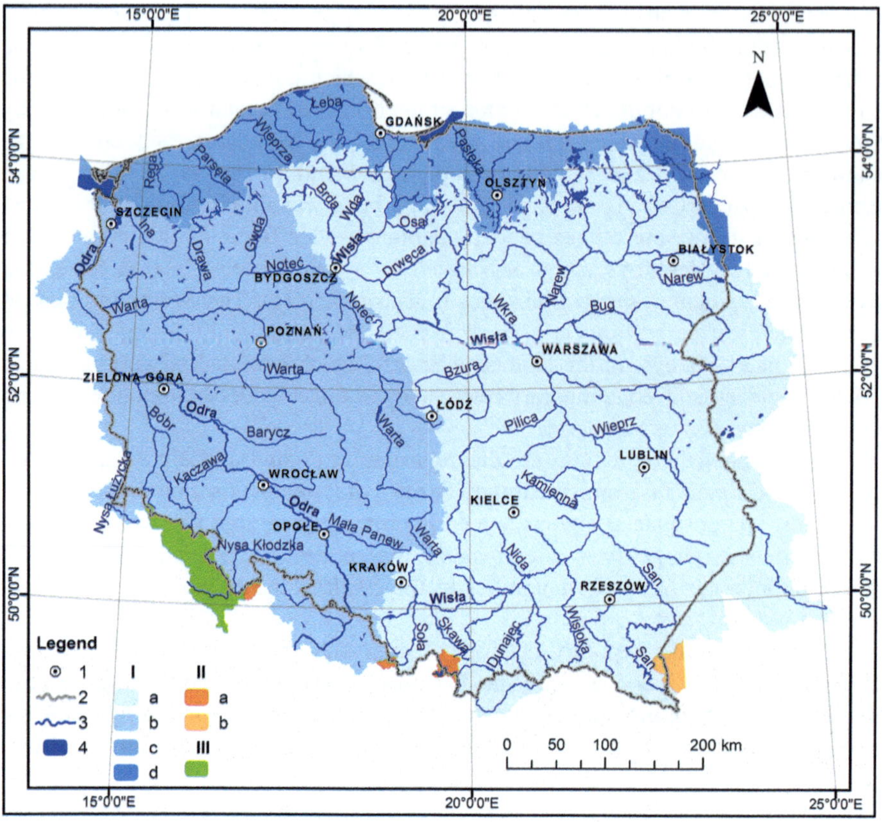

Fig. 2 Hydrographic division of Poland. Explanations: (1) Baltic Sea basin (a, Vistula basin; b, Oder basin; c, Niemen basin; d, basins of Przymorze rivers; e, Vistula Lagoon basin; f, Szczecin Lagoon basin), (2) Black Sea basin (a, Danube basin; b, Dniester basin), (3) North Sea basin (Elbe basin), (4) streams and reservoirs, (5) state borders, (6) provincial cities (Source: Map of the Hydrographic Division of Poland)

the Black Sea (the upper courses of the Orawa and the Skalniczanka in the catchment of the Danube and the upper course of the Strwiąż in the Dniester river basin), while nearly 0.1% in the catchment basin of the North Sea (the upper courses of the Jizera and the Orlica in the Elbe river basin).

2 Natural Determinants of Poland's River Network

The downward slope of Poland's surface to the northwest determines the near-meridional course of the rivers of Poland and the asymmetry of their basins. The occasional latitudinal course of the rivers is a consequence of marginal paleo-valleys, which, during its glaciation, latitudinally drained, along the snout of the melting ice sheet, huge amounts of water and the waters of the rivers that flew from the south towards the ice sheets in the north.

Polish rivers most commonly rise from sources located in the mountainous part in the south of the country, namely, in the Carpathians and the Sudeten. Source regions also include the Middle Polish highlands and lakelands in the north of Poland. In relation to the rivers, the lowlands of central Poland mainly play the role of a transit area and to a lesser degree function as a feeding area.

The latitudinal arrangement of landform belts – with the mountains in the south, followed northwards by basins, highlands, lowlands, lakelands and the coastal region of the Baltic Sea (Fig. 1) – resulted in the formation of gorges along the course of the rivers (e.g. the Dunajec River Gorge, the Vistula River Gorge of Lesser Poland, the Vistula River Gorge near Fordon in Bydgoszcz) and waterfalls in the mountains, which was mainly due to the tectonics of the deposits that formed fold mountain structures (e.g. Siklawa in the Tatras) or fault-block mountain structures (e.g. Szklarka, Kamieńczyk in the Sudeten).

In the mountains and highlands (south of Poland), the rivers generally adapted to the geological forms, especially to the lithological differences of bedrocks. Only some of the rivers transected mountain chains and hill chains across the river valleys. Such an arrangement of the river network is a consequence of the following: (a) persistence of former directions of outflow from the period preceding the slow uplift of land; (b) formation of a river network on the accumulation surface overlain on the previous geomorphological forms, which, upon the truncation of the overlaying layers, were exposed, revealing unconformity with the direction of the rivers; and (c) headward erosion, which led to the capture of waters from the neighbouring catchment, which in turn caused the direction of the stream outflow to reverse [7].

Five principal types of river channels are distinguished in the mountain and highland areas [6]:

1. High mountain river channels: bending, locally braided, cut out in pebbles and boulders and in solid rock, largely regulated.
2. Middle mountain river channels: bending with a system of riffles and pools, or braided or cut out in solid rock or pebbles, largely regulated.
3. Piedmont river channels: bending and meandering, with a system of chutes and pools, cut out in pebbles. Large allochthonous river channels are largely regulated, while the small ones are regulated to a lesser degree.
4. Foreland basin river channels: cut out in beds made up of fine pebbles and sands, usually meandering with a small gradient and with a low width-to-depth ratio.
5. Highland river channels: bending and meandering, cut out in alluvium, mostly sandy; during surges that exceed the riverbank water stage, the floodplain is remodelled. Transport of dissolved material overbalances that of clastic material. These include seminatural and regulated channels.

The arrangement of river valleys in the Polish Lowlands is associated with the course of the marginal dewatering pathways during glaciations and with climate changes. The impact of the latter is evident in valleys in the form of consecutive phases of erosional channel formation and accumulation and in the form of changes in the type and size of the river channels. Extreme water surges in subsequent millenniums contributed to the transformation of valley beds, whose direction of dewatering in individual segments of the river course often differed. In the southern part of the Polish Lowlands, late postglacial terrains (between the limit of maximum extent of the last glaciation in the north and the old mountains and highlands in the south) are dominated by autochthonous rivers whose basins lie in naturally uniform regions (e.g. the rivers Barycz, Prosna, Widawka, Bzura, Wkra, Liwiec, Narew). They are accompanied by allochthonous (transit) rivers, whose upper catchment parts are found in the highlands that are adjacent to them from the south (e.g. the rivers Bug, Wieprz, Pilica, Warta), or in the Sudeten (e.g. the rivers Bóbr, Kaczawa, Eastern Neisse). The river channels in this zone are horizontally stable, generally bending or meandering, and anastomosing within the widenings of the basins. Braided river channels are mostly found in the Middle Vistula Valley and in the lower courses of some of its tributaries. Rivers in early postglacial regions flow down polygenic valleys (tunnel valley-like segments, marginal paleo-valley-like segments, canyon-like segments, segments formed on outwash plains and within moraine plateaus). The river channels and floodplains here are largely inherited from the late Weichsel [Vistulian] and Holocene fluvial transformation [6]. The river network is dispersed and of the initial nature. The process of its consolidation and development in the lakelands is still ongoing. The contemporary river network of the early postglacial zone shows various directions of dewatering, which is mainly due to the morphogenesis of the relief of the Polish Lowlands. Five types of river channels are distinguished here [6]:

1. River channels of the southern slope of the lakeland hump (Fig. 1) developed in the beds of valleys shaped on outwash plains with lake segments, e.g. Drawa, Gwda, Brda, Wda, Wierzyca, Drwęca, Omulew, Pisa, Lega, Rospuda and Czarna Hańcza.
2. River channels in the northern slope of the lakeland hump with lake segments characterised by uneven longitudinal profiles, whose development was determined by the changes in the water level of the Baltic Sea, e.g. Ina, Rega, Parsęta, Wieprza, Słupia, Łupawa, Reda, Radunia, Pasłęka, Łyna and Węgorapa.
3. River channels shaped within the latitudinal marginal valleys or Urstromtäler, having a bending or an irregular broken pattern, e.g. the lower Noteć, the lower Warta and the upper Obra.
4. River channels in gorges, most commonly characterised by a meridional direction of flow (the Vistula River Gorge, the Warta River Gorge, the Obra River Gorge).
5. Channels of tributaries of large allochthonous rivers (especially of the Vistula and the Oder). The channels of these rivers in the plateaus differ from those in the main river bed.

3 The River Network of Poland

The river network of Poland consists of streams of various sizes. The upper courses of the rivers are made up of rapidly flowing mountain streams and brooks and, in the lowlands, of rapidly flowing streams and brooks and the slowly flowing brooks. These are all small water streams flowing from springs and mires of small topographical catchment areas (ranging from several to several dozen km^2), usually without tributaries [8]. They meet with each other to form higher-order streams. These are rivers of various sizes with well-defined channels (bending, meandering, braided, in some cases anastomosing). They flow along a valley in a bed that has been distinctly shaped by fluvial erosion and are fed from surface and underground sources with water from precipitation in their basins [8].

As a result of the belt-like arrangement of landforms, three principal segments differing in river gradient and activity are distinguished in the course of large Polish allochthonous (transit) rivers: the upper, middle and lower course. In its upper course, a river often flows on bedrocks, deepening its channel, drains excess precipitation and collections of underground water and has numerous tributaries which also feed it with bed load. In its middle course of a river, the slope of the valley decreases; the river begins to meander, cutting the banks in a wide floodplain on which it deposits most of its load in the form of point bars and alluvial soils. In its lower course with a very gentle slope, a river sometimes flows along numerous branches, and when it flows into the sea, it forms a delta (Fig. 3a). The longitudinal profiles of autochthonous lowland rivers are generally characterised by an even slope along its entire course (Fig. 3b). In early postglacial regions, lake segments and river segments are distinguished in the longitudinal profile of lake rivers and lake-river systems, which consist of lakes interconnected by short river segments (mostly overflows) [9–13]. The river segments are characterised by a greater slope if they have the form of gorges (Fig. 3c). The behaviour of the tributaries of the principal recipient (the main river of the system or lake), especially in their upper courses, depends on the catchment retention status [14, 15].

The density of the river network of Poland varies. The highest density is found in the Carpathians (an average of 1.4 km km^2; 2.2–2.5 km km^2 in the Pieniny, 3.3 km km^2 in the Polish Tatra Mountains, approximately 1.5 km km^2 in the Bieszczady) and the Sudeten (an average of 1.2 km km^2) and is due to the high precipitation, poorly permeable ground and steep gradient of the land. The density of the river network is much lower in highland areas formed by carbonate rocks (an average of 0.6 km km^2) and in the lowlands of central Poland (from 0.7 to 0.96 km km^2). The density ranges from 0.2 to 1.1 km km^2 in lakeland areas and from 0.8 to 1.3 km km^2 in coastal plains [16]. The eastern part of the Vistula delta is characterised by a very high density of streams (up to 20 km km^2) [17].

The largest river systems in Poland are formed by the Vistula and the Oder (Table 1). These rivers flow through the entire Poland, transporting water from the mountains in the south to the Baltic Sea in the north (Fig. 1). The Vistula, a river of intermediate length among the European rivers (11th longest river in Europe), is the

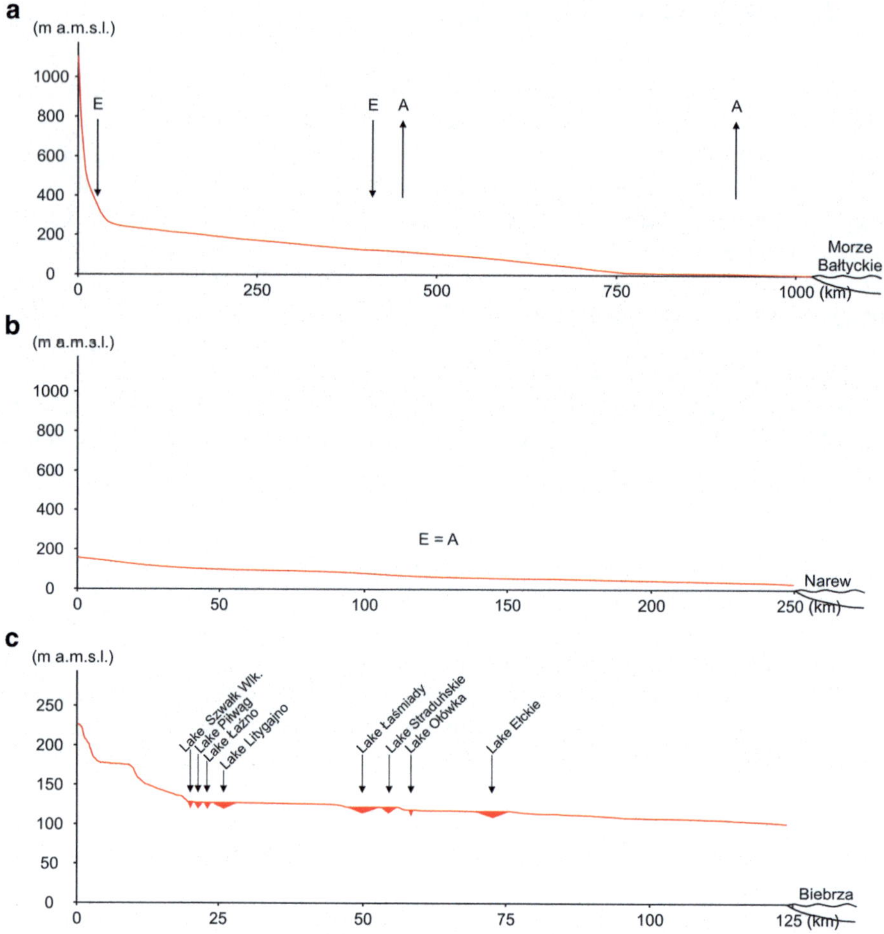

Fig. 3 Longitudinal profiles of rivers: (**a**) profile of an allochthonous river, (**b**) profile of an autochthonous lowland river, (**c**) profile of a lake-river system

largest and longest (1,022 km) [18] Polish river. It is also the river with the largest basin area (193.9 thousand km^2) [18] within the drainage basin of the Baltic Sea. Its average gradient is 1.08‰, although in 75% of its course, the gradient is below 0.3‰. The sinuosity index of the river[1] is low at 1.13. The tortuosity index[2] is 1.9. From its source in the Silesian Beskids, the Vistula flows from south to north, across a belt of highlands and lakelands, and empties into the Gulf of Gdańsk, forming a delta area at its mouth. The present estuary of the Vistula is an artificial cut excavated

[1]Sinuosity ratio is the ratio of the river length to the length of a straight line connecting the river source with the river-mouth [8].

[2]Tortuosity ratio is the ratio of the river length to the length of its valley [8].

Table 1 River systems of Poland

River	Recipient	Drainage basin area (km^2)		Length (km)		Average flow 1951–2000 (m^3/s)
		Total	Within borders of Poland	Total	Within borders of Poland	
Wisła river system						
Wisła	Baltic Sea	193,960	168,868[a]	1,022	1,022	1,080.0
Przemsza	Wisła	2,125	2,125	87	87	20.0
Dunajec	Wisła	6,796	4,838	249	249	85.5
Nida	Wisła	3,844	3,844	154	154	21.1
Wisłoka	Wisła	4,109	4,100	173	173	35.5
San	Wisła	16,877	14,426	458	457	129.0
Wisłok	San	3,538	3,538	220	220	24.5
Kamienna	Wisła	2,020	2,020	149	149	8.9
Wieprz	Wisła	10,497	10,497	349	349	36.4
Pilica	Wisła	9,258	9,258	333	343	47.4
Narew	Wisła	74,527	53,846	499	443	313.0
Biebrza	Narew	7,092	7,067	164	164	35.3
Pisa	Narew	4,510	4,510	82[b]	82	26.8
Orzyc	Narew	2,134	2,134	142	142	9.3
Bug	Narew	38,712[c]	19,239[c]	774	590	155.0
Krzna	Bug	3,273	3,273	107	107	11.4
Liwiec	Bug	2,763	2,763	142	142	12.1
Wkra	Narew	5,348	5,348	255	255	22.3
Bzura	Wisła	7,764	7,764	173	173	28.6
Drwęca	Wisła	5,697	5,697	231	231	30.0
Brda	Wisła	4,665	4,655	245	245	28.0
Wda	Wisła	2,324	2,324	198	198	14.3
Odra river system						
Odra	Baltic Sea	119,074	106,043	840	726	567.0
Nysa Kłodzka	Odra	4,570	3,742	189	189	37.7
Barycz	Odra	5,547	5,547	136	136	18.8
Bóbr	Odra	5,874	5,830	279	276	44.8
Nysa Łużycka	Odra	4,403	2,201	246	197	31.0
Warta	Odra	54,520	54,520	795	795	216.0
Prosna	Warta	4,917	4,917	227	227	17.4
Wełna	Warta	2,635	2,635	118	118	9.2
Noteć	Warta	17,302	17,302	391	391	76.6
Gwda	Noteć	4,947	4,947	140	140	27.9
Drawa	Noteć	3,291	3,291	192	192	21.3
Ina	Odra	2,151	2,151	125	125	13.0

(continued)

Table 1 (continued)

River	Recipient	Drainage basin area (km^2)		Length (km)		Average flow 1951–2000 (m^3/s)
		Total	Within borders of Poland	Total	Within borders of Poland	
Przymorze rivers						
Rega	Baltic Sea	2,767	2,767	188	188	21.1
Parsęta	Baltic Sea	3,084	3,084	143	143	29.1
Wieprza	Baltic Sea	2,313	2,313	133	133	23.8
Słupia	Baltic Sea	1,623	1,623	139	139	17.9
Łupawa	Baltic Sea	925	925	99	99	8.3
Łeba	Baltic Sea	1,801	1,801	117	117	11.7
Reda	Baltic Sea	639	639	51	51	4.3
Radunia	Motława[d]	837	837	103	103	6.3
Rivers of Vistula Lagoon basin						
Węgorapa	Pregoła	3,639	976	140	44	5.0[e]
Łyna	Pregoła	7,126	5,298	264	207	34.7[e]
Pasłęka	Baltic Bay	2,321	2,319	187	160	18.6
Nogat	Baltic Bay	1,330	1,330	62	62	30
Niemen river system						
Czarna Hańcza	Niemen	1,916	1,612	142	108	6.8[e]
Szeszupa	Niemen	6,105	184	297.6	24	1.62[e]

[a]Excluding the delta
[b]Calculated from Lake Roś
[c]Up to Lake Zegrzyńskie
[d]Martwa Wisła tributary
[e]At the border enclosing the drainage area

in 1891–1895 near the village of Świbno. Through its main mouth, the Vistula provides the Baltic Sea with one of the largest inflows of freshwater (the Newa, 2,488 m^3/s; the Vistula, 1,081 m^3/s according to HELCOM) [19].

The Vistula river system is asymmetrical. The ratio of left-bank basin to right-bank basin is 27:73. The main right-bank tributaries of the Vistula are Soła, Skawa, Raba, Dunajec, Wisłoka, San (with Wisłok), Wieprz, Świder, Narew (with Biebrza, Pisa, Bug and Wkra), Skrwa and Drwęca, and the main left-bank tributaries include Przemsza, Nida, Kamienna, Radomka, Pilica, Bzura, Brda, Wda and Wierzyca (Table 1 and Fig. 4).

The Vistula flows from south to north throughout Poland in a valley made up of segments of varying ages [20, 21]. As the valley travels across tectonic and geomorphological units arranged in belts and, in lowland areas, crosses the latitudinal courses of ice sheet overthrusts and their recession phases, individual segments of this valley vary in length, depth and width of the valley bed. During each major

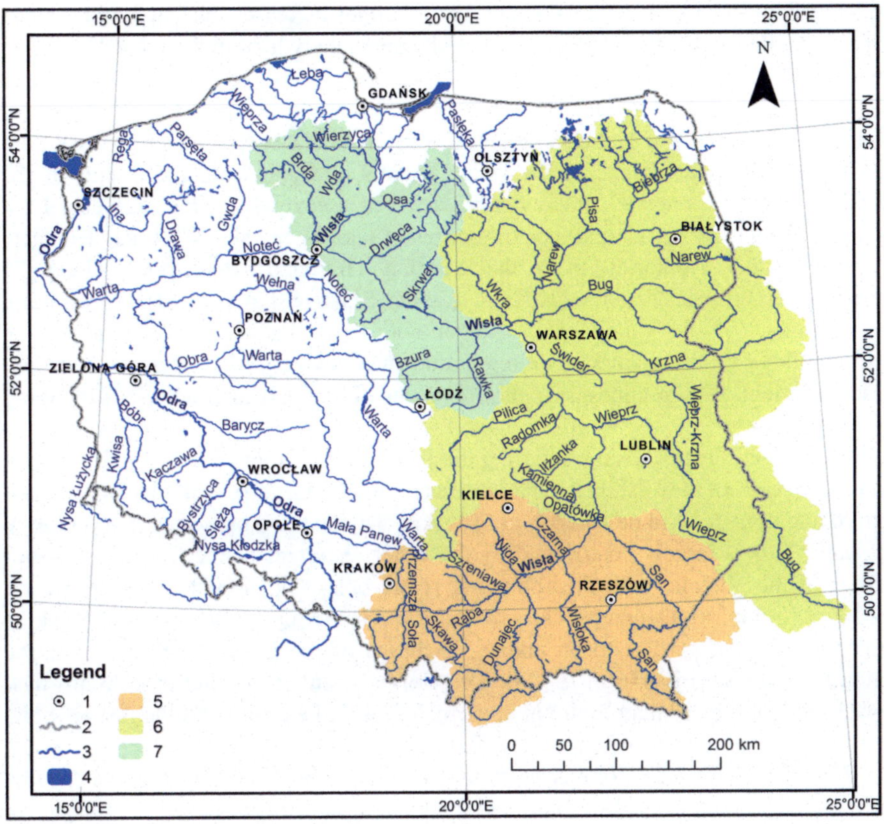

Fig. 4 The Vistula river basin with division into the Upper Vistula, the Middle Vistula and the Lower Vistula according to the Map of Hydrographic Division of Poland

water surge, this valley bed is shaped by the Vistula. According to hydrographic characteristics (mainly longitudinal slopes along its course) and water resources of its basin (mouths of tributaries that change the flow rate of the river), the Vistula is divided into three sections: upper, middle and lower. *The Upper Vistula* extends from its sources to the mouth of the San and is 385.3 km long [22]. The sources of the Vistula are in the Silesian Beskids, on the western slope of the Barania Góra. The Czarna Wisełka (1,107 m.a.s.l.) and the Biała Wisełka (1,080 m.a.s.l.) meet to form the Wisełka, which meets with the Malinka to form the Vistula. In its upper course, along a short segment in the mountains, the Vistula is a mountain stream with a steep gradient (several dozen ‰), an uneven longitudinal profile (the bed of the channel contains numerous cascades and natural sills), variable flow and large transport of rock load. Upon entering the foothills, the Vistula loses the characteristics of a mountain river (the longitudinal gradient of the river channel and its transport capacity decrease), gently turns eastwards and, as it approaches Kraków, assumes the character of a lowland river flowing through the horsts of Jurassic limestone. The river valley widens, the width of the river channel increases and numerous ponds appear in the bed of the valley. The river channel here, which in the preceding reach was meandering and unregulated, is now embanked and regulated. Two storage reservoirs (Wisła Czarne and Goczałkowicki) and spillway steps (Dwory, Smolice, Łączany, Kościuszko, Dąbie, Przewóz), forming the Upper Vistula Cascade, are found in the upper course of the river. Thanks to them, from the mouth of the Przemsza to the Przewóz spillway step, the river is navigable. This segment of the river is joined by its Carpathian tributaries (Biała, Soła, Skawa, Raba, Dunajec, Wisłoka) and tributaries from the Silesian Highland (Pszczynka, Gostynia, Przemsza, Dłubnia, Szreniawa, Nida, Czarna Staszowska) (Fig. 4). The natural environment of the valley of the lower course of the Middle Vistula and its surroundings were considerably altered after the discovery of sulphur deposits and establishment of strip mines, sulphur ore purification plants and factories producing sulphuric acid.

The Upper Vistula basin (including the San basin) accounts for 27% of the entire drainage basin of the Vistula and has an area of 50,655 km^2, 91% of which is situated in Poland, 5% in Ukraine and 4% in Slovakia. This is the most water-rich region in Poland. This is largely determined by the water resources of the Carpathian tributaries of the Vistula, with the following rivers being the richest in water: Skawa (MQ = 16.5 m^3/s), Dunajec (MQ = 88.5 m^3/s), San (MQ = 134 m^3/s) and Wisłoka (MQ = 38.9 m^3/s). The mean annual discharge of the Upper Vistula (below the mouth of the San) is 436 m^3/s [23]. The Upper Vistula basin features 26 artificial water bodies greater than 50 ha in area, including Lake Solina, the largest artificial lake in Poland in terms of volume (472.4 million m^3) with an area of 22 km^2 [22].

The Middle Vistula extends from the mouth of the San to that of the Narew (Fig. 4). The course of the Middle Vistula starts with the so-called Vistula River Gorge of Lesser Poland [24], a segment that passes through highlands (from the mouth of the San to that of the Wieprz). In this reach of the Vistula, the bed of its valley lowers from 135 to 115 m. The river breaks through from the south to the north, transecting rock layers of various resistance. Over about 75 km, the width of the Vistula changes from about 1 to 6 km and occasionally to up to 15 km. The slopes of the valley are steep and high, with an elevation of 30–80 m above river level. The bed of the river valley is flat and filled with alluvia. The floodplain features oxbow lakes and lateral canals which do not connect with the river from one end and referred to here as the Wisełki. The river flows in a channel filled with sandbanks and islands stabilised by vegetation. The following rivers join the Vistula in its gorge segment: Kamienna, Krępianka, Iłżanka and Zwoleńka from the west and Wyżnica, Chodelka and Bystra from the east (Fig. 4). Upon exiting the highlands, the Vistula enters the forefield of the last glaciation – the Mazovian Lowland – and flows down a valley which is a natural northwards continuation of the Vistula River Gorge of Lesser Poland [24]. This long (more than 200 km) segment of the Vistula river valley (from the mouth of the Wieprz to that of the Narew) cuts into the surrounding moraine plateaus at the depth of 25–55 m. As the Mazovian Vistula, the river receives the following tributaries: the Wieprz, Wilga and Świder from the east and the Radomka and Pilica from the west (Fig. 4). This section has a preserved natural character and the Vistula valley bed is 12–15 km wide. The river shapes its bed within the confines of the relatively widely distributed levees. It flows in a wide (600–1,200 m) braided channel with sandbanks, islands and islets. The floodplain, whose width south of Warsaw is up to 7 km, features many river lakes (the largest ones being Czerniakowskie, Powsinkowskie, Sielanka, Wilanowskie), lateral canals and overgrown depressions at sites of former meanders of the river. The mean annual discharge of the Vistula in Warsaw is 575 m^3/s [23].

Along its Warsaw section, the Vistula valley and bed naturally narrow due to the vicinity of a postglacial plateau. This natural narrowing was augmented after the Vistula channel was regulated following the flood in 1884. The regulation works carried out for dozens of years straightened, narrowed and deepened the Warsaw section of the river channel to about 350 m, as a result of which the velocity of the river increased twofold to threefold and erosion of the Vistula channel bed intensified, leading to bed depression by an average of over 2 m in the twentieth century [25].

Upon exiting Warsaw, the Vistula, which is sometimes called the queen of sand, resumes its braided character. Its channel is filled with point bars formed by the material carried by water and has many islands that divide the channel into several branches. The annual quantity of sand carried with the water of the Vistula has been estimated at approximately 1 million m^3 [25].

Upon entering the Warsaw Basin, the Vistula absorbs the Narew along with the Bug and the Wkra (the mean annual discharge of the Narew at its mouth is 313 m^3/s [23]) which force the course of the river westwards. In this latitudinal course, the bed of the Middle Vistula valley is up to 20 km wide. Also here, the Vistula displays the natural character of a braided river with numerous islands, oxbow lakes and lateral canals. The islands vary in form and succession stage, ranging from sandbanks to islands covered by herbaceous plants and willow and poplar communities. Adjacent to the river channel lies a floodplain dominated by meadows and pastures and, on the left bank, a sandy fluvial terrace covered with parabolic dunes arranged in belts and separated by peat bogs and marshes. This area features the Kampinos Forest protected within the confines of the Kampinos National Park.

The Middle Vistula basin (including the Narew basin) accounts for 53% of the entire drainage basin of the Vistula and has an area of 88,865 km^2, 81% of which is situated in Poland, 10% in Ukraine and 9% in Belarus [22]. The mean annual discharge of the Middle Vistula (below the mouth of the Narew) is 888 m^3/s [23]. The Middle Vistula basin features 18 storage reservoirs of more than 1 hm^3 in capacity. In terms of capacity, the two largest reservoirs are the Sulejowski Reservoir on the Pilica (86.6 hm^3) with an area of 23.8 km^2 and the Siemianówka Reservoir on the Narew (79.5 hm^3) with an area of 32.5 km^2 [22].

The mouth of the Narew marks the beginning of the *Lower Vistula* (Fig. 4). The length of this section to its entry into the Baltic Sea is 390.2 km. The initial fragment is similar to the Mazovian section and receives its left-bank tributary, the Bzura. The Vistula then changes its direction to northwestern, flows through the Thorn-Eberswalder marginal paleo-valley and enters the area of the last glaciation, separating the Greater Poland lakeland from the Chełmińsko-Dobrzyńskie lakeland

[24]. Here, the Kujawsko-Pomorski section of its course begins. The first tributaries of this section are the Skrwa Prawa and the Skrwa Lewa. The mean annual discharge of the Vistula below Płock is 932 m^3/s [23]. At Płock, the Vistula dramatically changes its character due to the presence of a spillway step constructed in Włocławek in 1970 (length, 670 m; maximum height from the bed of the river, approximately 20 m). Between 618 and 675 km, the river transforms into a river lake (Włocławskie), the largest artificial water body in Poland. It encompasses the ancient channel of high water of the Vistula along with low floodplains of the left bank. The average width of the resulting lake is 1,210 m and ranges from 500 to 2,500 m. The spillway step in Włocławek is the first and so far the only spillway step of the Lower Vistula Cascade that has been designed. Downstream this step, over a length of 9 km, accelerated erosion of the Vistula bed is observed, while further down, accelerated accumulation is seen. From Włocławek to, nearly, Toruń, the Vistula retains the untamed character. It flows through a valley whose bed is from several to 20 m wide, along a braided channel of up to 1 km in width. The Vistula channel features multiple shoals and picturesque islands. This section of the river receives its right-bank tributary, the Drwęca. Near Bydgoszcz, the Lower Vistula turns from the western direction (marked by the fluvioglacial water outflow through the Thorn-Eberswalder marginal paleo-valley) to the northeastern direction, and, having received the Brda from the left, it enters a narrow valley separating a belt of early postglacial lakelands (the Vistula River Gorge near Fordon). In this gorge section, the width of the valley decreases from 4 to 2.5 km and widens to about 7 km at the mouths of the two Vistula tributaries: the right-bank tributary Fryba and the left-bank tributary Wda. Having crossed the belt of lakelands, the Vistula turns north. Its meandering valley widens to up to 18 km. Here, the Vistula receives the lakeland left-bank tributary Wierzyca.

Below Gniew, in Biała Góra at Mątowski Headland, the Vistula enters a vast delta plain called Żuławy Wiślane [26]. It was formed by two main branch channels of the river: the western branch channel Leniwka, which followed the true Vistula channel, and the eastern branch channel Nogat. Surrounded by moraine plateaus from the south, the Vistula delta area widens northwards in a fanlike fashion to about 50–60 km over a length of about 50 km. The delta region covers an area of about 1,700 km^2, of which 450 km^2 constitute areas of depression sunken below sea level (with the lowest datum at 2.2 m.b.s.l.) [26].

Having given off the Nogat, the channel of the Leniwka at Głowa Gdańska branches into Szkarpawa (the Elbląg Vistula), which empties into the Vistula Lagoon, and the Gdańsk Vistula (currently called the Dead Vistula), which flows into the Baltic Sea (the Gulf of Gdańsk). In 1840, a new mouth channel of the Vistula was formed in Górki Zachodnie after a flood. This new channel was named Wisła Śmiała (the Bald Vistula), and the former channel of the Leniwka was renamed the Martwa Wisła (the Dead Vistula). In 1890–1895, in order to avoid further floods, a cut was excavated near Świbno (the Vistula Cut near Świbno). In 1895, the Vistula waters were redirected through this artificial cut straight into the Baltic Sea, closing the lateral branches (the Nogat, Szkarpawa and Dead Vistula) with locks. As a result, at present, the Vistula has two estuaries: the Bold Vistula estuary (secondary) and the Vistula Cut estuary (main).

The Lower Vistula basin covers an area of 34,136 km^2, accounting for 20% of the entire drainage basin of the Vistula. The basins of the Dead Vistula (along with the Motława), the Szkarpawa and the Nogat are excluded from this area, as they are separated from the main Vistula channel by hydraulic structures and levees. The mean annual discharge of the Lower Vistula (at its mouth profile) is 1,081 m^3/s [23]. The Lower Vistula basin features 11 artificial lakes with a capacity of over 1.0 hm^3. The largest one, the Włocławski Reservoir on the Vistula (capacity: 453.6 million m^3), is also the largest artificial lake in Poland in terms of area (about 70 km^2) [22].

The Oder, a river of intermediate length among the European rivers (21st longest river in Europe), is the second longest (840 km) [18] Polish river. It is also the second largest river in terms of basin area (119,074 km^2) [18]. The average gradient of the river is 0.7‰. The Oder starts its course in the Oder Mountains in the Fidlův mountain range (634 m.a.s.l.) in the Czech Republic. The character of a mountain river is evident only along its initial 50 km. Upon reaching the Moravian Gate, the gradient of the river abruptly lessens. Having exited the Moravian Gate, the river turns north-westwards. It initially flows down the Breslau-Magdeburg marginal paleo-valley (Urstromtal) and then cuts through to the Barycko-Głogowska marginal paleo-valley, followed by the Warsaw-Berlin marginal paleo-valley and the Thorn-Eberswalder marginal paleo-valley (Fig. 5). As it travels down the Breslau-Magdeburg marginal paleo-valley, the Oder receives left-bank tributaries from the Sudeten (the rivers Osobłoga, Eastern Neisse, Oława, Strzegomka) and right-bank tributaries (the Olza and the Mała Panew). Along the Barycko-Głogowska marginal paleo-valley, it receives the Barycz from the right and the Bóbr from the left. The Nysa Łużycka, which flows from the south, joins the Oder along the Warsaw-Berlin marginal paleoromtal. The Oder then turns northwards and generally maintains this direction until it reaches its mouth. Along this section, it only receives right-bank tributaries: the rivers Warta, Płonia and Ina (Fig. 5). At 704.1 km, it splits into the Eastern Oder (the Regalica), which passes through Lake Dąbie, and the Western Oder (the main course of the river), which passes through Szczecin. It empties into the southern part of the Szczecin Lagoon called the Oder Bay (Polish: Roztoka Odrzańska). The mean annual discharge of the Oder (at its mouth profile) is 567 m^3/s [27].

Similar to the Vistula, the river system of the Oder is asymmetrical. The ratio of left-bank basin to right-bank basin is 30:70. The main right-bank tributaries of the Oder are the rivers Warta (along with the Prosna, Wełna, Noteć, Obra) and Barycz, and the main left-bank tributaries are the rivers Bóbr (along with the Kwisa), the Lusatian Neisse (Polish: *Nysa Łużycka*) and the Eastern Neisse (Table 1 and Fig. 5). Similarly to the Vistula, the course of the Oder is divided into three sections: upper (from its source to Koźle, just before receiving the Gliwice Canal), middle (from Koźle to the mouth of the Warta) and lower (from the mouth of the Warta to the entry into the Oder Bay).

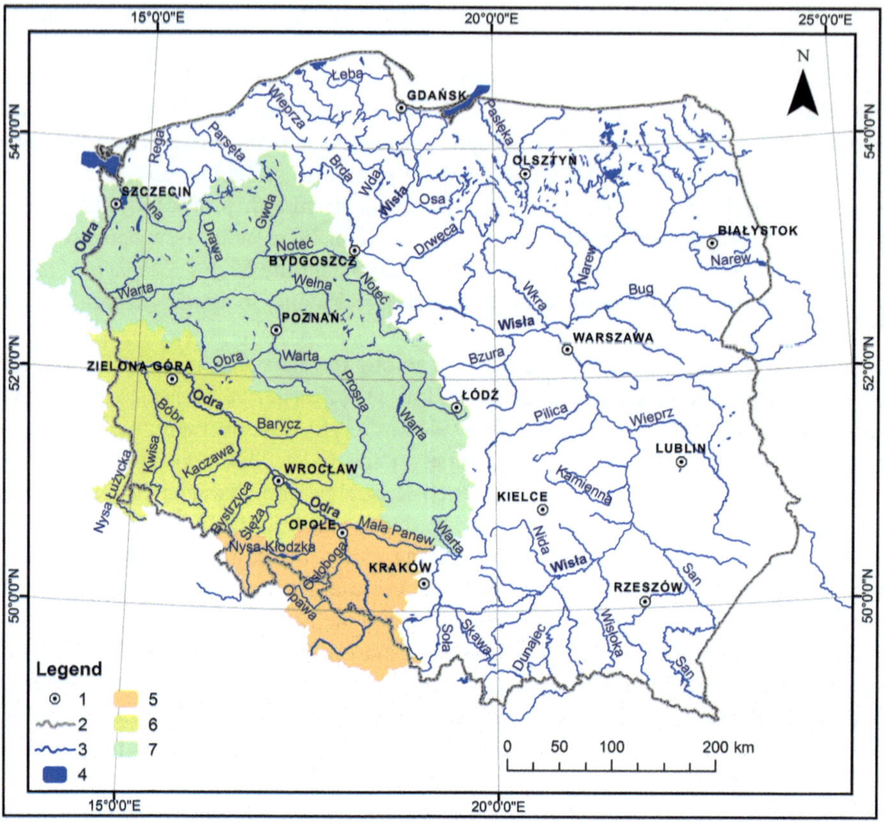

Fig. 5 The Oder river basin with division into the Upper Oder, the Middle Oder and the Lower Oder according to the Map of Hydrographic Division of Poland

The Oder is regulated almost along its entire course. From Koźle, it is channelled, and over a length of 711 km, it is an important waterway connecting Upper Silesia with the Baltic Sea. By the early nineteenth century, the river had been shortened by a total of 160 km as a result of channel straightening (through separation of the bends), which led to the rising of the channel in its lower course by about 4 m. In subsequent years, 21 spillway steps were constructed in the middle course of the Oder, which halted the channel bottom erosion and reduced the rate of floodplain rising. In the lower course of the Oder, the shallow gradient of the river along with the sparse quantity of the transported load results in its poor deposition on the floodplain and the basins between the groynes. The Oder drainage basin (mainly the basins of the Sudeten tributaries) contains 12 artificial lakes with a capacity of over 10 hm^3. The following artificial lakes have the highest total capacities: Jeziorsko on the Warta (203 hm^3 [8]; in 2014 the fill level was decreased, which changed its capacity to 114 hm^3 [28]) and Otmuchów (124 hm^3) and Nyski (114 hm^3) on the Eastern Neisse (Polish: Nysa Kłodzka) [8].

The largest rivers of the region called Przymorze are Rega, Parsęta, Wieprza, Słupia, Łupawa, Łeba and Reda (Fig. 2 and Table 1). Their upper courses, which drain the northern slope of the lakeland hump, are often river-lake systems. They often rise in boggy areas, postglacial ponds or lakes. Having exited the lakeland hump (Fig. 1), they enter the coastal lowlands and flow in the meridional direction towards the Baltic Sea. Their mouth section is often regulated, and its flow into the sea is directed by breakwaters, or it is shielded by breakwaters. Only the rivers Łupawa and Łeba pass through large coastal lakes (Lake Gardno and Lake Łebsko, respectively) before they enter the sea. The Reda, which flows into the Bay of Puck, is the only river in the mouth section of the Oder that forms a delta covering an area of 0.268 km^2 [29].

The largest tributaries of the Vistula Lagoon are the distributing channels of the Vistula within its delta region, namely, the Szkarpawa and the Nogat (permanently separated from the main Vistula channel by locks), and the rivers that drain the northern slope of the lakeland hump, namely, the rivers Elbląg, Bałda and Pasłęka, and, via the Pergoła, its tributaries Łyna and Węgorapa (Fig. 2 and Table 1).

The mouth sections of all the rivers of Przymorze and the Vistula Lagoon, and of the rivers Vistula and Oder, are under the influence of the Baltic Sea, whose range depends on the sea level.

The largest tributaries of the Niemen in Poland are the Czarna Hańcza and the Szeszupa (Fig. 2 and Table 1). The Czarna Hańcza (142 km long), the largest river in the Suwałki Region (Polish: Suwalszczyzna), flows as a small brook into Lake Hańcza, the deepest (108.5 m) lake in the Polish Lowland, and flows out as a rapid stream with a stony bed that flows down a narrow and deep (30 m) and very steep (18.5‰) valley with 13 elongated hills of the Turtulsko-Bachanowski Esker. Below this esker, the narrow valley of the Czarna Hańcza widens to over 1 km, and the river starts flowing down a meandering channel, maintaining the southeastern direction of course. It passes Lake Wigry, below which the middle course of the river commences, with an average gradient of 0.3‰. The river generates a mean annual outflow from this lake at 3.56 m^3/s. It slowly flows in the southeastern direction through pools and boggy meadows to subsequently accelerate and form meanders. Past 43 km, it joins the system of the Augustów Canal and 5 km further on enters Belarus, where it joins the Niemen [30].

The Szeszupa is 297.6 km long and 24 km of its upper course lies in Poland [30]. This river creates a river-lake system that drains, in the northeastern direction from the Niemen, an extensive kettle hole east of Lake Hańcza. It rises as a brook from marshland. It flows down a wide valley with a moderate midstream interrupted by rapids and receives small streams which drain water from the numerous lakes. In Poland, it passes through six shallow lakes: lakes Gulbin, Okrągłe, Krejwelek, Przechodnie, Postawelek and Pobondzie. The mean annual discharge of this river below Lake Pobondzie (near the border with Lithuania) is 1.61 m^3/s.

4 Hydrological Regimes of Polish Rivers

In Polish climate conditions, river discharge is the sum of underground and surface runoffs. The underground contributor to the total discharge of rivers (the underground runoff) originates from shallow or deeper rock strata containing free water which are drained by the river channels. As this drainage is generally permanent, the river network in Poland is stable. Only smaller streams may have limited hydraulic contact with underground waters, and given the lack of surface runoff, their channels are dry. The other contributor to river discharge, the surface runoff, is only generated during precipitation in those parts of the basin where the soil has low permeability, the slopes of the land are steep and the land is used for agricultural purposes. Surface runoff can also be produced in the direct vicinity of rivers. Surface runoff is also produced during snowmelt. As it flows into river channels, it generally causes a rapid increase in water stage and flow rate. Surface runoff is short-lived, rapid and irregular [8].

The underground contributor to river discharge plays an important role along the entire lengths of Polish rivers, all year long. Thanks to this component, rivers carry water all year long, even during long periods of no precipitation. River channels usually drain underground waters (draining channels). In some sections of their course, rivers can feed underground waters (infiltrating channels) or remain in hydraulic balance with underground waters (transit channels). The contribution of the underground component of the total discharge of Polish rivers is generally stable over the year and averages 55% [31–34]. Underground runoff accounts for 20–30% of the total discharge of Carpathian rivers, 40–50% of the total discharge of Sudeten rivers, 50–70% of the total discharge of highland rivers, 35–50% of the total discharge of lowland rivers and 60–80% of the total discharge of lakeland rivers, Przymorze rivers and marshland rivers (Table 1) [35].

Flow rates in Polish rivers vary greatly and, in general, gradually increase as the basin grows, i.e. from source to mouth. This flow variability is determined by the amount and distribution of precipitation that feeds the river basin, air temperature (which determines evapotranspiration, duration of snow cover and the rate of snow cover disappearance) and the sources of water inflow into the river (precipitation, underground water, snowmelt).

Discharge of Polish rivers varies over 1-year periods and multiple-year periods. Some years or series of years may be characterised by high flow rates while others by low flow rates. Seasonal variation of flow rates is also observed. About 55% of the total volume of water drained by the Polish rivers is drained by the Vistula river system, 25% by the Oder river system, 9.5% by the Przymorze rivers and 5.9% by the Vistula Lagoon river systems [36]. The mean annual discharge data on Polish rivers vary greatly with the river size and river basin storage (Table 1).

The discharge of the Polish rivers shows a seasonal variation, which is directly associated with their sources of water inflow (precipitation, snowmelt). With the exception of the mountains, the highest discharges are observed in the spring (after snowmelt). In the mountains, on the other hand, river discharges peak after heavy precipitation, which usually occurs early in the summer. Discharges of lakeland rivers show the smallest seasonal variation. In Poland, discharge in the cold half-year (November–April) is generally higher than that in the warm half-year (May–October): 57% vs 43% [36].

The temporal variability of the stage and discharge of a river over the course of a year that results from the sources of inflow is reflected by the river's hydrological regime. Hydrological regimes of Polish rivers have been determined based on the origin of the water (underground water, rainfall, snowfall) and contribution of each source of water (as a percentage of the annual total), discharge variability characteristics (based on the coefficient of variability of mean monthly flows), timing of the maximum(s), variability and irregularity of flows and stability of the outflow regime phases, i.e. regularity of high and low water periods during the water year [35]. Five basic types of river regime are distinguished in Poland [32–35]:

- *Type 1: Nival regime, poorly developed*, if the mean flow in a spring month does not exceed 130% of the mean annual flow. This regime characterises the rives of Przymorze (east of the Parsęta, i.e. the rivers Wieprza, Słupia, Łupawa and Łeba), the rivers of the Pomeranian Lakeland (Brda, Wda, upper Wierzyca and upper Radunia) and some of the rivers of the Masurian Lakeland (Pisa, upper Pasłęka and upper Łyna), the Lubusz Land (Ilanka and Pliszka) and the Silesian-Cracovian Upland (upper Warta, Kłodnica, Przemsza and Biała Nida) (Fig. 3). Rivers with this regime are characterised by the least variable flows and the greatest percentage contribution of underground water to the total discharge in Poland, which in many rivers exceeds 80% (Fig. 7 and Table 2). The rivers of Przymorze and lakelands show the lowest variation in monthly flows in Poland, as measured by the coefficient of variability (CV), and the smallest irregularity of flow among Polish rivers, as measured by the ratio of Q_{max} to Q_{min} (below 10) (Table 2). These rivers are characterised a high total discharge, which exceeds 200 mm and, in the case of the rivers of Przymorze, 300 mm. High water periods in the rivers of Przymorze (December–March, February–April) and some of the rivers of the Masurian Lakeland are relatively stable in terms of timing and occur during the winter and spring period. Low water periods, which are generally not very pronounced, generally occur in the summer (June–August) or during the summer and autumn period (Table 2 and Fig. 7).
- *Type 2: Nival regime, moderately developed*, if the mean flow in a spring month is 130–180% of the mean annual flow. The largest number of Polish rivers is characterised by this regime (Fig. 6), including rivers that flow in the north of Poland, e.g. in coastal lowlands (Ina, Płonia, Reda, Motława), the Pomeranian Lakeland (Rega and Parsęta, Drawa, Gwda with Piława, lower courses of the Radunia and the Wierzyca), most rivers of the Masurian Lakeland (except for the Rospuda, Pisa, upper Łyna and Pasłęka and Gubr) and Żuławy Wiślane (Szkarpawa, Nogat, Elbląg). In central Poland, this regime characterises transit rivers:

Table 2 Regime characteristics of selected Polish rivers

River	Gauge	Basin (km^2)	Total runoff (mm)	Groundwater inflow ratio[a] H_{ground} (%)	Discharge variability characteristics Cv[b]	Wn[c]	High water periods	Low water periods	Regime type
The Carpathians									
Soła	Oświęcim	1,357	447	22.6	1.963	908	Mar–Jul	Sep–Nov	5
Raba	Proszówki	1,473	365	31.8	1.953	2,652	Mar–Apr	Sep–Nov	4
Dunajec	Nowy Targ – Kowaniec	687	675	42.8	1.171	239	Apr–Jul	Jan–Feb	5
Wisłoka	Żółków	582	419	21.4	1.978	3,300	Mar–Apr	Aug–Oct	4
San	Lesko	1,617	581	33.5	1.063	323	Apr–Jul	Sep–Dec	4
The Sudeten									
Nysa Kłodzka	Nysa	3,276	281	39.4	1.089	475	Apr–Jul	Oct–Dec	5
Bystrzyca	Krasków	683	211	30.0	1.714	668	Mar–Jun	Sep–Nov	4
Kaczawa	Świerzawa	134	280	46.2	1.777	927	Mar–May	Aug–Oct	4
Kwisa	Nowogrodziec	736	311	48.2	1.217	382	Mar–May	Sep–Dec	4
Highlands (carbonates)									
Warta	Kręciwilk	66	384	68.2	0.610	44	Feb–Apr	Jul–Aug	1
Wieprz	Krasnystaw	3,010	133	66.9	0.553	36	Mar–Apr	Jul–Sep	2
Bystrzyca	Sobianowice	1,262	128	65.9	0.624	58	Feb–Apr	Jul–Oct	2
Kielecko-Sandomierska highland									
Czarna Nida	Tokarnia	1,211	173	50.3	1.193	160	Feb–Apr	Aug–Oct	
Koprzywianka	Koprzywnica	502	118	44.8	1.815	1,072	Feb–Apr	Jul–Sep	
Lowlands									
Mała Panew	Staniszcze Wielkie	1,107	203	51.8	1.053	144	Feb–Apr	Aug–Nov	
Mogilnica	Konojad	663	79	33.1	1.440	1,325	Feb–Apr	Jul–Oct	

Liwiec	Łochów	2,471	137	47.1	1.148	240	Feb–Apr	Jul–Oct	
Wkra	Trzciniec	1,955	168	53.0	0.746	94	Feb–Apr	Jul–Sep	
Lakelands									
Drawa	Drawsko Pomorskie	592	226	71.1	0.554	22	Jan–Apr	Jul–Oct	
Rega	Łobez	616	239	72.5	0.462	12	Jan–Mar	Jul–Sep	
Słupia	Słupsk	1,452	348	72.1	0.293	8	Jan–Mar	Jun–Aug	
Pisa	Ptaki	3,562	186	81.8	0.383	13	Feb–Apr	Jul–Oct	
Łyna	Sępopol	3,640	216	62.9	0.630	28	Feb–Apr	Jun–Aug	
Guber	Prosna	1,565	170	36.5	1.149	223	Feb–Apr	Aug–Nov	

[a]Groundwater inflow ratio is a contribution of groundwater inflow in total runoff
[b]Cv coefficient of variability of daily flows
[c]River regime coefficient (Q_{max}/Q_{min})

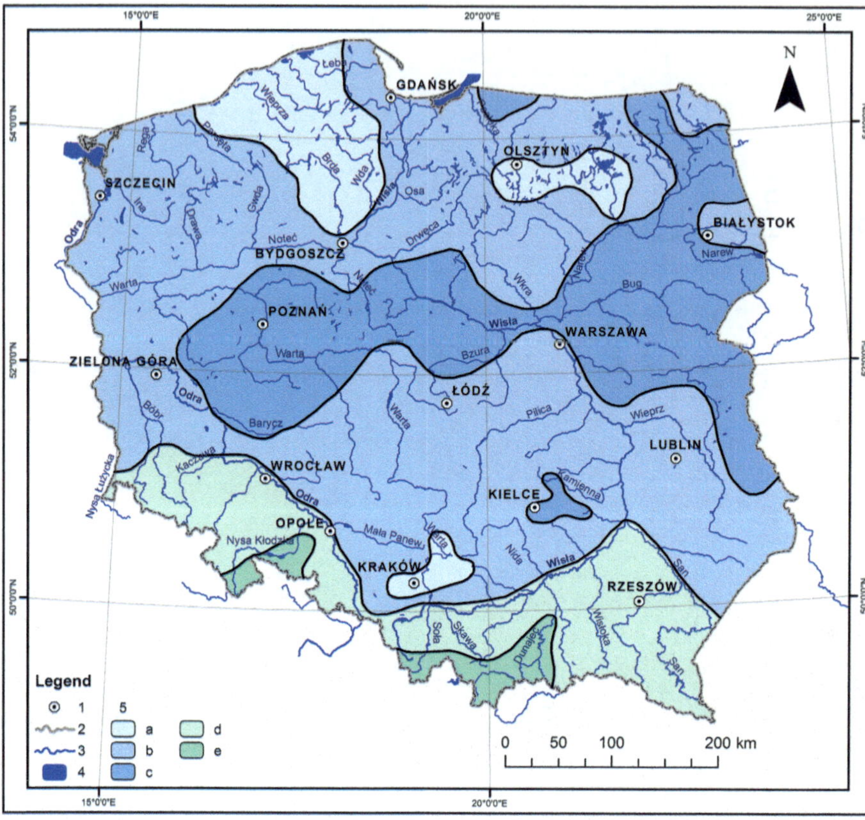

Fig. 6 Types of river regimes in Poland. Explanations: (1) cities, (2) Poland's border, (3) rivers, (4) lakes, (5) regimes. a, type 1 (poorly developed nival regime); b, type 2 (moderately developed nival regime); c, type 3 (well-developed nival regime); d, type 4 (nivo-pluvial regime); e, type 5 (pluvio-nival regime) (adapted from [35])

- The Middle Vistula and its tributaries (except for the lakeland tributaries of the Narew, except for the Pisa, Rospuda and Biebrza); the Lower Vistula from Toruń and its two tributaries, the Drwęca and the Osa; and the left-bank tributaries of the Upper Vistula and the right-bank tributaries of the Sam which drain the highlands.
- The Oder from the mouth of Kaczawa to the Szczecin Lagoon and its tributaries (except for the tributaries from the Sudeten and the upper courses of the rivers Bóbr, Kwisa and Nysa Łużycka and except for the Barycza river system).
- The Warta with the Noteć (except for the segment between the mouths of the Ner and Wełna and the upper Noteć). In the belt of highlands, this regime characterises rivers located between the Warta and the Middle Vistula and the Wieprz river system, which drains the Lublin Upland.

- The annual discharge of rivers with type 2 regime exceeds 200 mm (or 300 mm in the case of the Parsęta and the Szeszupa). The contribution of underground runoff to the total discharge of rivers characterised by this regime in the north of Poland and in the Wieprz river basin is high (60–80%). In the remaining rivers with this regime, the contribution of underground runoff is lower (40–60%). High water periods occur during the winter and spring period (January–April) or during the spring (March–April). In most of these rivers, low water periods, which are generally not very pronounced, occur during the summer and autumn period (July–September) or during the summer (June–August). Rivers having this regime are characterised by a greater variability of flow than rivers with type 1 regime. High and low water periods, however, occur in the same months of the water year (Fig. 7 and Table 2).

- *Type 3: Nival regime, well developed*, if the mean flow in a spring month exceeds 180% of the mean annual flow. This regime characterises the river systems of the Jarka, Rospuda and Biebrza (in Northeastern Poland); the Narew and its left-bank tributaries and the Bug river system (in the eastern part of the Polish Lowland); and the Bzura river system, the lower Prosna, upper Noteć and the Barycza river system (in the central part of the Polish Lowland) (Fig. 6). This regime also characterises rivers that drain the highland belt: the Kamienna with the Swiślina, the Koprzywianka and the Czarna Nida (Fig. 6). These rivers show the greatest flow variability over the course of a year. Their total discharge does not generally exceed 150 mm, although in some cases may exceed 200 mm (rivers in the northeast of Poland), while in others it can be as low as 60 mm (the Tążyna, a left-bank tributary of the Vistula in the Kuyavian region). The contribution of underground runoff to the total discharge of rivers with this regime is relatively low (below 40%). Periods of spring inflow of melted winter snow are brief and distinct. In Northeastern Poland (the rivers Narew, Bug, Biebrza, Rospuda, Szczeberka, Marycha), the high water period occurs from March to April. In the remaining rivers characterised by this regime, high water periods occur earlier, from January to March or from February to April. Low water periods occur during the summer and autumn period (July–September) and are generally profound (Fig. 7 and Table 2).
- *Type 4: Nivo-pluvial regime*, if the mean flow in a spring month is 130–180% of the mean annual flow in the summer months and there is a marked increase in flow in the summer months of at least 100% of the mean annual flow. This regime characterises the rivers of the Sudeten, the Oder from the mouth of the Kaczawa, most Carpathian tributaries of the Vistula (except for the upper courses of the rivers in the High Beskids and the Silesian Beskids and the upper and lower Dunajec) and the Upper Vistula (Fig. 6). The annual discharge of rivers with this regime ranges from 100 mm (the following tributaries of the Oder: Psina, Biała, Stradunia, Oława, Ślęza) to over 800 mm (the Upper Vistula, Żylica, Solinka). The contribution of underground runoff to the total discharge of rivers characterised by this regime is small (20–40%). The daily variability of flows is considerable, with the CV ranging from <1.0 in the lower course to >2.0 in the

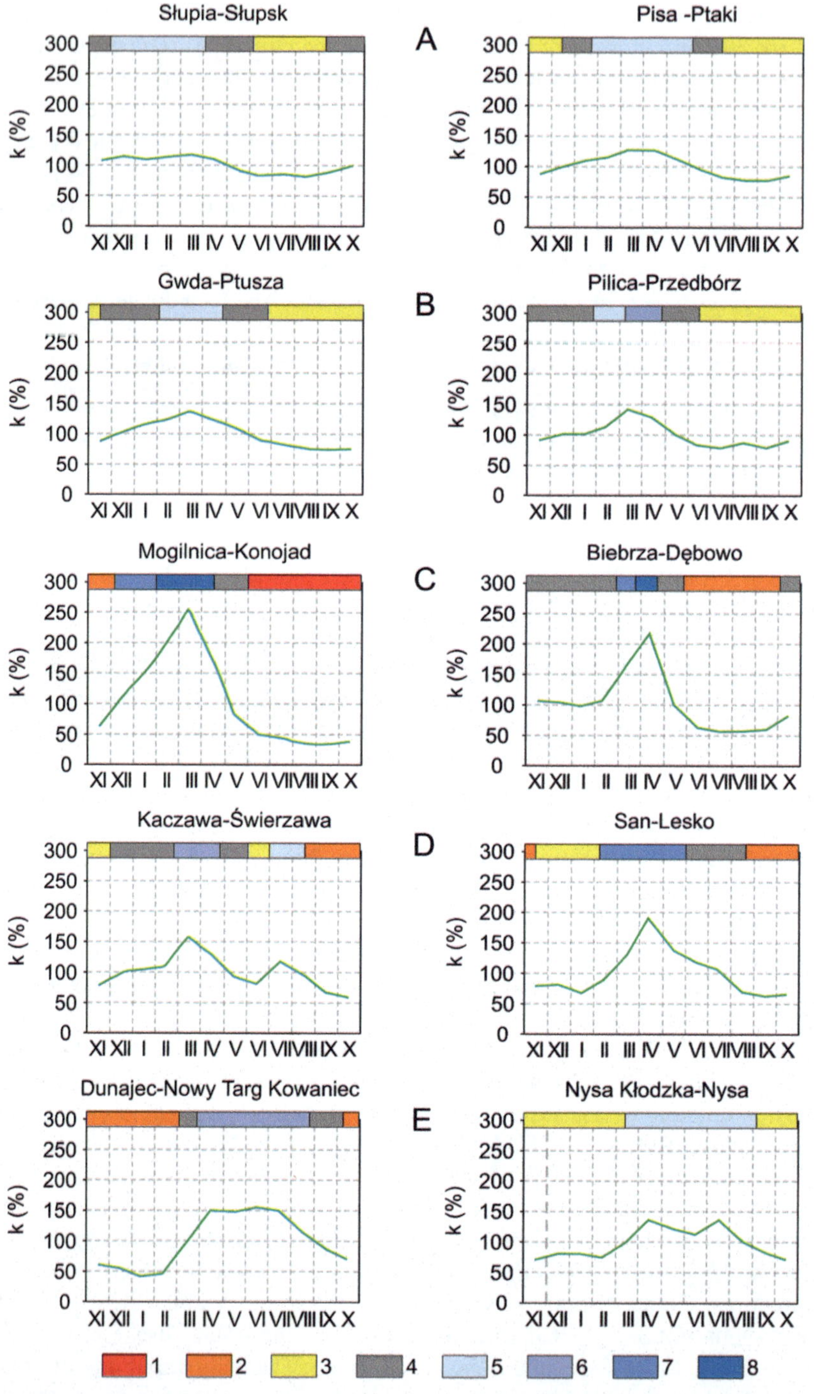

Fig. 7 Monthly flow ratios in selected rivers with a given type of regime. Explanations: k, monthly flow rate, MQ_m/MQ, regime type: (**a**) poorly developed nival regime, (**b**) moderately developed

upper course. High water periods occur in the spring (March–April) or during the spring and summer period (April–July). Low water periods most commonly occur in the autumn (September–November) or during the autumn and winter period (October–January) (Fig. 7 and Table 2).

- *Type 5: Pluvio-nival regime*, if the mean flow in a summer month exceeds or equals the mean flow in a spring month and in both cases it generally accounts for 130–180% of the mean annual flow. This regime characterises the Easter Neisse in the Sudeten and the Upper Vistula, Soła, Uszwica and the Dunajec river system (beyond its lower course) in the Carpathians (Fig. 6). These rivers have the highest discharge in Poland, averaging about 500 mm but much higher in mountain streams (over 1,000 mm, e.g. in the Czarny Dunajec and the Wołosaty, and over 1,500 mm in the Potok Kościeliski and the Białka). Surface runoff predominates in the total discharge of these rivers, and the contribution of underground runoff is generally lower than 40%. Rivers with this regime are characterised by the greatest variability of daily flow (CV > 1.5). High water periods occur in the spring (March–May) or during the spring and summer period (April–July). Low water periods, which are generally not very pronounced, most commonly occur in the autumn (September–November), during the autumn and winter period (October–January) and, in the case of Tatra streams, in the winter (December–February, January–February) (Fig. 7).

It follows from the above that most Polish rivers have simple regimes, only mountain rivers are characterised by complex regimes and the largest allochthonous (transit) rivers (the Vistula, Oder, Warta) have different regimes in different sections of their courses (Fig. 6). The Upper Vistula, in the section that receives the Carpathian tributaries, has maximum flows in the summer. Below the mouth of the Nida the maximum flows predominate in the spring, although they also still occur in July. Below the mouth of the San, the springtime flow cumulations are even more distinct and below the gap through the upland area, they become predominant. In terms of the major source of runoff, the Vistula along most of its length (the middle course and most of the lower course) is characterised by the simple, moderately developed nival regime. Only the segment between the mouth of the Pilica and the Włocławek Reservoir is typified by the simple well-developed nival regime. In its upper course, the Vistula has the complex nivo-pluvial regime with the exception of the source section, where the Carpathian Vistula also has a complex regime, but in this case it is the pluvio-nival regime. The Oder, from the mouth of the Kaczawa, i.e. along the section that receives the Sudeten tributaries, is characterised by the complex nivo-pluvial regime and, in its further course, the simple moderately developed nival regime. The Warta is characterised by the moderately developed nival regime almost along its entire length. Only the source section has the poorly developed nival regime, while the section between the mouth of the Ner and that of the Wełna is characterised by the well-developed nival regime.

Fig. 7 (continued) nival regime, (**c**) well-developed nival regime, (**d**) nivo-pluvial regime, (**e**) pluvio-nival regime (adapted from [35])

5 Transport of Sediments by the Rivers

Rivers transport with their waters material from denudation of the their basins, which reaches the river with surface runoff and also originating from river channel erosion. The whole of the matter transported in steams is referred to by hydrologists as the river bed load and can be subdivided into material moved by rolling (large rocks), material moved by sliding (cobble, pebble, sand), material moved by saltation, suspensions (equivalent to seston), and solutions. In a stream, the bed load moves from the source to the mouth, depending on the quantity and velocity of the flowing water. The largest amounts of load is transported by the Vistula, which is sometimes called the queen of sand. The mean annual transport of load carried on the bed by the Vistula at its mouth is 1.2 million tonnes, over 2 million tonnes in a wet year, 0.6 million tonnes in a dry year, while the annual transport of suspended load is about 1.8 million tonnes [37]. The load deposited by the Vistula undergoes partial dispersion by the currents along the shores of the Bay of Gdańsk and by waves, to form shoals and islets at its new mouth. In order to maintain the patency of this river-mouth, the directing breakwaters are gradually extended. They are currently about 2,370 m long [38]. Failure to construct these breakwaters would allow the Vistula to form its new delta in an uncontrolled manner, as it had done until the end of the nineteenth century. The Vistula provides the Baltic Sea with the largest input of biogenic elements. In 2014, the total nitrogen load was 64.87 thousand tonnes and the total phosphorus load amounted to 7.89 thousand tonnes [37].

The Oder, given the presence of spillway steps and regulated channel, has an almost twice lower transport potential than the Vistula. The greater supplier of bed load to the Oder is the Eastern Neisse. The annual transport of the Oder river load carried on the bed at its mouth is estimated at 0.35 million tonnes and that of the suspended load at 354 thousand tonnes. The annual supply of total nitrogen from the Oder to the Baltic Sea is 39.97 thousand tonnes and that of total phosphorus 2.29 thousand tonnes [37].

6 Conclusions

Rivers are the major hydrographic objects in Poland. The present arrangement of Poland's river network has been shaped as a result of complex geological and landscaping processes. The river network in the south of Poland assumed a shape similar to the present one in the Pliocene Epoch. The river network of the Polish Lowland formed in a step-wise manner, as the snout of the ice sheet moved and deglaciation ensued. Hence the course of the larger river valleys in the Polish Lowland consists of latitudinally arranged marginal paleo-valley sections and gorge segments arranged in the meridional direction. However, the arrangement of the contemporary river network has to the largest degree been shaped by climate elements which determine the changes in the hydrological regime of rivers.

Many Polish rivers, including the largest ones, flow down inherited valleys. The largest Polish river, the Vistula, with the exception of the Carpathian section and the gorges, failed to form its own valley but instead used the pre-existing concave landforms of different genesis and age, cutting through them. The Vistula flows in its own river valley only in its Carpathian section. The Vistula also contains gorges and used to include Żuławy.

While the river network in southern and central Poland is stabilised, it is still in the initial stage in northern Poland, on early postglacial terrains.

Due to climate conditions, most Polish rivers are characterised by simple nival regimes with high snowmelt flows in the springtime (March–April). Complex regimes are only found in mountain rivers. The nivo-pluvial regime, with the main maximum in the spring and a secondary one in the summer, prevails. Only some of the Carpathian rivers have the pluvio-nival regime with the main maximum in the summer and a secondary one in the spring. The Vistula has a large potential for bed load transport, especially the load carried on the bed and the suspended load, while the Oder's potential is nearly twice as low.

References

1. Starkel L (2008) Paleogeograficzne korzenie współczesnej Polski. In: Starkel L, Kostrzewski A, Kotarba A, Krzemień K (eds) Współczesne przemiany rzeźby Polski. IGiGP UJ, Kraków, pp 13–20
2. Gilewska S (1991) Współczesne środowisko przyrodnicze. Rzeźba. In: Starkel L (ed) Geografia Polski. Środowisko przyrodnicze. Wydawnictwo Naukowe PWN, Warsaw, pp 248–296
3. Michno A (2017) Rzeźba. In: Jokiel P, Marszelewski W, Pociask-Karteczka J (eds) Hydrologia Polski. Wydawnictwo Naukowe PWN, Warsaw, pp 31–36
4. Klimaszewski M, Starkel L (1972) Karpaty Polskie. In: Klimaszewski M (ed) Geomorfologia Polski. I. Polska Południowa. Góry i wyżyny. PWN, Warsaw, pp 167–231
5. Klimek K, Starkel L (1972) Kotliny Podkarpackie. In: Klimaszewski M (ed) Geomorfologia Polski. I. Polska Południowa. Góry i wyżyny. PWN, Warsaw, pp 117–138
6. Andrzejewski L, Krzemień K (2017) Doliny i koryta rzek. In: Jokiel J, Marszelewski W, Pociask-Karteczka J (eds) Hydrologia Polski. Wydawnictwo Naukowe PWN, Warsaw, pp 121–127
7. Gutry-Korycka M (2005) Rzeki. In: Richling A, Ostaszewska K (eds) Geografia fizyczna Polski. Wydawnictwo Naukowe PWN, Warsaw, pp 151–156
8. Bajkiewicz-Grabowska E, Mikulski Z (2006) Hydrologia ogólna. Wydawnictwo Naukowe PWN, Warsaw, p 339
9. Hillbricht-Ilkowska A, Wiśniewski RJ (eds) (1994) Jeziora Suwalskiego Parku Krajobrazowego. Związki z krajobrazem, stan eutroficzny i kierunki ochrony. Zesz. Nauk. „Człowiek i Środowisko" PAN, 7, Zakład Narodowy Ossolińskich, s. 281
10. Hillbricht-Ilkowska A, Węgleńska T (1995) System rzeczno-jeziorny jako układ płatów krajobrazowych i stref granicznych. In: Dąbrowska-Prot E, Łuczak J (eds) Problemy ekologii krajobrazu pojeziernego Polski Północno-Wschodniej. Zesz. Nauk. „Człowiek i Środowisko" PAN, 12. Oficyna Wydawnicza, Instytut Ekologii PAN, pp 185–202
11. Bajkiewicz-Grabowska E (2002) Circulation of matter in the river-lake systems. Warsaw University, Faculty of Geography and Regional Studies, Warsaw, p 274

12. Bajkiewicz-Grabowska E (2007) Jeziorne ogniwa sieci hydrograficznej obszarów młodoglacjalnych i ich rola w kształtowaniu odpływu rzeki pojeziernej. Obieg wody w środowisku naturalnym i przekształconym. Wyd. UMCS, Lublin, pp 39–46
13. Bajkiewicz-Grabowska E (2017) Sieć rzeczna pojezierzy. In: Czerniawski R, Bilski P (eds) Funkcjonowanie i ochrona wód płynących. Szczecin, pp 297–306
14. Bajkiewicz-Grabowska E, Golus W (2009) Organizacja sieci hydrograficznej w zlewni pojeziernej przy różnym stanie jej retencji. In: Bogdanowicz R, Fac-Beneda J (eds) Zasoby i ochrona wód. Obieg wody i materii w zlewniach rzecznych. Fundacja Rozwoju Uniwersytetu Gdańskiego. Gdańsk, pp 159–166
15. Kwidzińska M (2010) Rola oczek w uaktywnianiu odpływu powierzchniowego. Praca magisterska wykonana w Katedrze Limnologii Uniwersytetu Gdańskiego (typescript)
16. Fac-Beneda J (2017) Struktura sieci hydrograficznej. In: Jokiel P, Marszelewski W, Pociask-Karteczka J (eds) Hydrologia Polski. Wydawnictwo Naukowe PWN, Warsaw, pp 116–127
17. Błaszkowski J (1992) mapa gęstości sieci wodnej na Żuławach Elbląskich. Zeszyty Naukowe UG, Geografia 18, Gdańsk
18. Rocznik statystyczny Rzczypospolitej Polskiej 2015, Rok LXXV. GUS, Warsaw
19. HELCOM (2004) 30 years of protecting the Baltic Sea-HELCOM1974-2003. http://www.helcom.fi
20. Starkel L (1996) Rivers in space and time. In: Kułtuniak J (ed) Rzeki, Kultura-Cywilizacja-Historia, vol 5, pp 43–60
21. Starkel L (2001) Evaluation of the Vistula river valley since the last glaciations till present (Historia doliny Wisły od ostatniego zlodowacenia do dziś). IGiPZ PAN, seria Monografie 2, p 263
22. Majewski W (2010) Perspektywiczne zagospodarowanie dorzecza Wisły wraz z systemem ocen wpływu inwestycji hydrotechnicznych na środowisko. Sprawozdanie z realizacji zadania 9 projektu „Klimat. Wpływ zmian klimatu na środowisko, gospodarkę i społeczeństwo" współfinansowanego przez UE z Europejskiego Funduszu Rozwoju Regionalnego. Warsaw, p 31
23. Majewski W (2016) Monografia dolnej Wisły. IMGW-PIB, Warsaw, p 173
24. Kondracki J (2009) Geografia regionalna Polski. Wydawnictwo Naukowe PWN, p 450
25. Biernacki Z (2000) Geomorfologia i wody powierzchniowe. [w}: Wisła w Warszawie. Biuro Zarządu m.st. Warszawy, Wydział Planowania Przestrzennego i Architektury, Opracowanie graficzne, druk i oprawa: Dom Wydawniczy ELIPSA 2000:22–70
26. Cebulak K (2010) Delta Wisły powyżej i poniżej morza. Stowarzyszenie Żuławy i Lokalna Grupa Działania Żuławy i Mierzeja. Nowy Dwór Gdański, p 59
27. Pociask-Karteczka J (2017) Położenie hydrograficzne Polski na tle Europy. In: Jokiel P, Marszelewski W, Pociask-Karteczka J (eds) Hydrologia Polski. Wydawnictwo Naukowe PWN, Warsaw, pp 15–21
28. Walczak Z, Sroka Z, Tschuschke W, Stefanek P, Wrzosek K (2017) Straty filtracyjne zrzutów ze zbiornika Jeziorsko zasilających przepływy niżówkowe Warty. Acta Sci Pol Form Circum 16(4):217–232
29. Rudyk S (2013) Zmiany sieci wodnej delty Redy i jej linii brzegowej w świetle dokumentacji kartograficznej oraz zobrazowań lotniczych i satelitarnych. PhD thesis typescript, s. 266
30. Mapa Podziału Hydrograficznego Polski w skali 1 : 50 000 (2005) IMGW-PIB, Warsaw. http://www.kzgw.gov.pl. Accessed 8 July 2018
31. Dynowska I (1971) Typy reżimów rzecznych w Polsce. Zeszyty Naukowe UJ, CCLXVIII, Prace Geograficzne 28, Kraków, pp 67–81
32. Dynowska I (1991) Obieg wody. In: Starkel L (ed) Geografia Polski. Środowisko przyrodnicze. Wydawnictwo Naukowe PWN, Warsaw, pp 355–387
33. Dynowska I (1994) Reżim odpływu rzecznego, plansza 32.3 Odpływ rzeczny. In: Atlas Rzeczypospolitej Polskiej. IGiPZ PAN, Główny Geodeta Kraju, PPWK im. E. Romera S.A., Warsaw

34. Somorowska U (2005) Ustroje hydrologiczne. In: Richling A, Ostaszewska K (eds) Geografia fizyczna Polski. Wydawnictwo Naukowe PWN, Warsaw, pp 165–168
35. Wrzesiński D (2017) Reżimy rzek Polski. In: Jokiel P, Marszelewski W, Pociask-Karteczka J (eds) Hydrologia Polski. Wydawnictwo Naukowe PWN, Warsaw, pp 215–221
36. Michalczyk Z (2017) Odpływ rzeczny w Polsce. In: Jokiel P, Marszelewski W, Pociask-Karteczka J (eds) Hydrologia Polski. Wydawnictwo Naukowe PWN, Warsaw, pp 153–160
37. Ciupa T, Suligowski R, Łajczak A, Babiński Z (2017) Transport fluwialny w rzekach Polski. In: Jokiel P, Marszelewski W, Pociask-Karteczka J (eds) Hydrologia Polski. Wydawnictwo Naukowe PWN, Warsaw, pp 201–213
38. Szymkiewicz R (2018) Dolna Wisła rzeka niewykorzystanych możliwości. Wyd. Politechniki Gdańskiej, Gdańsk, p 132

Geoecosystems of Polish Lakes

Elżbieta Bajkiewicz-Grabowska

Contents

Abstract This chapter contains an analysis of the geoecosystem types of 388 lakes of northern Poland determined by the method elaborated by E. Bajkiewicz-Grabowska, on the basis of data from the literature. The concept of lake geoecosystem is understood as a natural landscape system, composed of a lake with its catchment. The type of geoecosystem determines the rate of natural eutrophication of the reservoir's water, and the rate is usually strongly modified by human activity. The method used to determine the geoecosystem type of a lake involves two elements: (1) the catchment as a supplier of matter to the lake and (2) the lake as a recipient of matter. Therefore, it considers the relationship between the lake and its alimentation area. It can therefore be used in different types of natural landscapes, and its core is the determination of (a) the vulnerability of the catchment to supplying matter into the lake, based on the properties of the physico-geographical environment of the catchment, and (b) lake resistance to the catchment impact determined on the basis of the limnological features of the lake. The combination of these two characteristics allows to distinguish four lake geoecosystem types with a specific natural eutrophication rate. It was shown that the majority (39%) of the analysed lakes have a geoecosystem of type 4 (low lake resistance to external pressure and high catchment vulnerability to supplying matter into the lake). The smallest group (18%) is composed of lakes representing geoecosystem of type 1, i.e. those characterized by low activity of the catchment in

E. Bajkiewicz-Grabowska (✉)
Department of Limnology, Institute of Geography, University of Gdańsk, Gdańsk, Poland
e-mail: bajka37@wp.pl

E. Korzeniewska, M. Harnisz (eds.), *Polish River Basins and Lakes – Part I*,
The Handbook of Environmental Chemistry 86,
https://doi.org/10.1007/978-3-030-12123-5_3

the supply of areal loads, as well as high lake resistance to external impact. The remaining lakes are characterized by a moderate eutrophication rate (geoecosystem of types 2 and 3).

Keywords Eutrophication · Geoecosystem types of Polish lakes · Lake geoecosystem · "Natural" indices of environmental pressure and lake resistance

1 Introduction

In lakelands, lakes are the main component of the hydrographic network. They shape the circulation of energy and matter in this young glacial landscape. One of the most important features of these water reservoirs is a high capacity of energy and mass accumulation. It manifests itself by unbalanced exchange balances with the environment, mainly streams of radiation energy and solid matter. Annual energy surpluses of lakes are indicated by high heat resources stored in their waters, with mean temperature values higher by 2–4°C in relation to the temperature of the surroundings. For matter migrating from the catchment, a lake is a trap that retains dragged material and most of floating material in its basin. In addition, lakes have the ability to generate large amounts of organic matter.

The surplus nature of the exchange of energy and matter with the environment causes a continuous disruption of the physical balance in lakes, which naturally triggers processes successively changing the water environment properties. The mechanism of evolutionary changes of most lakes, whose main cause is the eutrophication of their waters, is similar. Slowly and continuously it leads to the transformation of their limnological properties (changes in water fertility, changes in morphometry caused by accumulation of bottom sediments, overgrowth of the coastal zone, changes in water level caused by changes in supply) and, consequently, to their disappearance. However, the rate and character of these transformations are quite diverse. Hence, despite the fact that a similar period has elapsed since the formation of lakes in lakeland areas, reservoirs (even not far apart from each other) may be on various stages of development. This also depends on the development of their internal structure which determines the natural ability to accumulate energy and matter in lake basins [1, 2] and on the physico-geographic characteristics of the area supplying the lake with matter and energy (catchment). For the lake and the basin constitute a natural landscape system called the lake geoecosystem [3–5]. Therefore, the rate of natural eutrophication of reservoir water, which determines its evolutionary transformations – the slow and natural process of ageing and disappearance – depends on the lake geoecosystem type [5].

The rate of natural eutrophication of lakes is strongly modified by anthropogenic activities, which usually take place quickly and almost all of them are disadvantageous for lakes, as they generally accelerate the rate of their evolutionary development and eventual disappearance. Each lake has a certain level of tolerance (resistance) to environmental pressure. Exceeding this level triggers an entire set of transformation processes (often degradative) that change the properties of lake ecosystems. The current state of the lake trophy is thus a resultant of natural and anthropogenic pressure as well as the reaction of the lake itself to these impacts. Determined on the basis of actual measurements, it is the result of the quantity of biogenic compounds found in the lake waters (from external and internal supply) depending on environmental pressure (determined by the type of the lake geoecosystem) and anthropogenic pressure (determined as the category of threat to the lake).

2 Lake Geoecosystem

The functioning of the lake geoecosystem depends on the following: (1) climatic conditions (solar radiation intensity, air temperature, wind direction and speed, quantity and intensity of atmospheric precipitation) that shape the abiotic conditions of the lake (lighting, water temperature, oxygen conditions, water dynamics), (2) hydrological conditions (streams of water exchange in the lake, water retention time), (3) physico-geographical features of the catchment environment (the amount of matter flowing from the catchment to the lake, chemical composition of lake water) and (4) limnological (abiotic and biotic) features of the lake describing its (hydrological, balance, thermal, oxygen, fishing, trophic) type. It is based on the constant transport of matter from the catchment (whose physico-geographic structure may favour or restrict areal runoffs) and its accumulation in the lake. The basic means of transporting various forms of matter in this system is water. Therefore, the relative durability of the lake geoecosystem depends on the maintenance of the equilibrium between income and outflow of water.

In the geographical environment, there are lake geoecosystems which differ in the rate of natural eutrophication resulting from the physico-geographic structure of the alimentation area of the lake (catchment) and morphometry of the lake basin and the hydrological regime of the water reservoir [5]. Each lake geoecosystem, regardless of the character of the lake (small and large, shallow and deep, lobelia and Charophyceae, humus, without outflow, with outflow and flow-through, permanent and periodical, natural and artificial) and its location in the physico-geographical mesoregion (lakelands, outwash plains, coastal lowlands, alluvial valleys, karst highlands, mountains), determines the rate of natural eutrophication of lake waters. The lake geoecosystem thus represents the expected state in reference conditions, i.e. a state that reflects conditions close to natural and does not show or shows only minimal disturbances as a result of human activity, which is entirely in line with the Water Framework Directive (Directive 2000/60/EC).

3 Lake Geoecosystem Determination Method

The geoecosystem of each lake can be determined by the method developed by E. Bajkiewicz-Grabowska [5–11]. This method assesses both the catchment as the supplier of matter (including biogenic) to the lake, as well as the lake – the recipient of this matter, taking into account the relations binding the lake with its alimentation area (Fig. 1). The assumption of this method is that the resultant of all factors favouring eutrophication has a different value for each lake. The same degree of the catchment impact, expressed as the amount of biogenic substances flowing from the catchment into the lake, yields a different effect in the case when the lake's natural resistance to degradation is high and different when the lake is susceptible to external impact. And reversely, with the same resistance of lakes to external impact, one should expect differences in their trophic state if their catchments affect them to a different extent [11].

The degree of the impact of the catchment on the lake is assessed on the basis of features characterizing both the total catchment of the lake and its immediate catchment. The features describing the impact degree of the total catchment on the lake are as follows: (1) the lake index (in hydrobiology Ohle index; quotient of total lake catchment and reservoir area), informing about the location of the lake in the hydrographic system (the lower its value, the lower the pressure exerted by the catchment on the reservoir) and balance type of the lake (flow-through, outflow, without outflow), which determines the potential of matter supply to the lake from

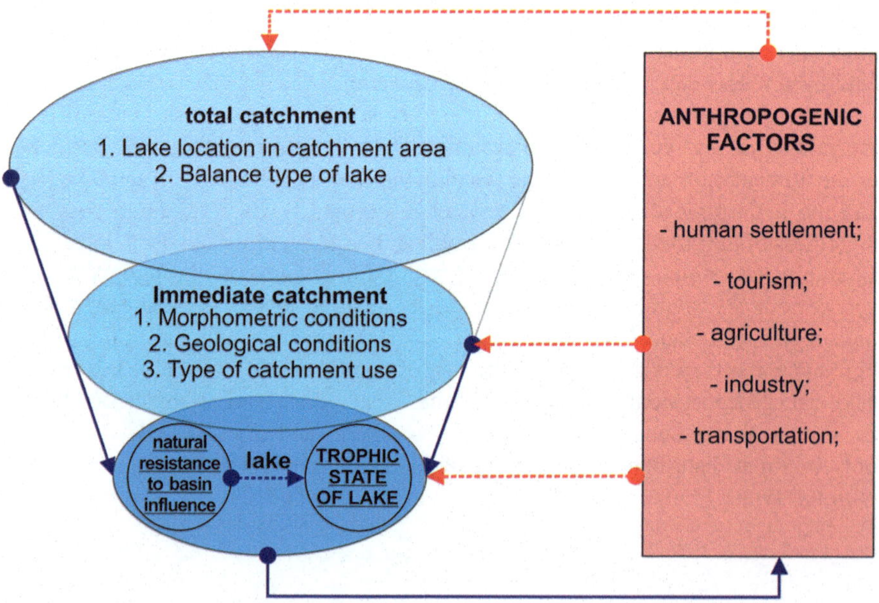

Fig. 1 Interactions scheme of the loop: catchment – lake – anthropopressure [12]

point sources and its "exit" from the lake via a watercourse. Lakes with no outflow are a trap for matter, outflow lakes take matter out of the catchment via a watercourse flowing out of them, while the role of flow-through lakes may vary depending on local conditions. Nevertheless, flow-through lakes are subjected to the "flushing" process, which facilitates the transport of matter outside the catchment.

The volume of the areal load entering the lake also depends on the physico-geographic environment of its immediate catchment. In young glacial landscape, the volume of supply is determined by the following: (1) the size of the catchment effectively involved in the matter supply – its measure is the degree of the catchment's lack of outflow[1] – (2) the average catchment slope[2], whose value conditions the areal runoff and the intensity of water erosion; (3) river network density[3] which is a measure of direct and rapid transport of matter to the lake; (4) surface formations and soil conditions, determining the permeability of the ground, which influences the possibility of transporting matter to groundwaters drained by the lake; and (5) type of land use, which determines the amount and form of biogenic elements in areal runoff [14].

According to the adopted method, the impact of the catchment on the matter supply to the lake is assessed using the bonitation of each of the above-mentioned features on a scale of 0 to 3 points, where 0 means a very low impact on the matter supply and inability to reach the lake and 3 is high impact and fast delivery of matter to the reservoir (Table 1). The final grade is the arithmetic mean of the points obtained from the assessment of particular features, and it qualifies the given lake catchment to one of the four vulnerability groups (Table 2). The assessment also indicates which of the considered features promotes the supply of matter to the lake (e.g. large slopes, dense drainage, the form of catchment use) and which has an inhibitory effect on the process (e.g. a high degree of lack of outflow, poor drainage, small slope).

In determining the type of the lake geoecosystem, the assessment of the natural resistance of the lake to the pressure from the alimentation area is also taken into account. The same impact of the catchment gives different results in the case of lakes whose morphometric and hydrological features shape their high natural resistance to eutrophication, than when the reservoir shows high susceptibility to external influences. The rate of natural evolution (eutrophication) will also be different if the lake with a certain degree of resistance to the catchment impact is affected by a catchment vulnerable to activation of areal load and its supply to the reservoir, than by a catchment of strongly limited matter supply possibilities.

---

[1]Degree of lack of outflow of catchment indicates the % of its area excluded from surface outflow [13].

[2]Mean slope of catchment is a quotient of catchment denivelation and square root of its area [13].

[3]River network density is a quotient of the length of the river network in the catchment and its area [13].

Table 1 Point criteria for the evaluation of a catchment as matter supplier to a lake [6]

	Number of points			
Characteristics	0	1	2	3
Ohle's index	<10	10–40	40–150	>150
Type of lake water balance	–	Outflow	Without outflow	Throughflow
River network density in direct catchment (km km^{-2})	<0.5	0.5–1.0	1.0–1.5	>1.5
Average slope of direct catchment (‰)	<5	5–10	10–20	>20
Endorheic areas in catchment (%)	>60	45–60	20–45	<20
Geological type of direct catchment	Clay	Sand-clay	Clay-sand	Sand
Land use type of direct catchment	Forest, swamp, farmland-forest, pasture-farmland-forest, pasture, pasture-forest	Forest-farmland, pasture-farmland	Farmland, pasture-forest-farmland with settlements, forest with settlements	Forest-farmland with settlements, pasture-farmland with settlements, farmland with settlements

Table 2 Groups of catchment susceptibility to activation and transport of biogenic matter according to [6]

Final score	Catchment description	Catchment susceptibility group
≤1	Catchment has a limited impact on activation and supply of biogenic matter into the lake	*Susceptibility group 1*
1.01–1.49	Catchment has a small impact on activation and supply of biogenic matter into the lake	*Susceptibility group 2*
1.50–1.99	Catchment has a medium ability to activate and supply biogenic matter into the lake	*Susceptibility group 3*
≥2	Catchment has a large ability to activate and supply biogenic matter into the lake	*Susceptibility group 4*

The lake resistance to the catchment impact, according to the adopted method, is indicated by (1) the lake's average depth, (2) the quotient of the lake's capacity and the length of its shoreline, (3) the quotient of active bottom area and epilimnion capacity, (4) percentage of water stratification, (5) intensity of water exchange and (6) Schindler index (quotient of the surface receiving the matter, i.e. the total catchment of the lake and the amount of water diluting it, i.e. the capacity of the lake basin) [5].

The depth ratios of the lake, thus also the shape of the lake basin, determine the water potential of a given reservoir, expressed as lake capacity. In turn, the quotient of the lake's capacity and the length of the shoreline is a measure of its resistance to the catchment impact. The greater its value, the more resistant the lake is to external influences [15]. The resistance of lakes to the catchment impact is also reduced by low intensity of "flushing", expressed as the quotient of the average annual outflow from the lake and the capacity of the lake [16]. Polymictic lakes without permanent stratification have lower resistance to influences of the catchment. The active bottom area in these lakes covers the entire bed of the lake basin, and thus favourable conditions for the recirculation of biogenic substances from sediments into the water are met. Both the low percentage of stratified waters and the high value of the index expressed as the quotient of active bottom area and lake capacity significantly reduce the resistance of these lakes to eutrophication. The occurrence of summer thermal stratification in deeper lakes reduces the scope of the active bottom to the epilimnion layer. Recirculation of biogenic substances from sediments into epilimnion waters is therefore limited, which increases the resistance of these lakes to external influences. The resistance of lakes is also influenced by the Schindler index, which links the lake to its catchment. As the value of this index increases, the lake's resistance decreases.

Similarly as in the assessment of the catchment vulnerability to the supply of matter to the lake, in the case of assessing the lake's resistance to catchment influences, each of the above-mentioned features is assigned points from 0 to 3, where 0 means high resistance and 3 lack of resistance to catchment impact (Table 3). The final grade is the arithmetic mean of points from the assessment of individual resistance features, and it defines one of the four categories of lake resistance (Table 4). The assessment also indicates which of the considered features decreases the lake resistance to external influences and which increases it.

The vulnerability group of the lake's catchment and the category of its resistance determine the type of the lake's geoecosystem with a given rate of natural eutrophication. *The first lake geoecosystem type* represents such a catchment – lake relationship in which both natural features of the reservoir (resistance category I or II) and catchment (vulnerability group 1 or 2) are not conducive to lake water eutrophication; the lake is resistant to external influence, and its catchment is not very active in the supply of areal load into the reservoir. Natural conditions are

Table 3 Criteria for evaluation of lake resistance to catchment impact [according to 6]

	Number of points			
Characteristics	0	1	2	3
Mean lake depth (m)	>10	5–10	3–5	<3
Lake volume (h m^3) to shoreline length (m)	>5	3–5	1–3	<1
Thermal stratification (%)	>35	20–35	10–20	<10
Active bottom surface (m^2) to epilimnion volume (m^3)	<0.10	0.10–0.15	0.15–0.30	>0.30
Rate of annual water exchange	>10	5–10	1–5	<1
Schindler's index (m^2 m^{-3})	<10	10–30	30–100	>100

Table 4 Categories of lake resistance to catchment impact, according to [6]

Final score	Lake description	Lake resistance category
≤0.89	High lake resistance to catchment impact	Category I
0.90–1.69	Medium lake resistance to catchment impact	Category II
1.70–2.40	Small lake resistance to catchment impact	Category III
≥2.41	Lake without resistance, heavily susceptible to catchment impact	Category IV

conducive to slow evolution of the lake, and such an arrangement has a chance to keep its trophy at a low level. *The second geoecosystem type* has unfavourable catchment conditions for the lake (high possibility of supplying matter to the reservoir – catchment vulnerability group 3 or 4), which are balanced by high resistance to external influence of the lake itself (resistance category I or II). As a result, the rate of natural eutrophication in lakes representing this geoecosystem type is moderate. *The third geoecosystem type* has favourable catchment conditions (the catchment is not very active in supplying matter to the reservoir – vulnerability group 1 or 2), but the lake itself is susceptible to external influence (resistance category III or IV). Eutrophication of lakes with this geoecosystem type proceeds moderately; however, an interference with catchment conditions (e.g. tourism development, land use change, melioration works) can lead rather quickly to an accelerated eutrophication rate of lake waters. *The fourth geoecosystem type* is such a catchment – lake relationship, whose natural conditions favour rapid eutrophication of lake waters. Lakes are susceptible to external influence (resistance category III or IV), and their catchments are active in supplying areal loads into them (vulnerability group 3 or 4).

4 Results

The presented method, recommended by the Polish Limnology Society [10], was applied to determine the geoecosystem types of 388 lakes of northern Poland [6–13, 17–28]. This group is dominated by flow-through lakes with an area of over 50 ha.

Lake geoecosystem type 1, i.e. revealing the chance to maintain eutrophication of lake waters at a low level, is represented by 71 lakes of the studied group, i.e. 18% (Fig. 2). These are generally deep, stratified, mostly flow-through lakes, usually with a low Schindler index (WS <2).

Geoecosystem type 2 indicating a moderate rate of eutrophication of lake waters was observed in 80 lakes of the studied group, i.e. 21%. Among them, apart from deep lakes, are already medium deep; these are also generally flow-through stratified lakes, usually with a low Schindler index (WS ≤2).

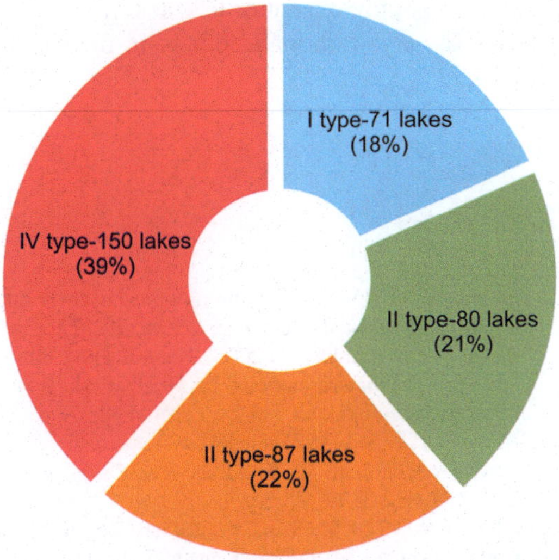

Fig. 2 Geoecosystem types of the analysed group of lakes of northern Poland

Geoecosystem type 3, also indicating a moderate rate of lake water eutrophication, was found in 22% of the lakes of the studied group (87). These are already shallower lakes, also generally flow-through, although there are lakes without outflow too. Among them are stratified and polymictic lakes. Lakes with this geoecosystem type have a high Schindler index (WS >30).

Geoecosystem type 4, whose natural conditions favour fast eutrophication of lake waters, was observed in as many as 150 lakes of the studied group (39%). Among them are stratified (deeper) and polymictic (shallower) lakes, without and with outflow, less frequently flow-through, usually with a high lake index (generally C >60) and a high Schindler index (mostly WS >80).

Therefore, in the conditions of Polish lakelands, the majority of analysed lakes have a geoecosystem type 4. The majority of Polish lakes have catchments, which exert considerable pressure on them, because their physico-geographical environment promotes the supply of matter to the reservoir. The lake itself is susceptible to the influence of the catchment.

5 Conclusion

The lake geoecosystem (catchment – lake ecological system) described by the vulnerability group (catchment) and the resistance category (lake) can be reproduced in various natural landscapes. While the selection of features determining the resistance of the lake does not depend on the type of natural landscape, the selection of features determining the vulnerability of the catchment to supplying matter to the lake is dependent on the specificity of the landscape. Other catchment features favour the supply of matter to lakes in young glacial landscapes; others in upland landscapes, e.g. carbonate; others in marshy landscapes; and others in mountain landscapes.

References

1. Lange W (1986) Fizycznolimnologiczne uwarunkowania tolerancji systemów jeziornych Pomorza. Zesz Nauk UG Rozpr i Mon. 79:177
2. Lange W, Maślanka W (2005) Natural and anthropogenic conditions of Pomorze selected lakes present transformations. In: Gołębiewski R (ed) Peribalticum. Z problematyki Pobrzeża i Pojezierzy Bałtyckich, vol IX, GTN Wydz. V Nauk o Ziemi, Gdańsk, pp 113–138
3. Kostrzewski A (1993) Geoekosystem obszarów nizinnych. Koncepcja metodologiczna. (The geoecosystem of lowlands areas a methodological conception). In: Kostrzewski A (ed) *Geoekosystem obszarów nizinnych*. Polska Akademia Nauk, Kom. Nauk. Prez. PAN *"Człowiek i Środowisko"*, Zakład Narodowy im. Ossolińskich, Wrocław, pp 11–17
4. Kostrzewski A (ed) (1991) Koncepcja programu: Monitoring obiegu materii, kompleksowy monitoring środowiska przyrodniczego w podstawowych typach geoekosystemów Polski (The concept of the program: Monitoring of the matter circulation, a comprehensive monitoring of the environment in the basic types of geoecosystems in Poland). Komitet Naukowy przy Prezydium PAN "Człowiek i Środowisko", Poznań
5. Bajkiewicz-Grabowska E (2008) Lake geoecosystem as an indicator of the rate of natural eutrophication of water reservoirs. Limnol Rev 8:1–2
6. Bajkiewicz-Grabowska E (1981) The influence of the physical geographic environment on the biogenous matter delivery to the Lake. J Hydrol Sci 8(1–2):63–73
7. Bajkiewicz-Grabowska E (1983) Ecological characteristics of lakes in North-Eastern Poland versus their trophic gradient. II. Lake catchment areas – phusico-geographical environment description of the region and 43 lakes. Ekol Pol 31(2):257–286
8. Bajkiewicz-Grabowska E (1985) Struktura fizycznogeograficzna zlewni jako podstawa oceny dostawy materii biogennej do jezior (Effect of the physical geographic environment on the biogenous matter delivery to the lakes). Prace i Studia Geogr 7:65–89
9. Bajkiewicz-Grabowska E (1987) Evaluation of natural susceptibility Shallow lakes to degradation. GeoJournal 14:3
10. Bajkiewicz-Grabowska E (2010) Assessment of the ecological state of lakes as proposed by the Polish Limnological Society. Limnol Rev (3–4):105–116
11. Bajkiewicz-Grabowska E (2002) Obieg materii w systemach rzeczno-jeziornych (Circulation of matter in the river-lake systems). Wyd. UW, Warszawa, p 274
12. Markowski M, Kwidzińska M (2015) Types of geoecosystems of the lobelia lakes of the Tricity area. Quaestiones Geographicae 34(1):15–25. Bogucki Wydawnictwo Naukowe, Poznań
13. Bajkiewicz-Grabowska E, Magnuszewski A (2012) Przewodnik do ćwiczeń z hydrologii ogólnej. Wyd. Nauk. PWN, Warszawa, p 196
14. Kajak Z (1998) Hydrobiologia – Limnologia. Ekosystemy wód śródlądowych. Wyd. Nauk. PWN, Warszawa, p 355
15. Sylwester H, Hitchins M, Gould L, Hall MW (1974) A quantitative classification of Maine Lakes. The Environmental Studies Center, University of Maine, Orno
16. Vollenweider (1968) Scientific fundament of the eutrophication of lakes and flowing waters with particular reference to nitrogen and phosphorus as factors in eutrophication. DAS/CSIO/68.27. OECD, Paris, p 192
17. Bajkiewicz-Grabowska E (1990) Stopień naturalnej podatności jezior na eutrofizację na przykładzie wybranych jezior Polski. Gosp Wodna 12:279–289

18. Bajkiewicz-Grabowska E, Hilbricht-Ilkowska A, Zdanowski B (1992) Ocena podatności na degradację, stan czystości wód i tempa eutrofizacji jezior. In: Zdanowski B (red.) Jeziora Wigierskiego Parku Narodowego. Stan eutrofizacji i kierunki ochrony. Zesz. Nauk. "Człowiek i Środowisko", vol 3, pp 163–190
19. Bajkiewicz-Grabowska E, Malczyk M (1993) Stan ekologiczny wybranych jezior obszaru pojezierzy. In: Dynowska I (ed) Przemiany stosunków wodnych w Polsce w wyniku procesów naturalnych o antropogenicznych. Kraków, pp 137–147
20. Bajkiewicz-Grabowska E (1994) Waloryzacja zlewni jezior Suwalskiego Parku Krajobrazowego i ich naturalnej podatności na degradację. In: Hillbricht-Ilkowska A, Wiśniewski RJ (eds) Jeziora Suwalskiego Parku Krajobrazowego. Związki z krajobrazem, stan eutrofizacji i kierunki ochrony. Zesz. Naukowe Kom. "Człowiek i Środowisko", vol 7, Ossolineum, pp 47–54
21. Bajkiewicz-Grabowska E (1999) Struktura fizycznogeograficzna układu krajobrazowego zlewnia-jezioro i jej wpływ na tempo naturalnej eutrofizacji jezior. In: Funkcjonowanie i ochrona ekosystemów wodnych na obszarach chronionych. Wyd. IRŚ, Olsztyn, pp 77–86
22. Bajkiewicz-Grabowska E, Nosal K (1999) Zastosowanie geograficznego systemu informacji (GIS) w ocenie tempa naturalnej eutrofizacji jezior Wigierskiego Parku Nardowego. Funkcjonowanie i ochrona ekosystemów wodnych na obszarach chronionych. Wyd. IRS, Olsztyn, pp 203–230
23. Bajkiewicz-Grabowska E (2004) Podatność jezior na eutrofizację. In: Zdanowski B, Hutorowicz A, Białokoz W (eds) Ekosystemy wodne Parku Narodowego "Bory Tucholskie", vol 17, Wyd. IRS Olsztyn, p 32
24. Bajkiewicz-Grabowska E (2007) Zróżnicowanie geoekosystemów jeziornych ze względu na tempo naturalnej eutrofizacji. (Diversity of lake geoecosystems in terms of natural eutrophicaton rate). In: Borowiak D (ed) Jeziora Kaszubskiego Parku Krajobrazowego. Badania Limnologiczne, vol 5, Katedra Limnologii UG, Gdańsk, pp 279–292
25. Bajkiewicz-Grabowska E (2008) Ocena tempa naturalnej eutrofizacji jeziora Hańcza. In: Kozłowski J, Poczyczyński P, Zdanowski B (eds) Środowisko i ichtiofauna jeziora Hańcza. Wyd. IRŚ, Olsztyn, pp 37–42
26. Bajkiewicz-Grabowska E (2008) Ocena zlewni górnego Ełku (Łaźnej Strugi) i podatność jezior na degradację. In: Białokoz W, Chybowski Ł (eds) Jeziora Puszczy Boreckiej – stan ekosystemów i kierunki ochrony. Wyd. IRŚ, Olsztyn, pp 19–28
27. Bajkiewicz-Grabowska E (2010) Czyste jeziora Pomorza – Diagnoza. Badania Limnologiczne, vol 7. Katedra Limnologii Uniwersytetu Gdańskiego, Gdańsk, p 74
28. Bajkiewicz-Grabowska E, Markowski M (2016) Geoekosystemy jezior Parku Narodowego "Bory Tucholskie" i stan ich zagrożenia. In: Choiński A, Kochanowska M, Marszelewski W (eds) Przyroda abiotyczna Parku Narodowego "Bory Tucholskie". Bogucki Wyd. Nauk, Poznań, pp 345–359

Occurrence, Genetic Types, and Evolution of Lake Basins in Poland

Adam Choiński and Mariusz Ptak

Contents

Abstract The territory of Poland includes more than 7,000 lakes (1 ha or larger) with a total surface area of 281,377.0 ha, translating into the lake density index of 0.90%. The territory of Poland is evidently divided into the northern part, including 95% of all lakes, and the southern part poor in lakes. The situation is related to the last Scandinavian glaciation. Due to this, in terms of genesis of basins, the majority of lakes are of postglacial character, where the occurrence and course of channel lakes is an indicator of the maximum range of the ice sheet. Lakes fulfil a number of important functions in the environment. The functions refer among others to the conditions of water circulation (suppressing extreme situations, both floods and draughts), affect the biodiversity, shape climatic conditions, and are of key importance for the development and functioning of many branches of the economy (agriculture, energy engineering, tourism), etc. Both environmental and anthropogenic effects resulting from the occurrence of lakes can soon be lost as a result of the progressing process of their disappearance. The situation is determined by the natural process of evolution, accelerated by human activity related to the regulation of water relations, and an increase in productivity of agriculture, and consequently an increase in supply of nitrogen and phosphorus compounds to water. It is estimated that from the moment of development of lakes until today, approximately 60% of the surface area of lakes disappeared.

A. Choiński (✉) · M. Ptak
Department of Hydrology and Water Management, Adam Mickiewicz University, Poznań, Poland
e-mail: choinski@amu.edu.pl

E. Korzeniewska, M. Harnisz (eds.), *Polish River Basins and Lakes – Part I*,
The Handbook of Environmental Chemistry 86,
https://doi.org/10.1007/978-3-030-12123-5_4

Keywords Decline of lakes · Evolution · Lakes · Morphometry

1 Introduction

Lakes through their parameters, namely, the ability to accumulate energy and collect and produce energy, constitute specific object in the natural environment. They fulfil very important functions in the geosystem, ascribed exclusively to them. They develop biodiversity (with unique flora and fauna), affect microclimatic conditions (change of the thermal regime in adjacent areas, change of humidity conditions, occurrence of breeze circulation), fulfil the protective function inhibiting movement of biogenic compounds [1] but particularly constitute a very important element of water circulation. This is manifested in the effect on both the volume of elements of the water balance and their dynamics. They constitute a factor increasing water retention in a given area. The presence of lakes in the catchment causes higher than average evaporation and a change of the regime of rivers flowing out and through lakes: in the dry period, the lakes exceed the water level in rivers, and in the humid period, they decrease it [2].

Lakes also considerably affect human existence. Already thousands of years ago, people would take advantage of the vicinity of lakes, locating their settlements there to provide a certain degree of safety and facility of obtaining food. Also today, lakes have been comprehensibly adapted by different branches of the economy (tourism, agriculture, industry, etc.). Therefore, in areas with their considerable abundance (lakelands), they have largely determined and still determine the economic development of a given region. Therefore, in areas with no lakes, artificial reservoirs are often built, aimed among others at filling the gap in the scope.

2 Occurrence of Lakes in Poland

The majority of currently existing lakes in the northern part of Europe and North America originate from the activity of ice sheets from the Pleistocene. The effects of the last period of continental glaciation are particularly evident. In the territory of Poland, its maximum range had a course exceptionally parallel to isolines, reflecting lake density. In comparison to the neighbouring countries, Poland is included to the zone with low lake density. It covers only 7,081 lakes with a surface area equal or higher than 1 ha, and the lake density amounts to only 0.90% of the area of the country [3–5]. Considerable variability in lake density occurs, however, between the north lakeland zone and the south zone poor in lakes (Fig. 1). The comparison of the number of lakes and their surface area is presented in Table 1. Table 2 presents ten largest and ten deepest lakes in Poland.

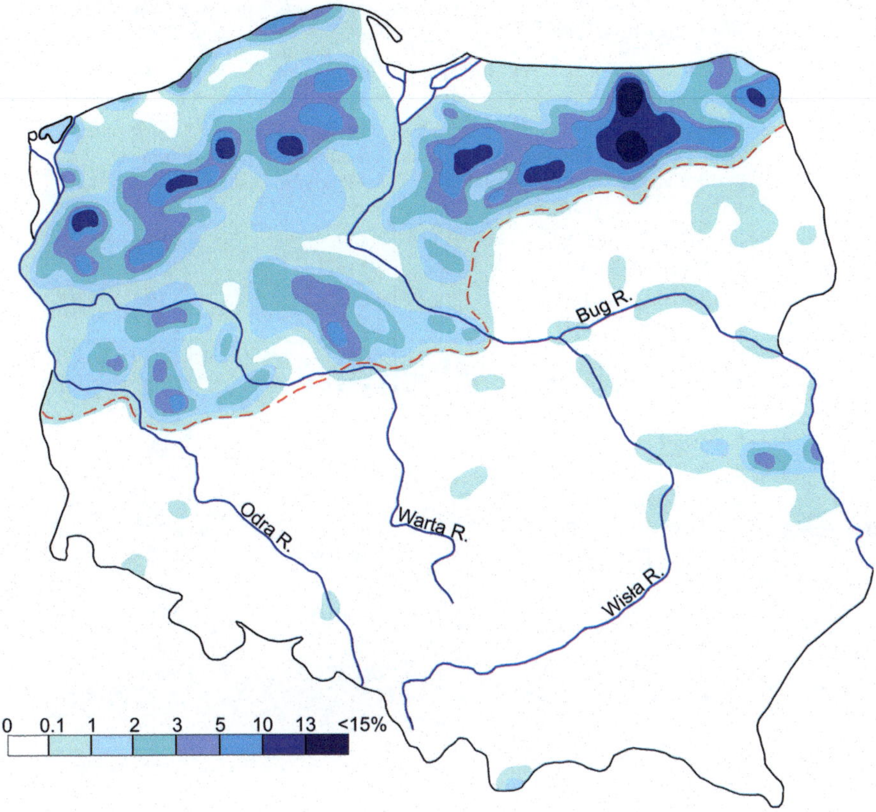

Fig. 1 Lake density in Poland (following [6], partially changed)

3 Genetic Types of Lake Basins

Basins of a large majority of Polish lakes originate from Pleistocene formations. It is therefore genetically related to the period of glaciation in spite of considerable modifications of their basins in the Holocene. Glacial formations of lake basins that survived to the modern times originated through melting of lumps of dead ice, gauging of depressions by the waters of the ice sheet, chaotic accumulation of glacial sediments, evorsive activity of the waters of the melting ice sheet, and ploughing of depressions by the mass of the mobile ice sheet. The development of lakes in the Holocene can be associated with the following processes: filling of inter-dune depressions with water, losing sea bays by spits, development of karst depressions, and development of oxbow lakes. The determination of the genesis of a lake is important, because it permits among others the determination of the age of the lake,

Table 1 Lakes in size classes according to the *Catalogue of Polish lakes* from 1954 and the *Catalogue of Polish lakes* from 2006

Size class in ha		*Catalogue of Polish lakes* (1954) a – number of lakes b – lake surface area in ha	a – % of total number b – % of total surface area	*Catalogue of Polish lakes* (1991a, b, 1992) a – number of lakes b – lake surface area in ha	a – % of total number b – % of total surface area
1–5	a	4,734	50.93	3,112	43.95
	b	10,387.8	3.28	7,116.7	2.53
5–10	a	1,316	14.16	945	13.34
	b	9,239.4	2.91	6,507.9	2.31
10–20	a	1,91	11.74	1,047	14.79
	b	15,263.6	4.82	14,287.4	5.08
20–50	a	1,043	11.22	981	13.85
	b	32,747.9	10.33	31,183.0	11.08
50–100	a	533	5.73	492	6.95
	b	36,783.2	11.61	33,875.0	12.04
100–1,000	a	545	5.86	476	6.72
	b	136,262.3	42.99	116,615.5	41.45
>1,000	a	34	0.36	28	0.40
	b	76,242.8	24.06	71,791.5	25.51
Total	a	9,296	100.00	7,081	100.00
	b	316,927.0	100.00	281,377.0	100.00

Table 2 List of the largest and deepest lakes in Poland

No.	Name of lake	Surface area [ha]	Name of lake	Max depth [m]
1	Śniardwy	11,487.5	Hańcza	106.1
2	Mamry	9,851.0	Drawsko	82.2
3	Łebsko	7,020.0	Wielki Staw	80.3
4	Miedwie	3,491.0	Czarny Staw pod Rysami	77.0
5	Jeziorak	3,152.5	Wigry	74.2
6	Niegocin	2,595.0	Wdzydze	69.5
7	Gardno	2,337.5	Wukśniki	67.3
8	Jamno	2,231.5	Babięty Wielkie	65.0
9	Gopło	2,121.5	Morzycko	60.7
10	Wigry	2,115.0	Trześniowskie	58.8

directions of its evolution, and frequently physical-chemical water parameters and the determination of the dependency between the lake and its catchment. Approximately 20 types (Fig. 2) of lake basins can be designated in Poland. It should be emphasised that the determination of the type of lake is defined here as the genetic type of its basin.

Channel Lakes It is the most characteristic type of lakes within the range of the Baltic glaciation. One of the most important features of channel lakes is their directional character. Both in Poland and in other Baltic lakelands, directions of

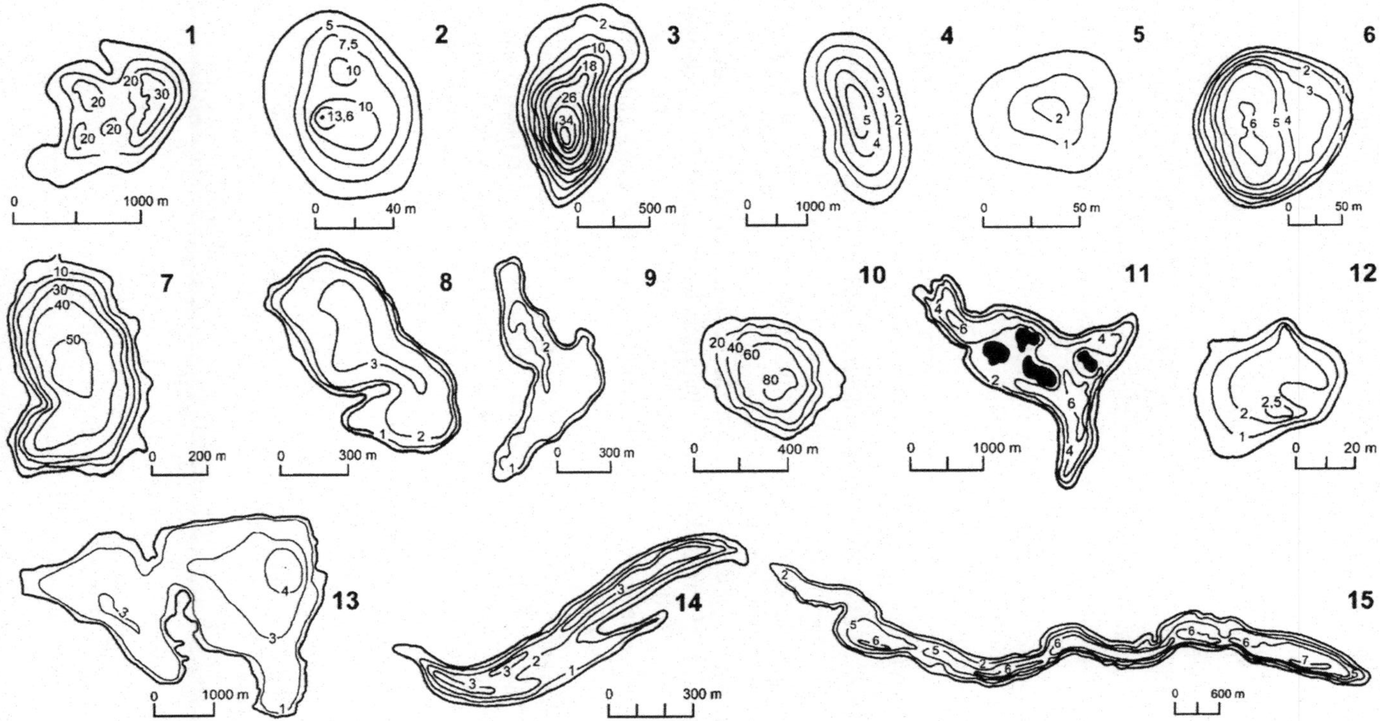

Fig. 2 Selected bathymetric plans of different genetic types of lake basins: 1 – evorsive lake (Gostynińskie near Międzychód – according to [14]), 2 – explosive lake (Tobellus – according to [18]), 3 – karst lake (Piaseczno – according to [11]), 4 – interdunal lake (Moczydło – according to [19]), 5 – waterhole (according to [19]), 6 – relic lake after pingo (Pietronajć – according to [20]), 7 – mountain moraine lake (Morskie Oko – according to [21]), 8 – sandur lake (Pniewo – according to [22]), 9 – terminal moraine lake (Kruchowskie – according to [22]), 10 – tarn (Czarny Staw at Morskie Oko – according to [23]), 11 – ground moraine lake (Bytyńskie – according to [22]), 12 – meteorite lake (Moraskie – according to [24]), 13 – coastal lake (Wicko – according to [25]), 14 – oxbow lake (Martwe – according to [26]), 15 – channel lake (Strykowskie – according to [22])

lake basins are closely related to the system of sequences of terminal moraine to which they are usually perpendicular. Other characteristic features of the discussed type of lake basins include considerable depths, bottoms with uneven configuration with frequently occurring thalwegs and shallow areas constituting steps, and high slope inclinations of the bottom. The lakes are narrow – this translates into high elongation indices. Channel lakes usually occur in characteristic sequences, delineating former ranges of channels that were divided into a number of smaller basins. Examples include channels of lakes near Kórnik, Łagów, and Chojnice.

Ground Moraine Lakes Are characterised by substantial surface areas and diverse shoreline, associated with variable deposition of sediments or melting of variable sizes of lumps of dead ice. Lakes of the type have bays and often also islands and peninsulas. Bottom slopes are relatively gentle, and depths are usually largely variable and unevenly distributed. In the conditions of lowering the water level, this causes an increase in the surface area of the already existing islands or the occurrence of new islands. Examples of ground moraine lakes are Lakes Śniardwy, Wielimie, Niegocin, and Bytyńskie.

Terminal Moraine Lakes Fill concave landforms developed after melting of sheets of dead ice and chaotic accumulation of clastic material in the marginal zone or through blocking outflow due to folding of material in front of the glacier terminus. The lakes are often located on the internal side of terminal moraines, and their longitudinal axis is usually parallel to moraine ramparts. Due to the high elevation of the zones of terminal moraines, the water level in the lakes is relatively high, often exceeding 100 m a.s.l. Examples of the discussed type of lakes occur in the vicinity of Trzemeszno (e.g. Lake Kruchowskie – 110 m a.s.l.) and in many places of the watershed zone of the Pomeranian and Mazurian Lakelands.

Kettle Lakes Are lakes with a small surface area and considerable depth, reaching up to 50 m. Usually without surface outflow, they have very steep bottom slopes, conical shape of the basin, and the outline of the shoreline is usually oval. This type of lakes usually has no islands. The genesis of their lake basins can be associated with melting out of thick lumps of dead ice or evorsive thalweg in the bottom formed by the waters of the melting ice sheet. An example of kettles can be lakes in the vicinity of Sierakowo [7] or Wałcz.

Waterholes Are small closed-drainage water bodies usually with a circular shape and depth not exceeding 3 m. Often in the summer period, as a result of a natural decrease in the water level of the first aquifer, the lakes can be left without water. The process of decline of waterholes is usually related to melioration works covering large areas. Their considerable part is covered with peat or boggy vegetation. Waterholes occur both north and south of the line of the maximum range of the Baltic glaciation. They are usually characteristic of areas of moraine plateaus and areas of the ground moraine. On the Polish Lowland, the number of waterholes is estimated for 100 thousand.

Sandur Lakes In the case of melting of lumps of dead ice, depressions developed in outwash plain areas. In the conditions of an appropriate arrangement of the depth of the basin and stagnation of groundwaters, they are filled with water. They are usually lakes with a length of several metres, however often with considerable surface areas. Notice that lakes located on outwash plains are often within the range of channels eroded in fluvioglacial formations. Such examples are encountered, e.g. in the south-west part of the Lubusz Plateau [8]. Lakes with a water level approximate to the surface of the outwash plain (without the channel character) occur among others in the vicinity of Człopa – outwash plain of the Drawa River (Lakes Pinow and Strzeleckie) and on the outwash plain Pliszki on the Lubusz Land – e.g. Lake Dobrosułowskie [9].

Esker and Drumlin Lakes Fill longitudinal depressions formed during the processes of development of eskers and drumlins. The lakes are usually shallow and filled with peat. Many lakes of the type are located in the zone of occurrence of eskers between Duszniki and Stęszewo in Wielkopolska or on the Dobrzyńskie Lakeland.

Oxbow Lakes [10] – occur in river valleys (proglacial stream valleys). They are remains of former river channels, hence their characteristic oval and considerably elongated shape. Their depths are small, and the maximum depth occurs in the vicinity of the concave shores. Due to the shallow occurrence of groundwaters in valley zones, shores of the lakes are often boggy. Their water level fluctuations are often related to the amplitudes of river water.

Interdunal Lakes Are often boggy and filled with peat. They develop in depressions between dunes, often located in valley zones and on outwash plain surfaces covered with dunes. The depressions are usually deflation basins. Lakes of the type occur, for example, in the area of the Middle Noteć River valley.

Karst Lakes Develop as a result of melting and collapsing of gypsum or lime substrate washed out by water. An example are some lakes of the Łęczna-Włodawa Lakeland – e.g. Lake Krasne [11] or small lakes in the vicinity of Busko [12]. The discussed type of lakes is characterised by depths of more than 30 m, oval outlines of the shoreline, relatively regular bathymetry – so-called centric system of isobaths, and the maximum depth usually located near the centre of the lake. A specific variety of karst lakes are cave lakes. Their genesis is associated with melting calcium carbonate. This type of lakes occurs, e.g. in the Bear Cave near Kletno [13].

Groundwater (Boggy) Lakes Develop as a result of obstructed inflow of groundwaters which "appear" on the surface in depressions. They are small and shallow lakes with overgrown and boggy shores. The areas of occurrence of this type of lakes are usually surrounded by bogs – they develop flat areas with largely impermeable ground. In Poland, such lakes are encountered near Włodawa in Polesie Lubelskie.

Wellhead Lakes Develop on powdery ground, where the succouring activity of water causes the development of a basin alimented with the waters of the spring or groundwaters. An example of this type of lakes is the lake in Miedzierza near Końsk [14].

Delta Lakes In Poland, this type is represented by only two lakes, namely, Lake Dąbie at the mouth of the Oder River to the Szczecin Reservoir (which can be currently treated as its southern part) and Lake Druzno in the Vistula delta. They are relic lakes with advanced overgrowing process, shallow, with low and boggy shores. Lake Dąbie constitutes a former bay of the Szczecin Reservoir, separated with the Ina delta, and Lake Druzno was separated with the Nogat delta [15].

Coastal Lakes The development of the basins of the lakes is associated with the separation of sea bays with spits developed by the deposition of clastic material transported by waves and littoral currents. They are lakes with considerable surface areas (usually more than 10 km^2), but their maximum depths do not exceed 6 m. They have largely inaccessible boggy shores. The relief of the original bottoms is obscured by a layer of sediments with a thickness of several or a dozen metres. They are separated from the sea with sandy spits covered with dunes. They are connected with the sea through narrow isthmuses with a width of several tens of metres. The lakes can be described as brackish. This type of lakes originates from lagoons – Szczecin and Vistula lagoons – constituting the intermediary stage between the sea and lake [16]. In the conditions of low water exchange, the lakes are subject to rapid terrestrialisation. An example of coastal lakes is Wicko, Sarbsko, Bukowo, and the westernmost Koprowo.

Dam Lakes Develop as a result of damming of valleys by landslides, originating from weathered rocks or loose near-surface rocks. An example of this kind of lakes are the Duszatyń Lakes near Komańcza on the slope of Chryszczata in the Olchowaty Stream valley in Beszczady. They developed as a result of sliding of a mass of weathered rock and damming of waters in the valley. Lake Duszatyńskie Górne with a surface area of approximately 1 ha has a maximum depth of 6.2 m, and Lake Duszatyńskie Dolne has a maximum depth of 6.6 m and surface area of approximately 0.8 ha. The lakes constitute a nature reserve.

Organic Lakes The lakes include phytogenic and zoogenic lakes. The former are developed among peatlands and are located in places with limited vegetation development. The latter constitute an effect of the activity of animals – e.g. beavers – and in this case have a character of dam reservoirs. Lakes of the type are shallow and small and have boggy shores [13].

Lakes with Polygenetic Origin of Basins The lakes include a high number of lakes with origin so complicated that it would be difficult to classify them as any of the aforementioned types. For example, oxbow lakes can be simultaneously interdunal lakes, and lakes located on the surface of the outwash plain can originate from a channel eroded in deeper ground. Finally, a perfect example can also be lakes commonly considered coastal, whose original bottom is covered with sediments of organogenic accumulation with a thickness of several metres. The course of isobaths shows that, e.g. Lake Jamno developed as a result of damming of outflows of several channels with a spit [17].

Meteorite Lakes In the northern part of the city of Poznań, in the area of the Moraska Mountain, seven depressions are located, including six periodically filled with water. The largest one has several tens of metres in diameter and more than 2.5 m in depth. Many fragments of meteorites have been found in the vicinity of the lakes, with maximum weight reaching 300 kg.

Mountain Lakes Developed as a Result of Activity of Glaciers They have a completely different character than lakes in the lowland part of Poland, and examples of this type of lakes occur both in the Carpathians and Sudetes. Basins of the lakes are partially or completely developed in bedrock. Other characteristic features are considerable depths, often exceeding 50 m, very substantial slope inclinations of basins, and their oval shape. In the group of lakes developed in basins eroded by mountain glaciers, the following two types can be designated:

- *Tarns* – filling the space of former firn fields. They are confined by steep slopes on three sides and blocked with a clastic rock bar. An example is Czarny Staw at Morskie Oko and Czarny Staw Gąsienicowy in the Tatra Mountains or Mały and Wielki Staw in Karkonosze.
- *Moraine lakes* – developed due to damming of a valley with a moraine rampart increasing the water level. Examples include Morskie Oko, Toporowy Staw, and Smreczyński Staw.

The last type of lake basins are *artificial lakes*, i.e. *anthropogenic reservoirs*. They are characterised by variability of all the previously mentioned parameters, determined by the function fulfilled by each reservoir. Two groups of anthropogenic reservoirs can be designated:

- *Post-exploitation reservoirs* with extensive surface area (e.g. in the Konin Coal-field), as well as very small (e.g. after exploitation of loam, peat, etc.) ones
- *Retention reservoirs*, fulfilling many different functions, e.g. agricultural, for irrigation; anti-flood, for reduction of flood waves; energy engineering, animal breeding, navigation, for increasing low water levels in rivers; recreation, water supply to industry, for open and closed cycles of circulation

4 Evolution of Lakes

From the moment of their development, lakes are subject to evolution process. Their rate and scale depend on a number of overlapping factors (location of the lake, climate conditions, morphometry, human pressure, etc.). The analysis of cartographic materials since the first half of the nineteenth century generally shows four situations suggesting stability of the surface area (Fig. 3a), an increase in the surface area (usually as a result of hydrotechnical works, Fig. 3b), a decrease in the surface area (Fig. 3c), and an increase in the number of lakes as a result of a

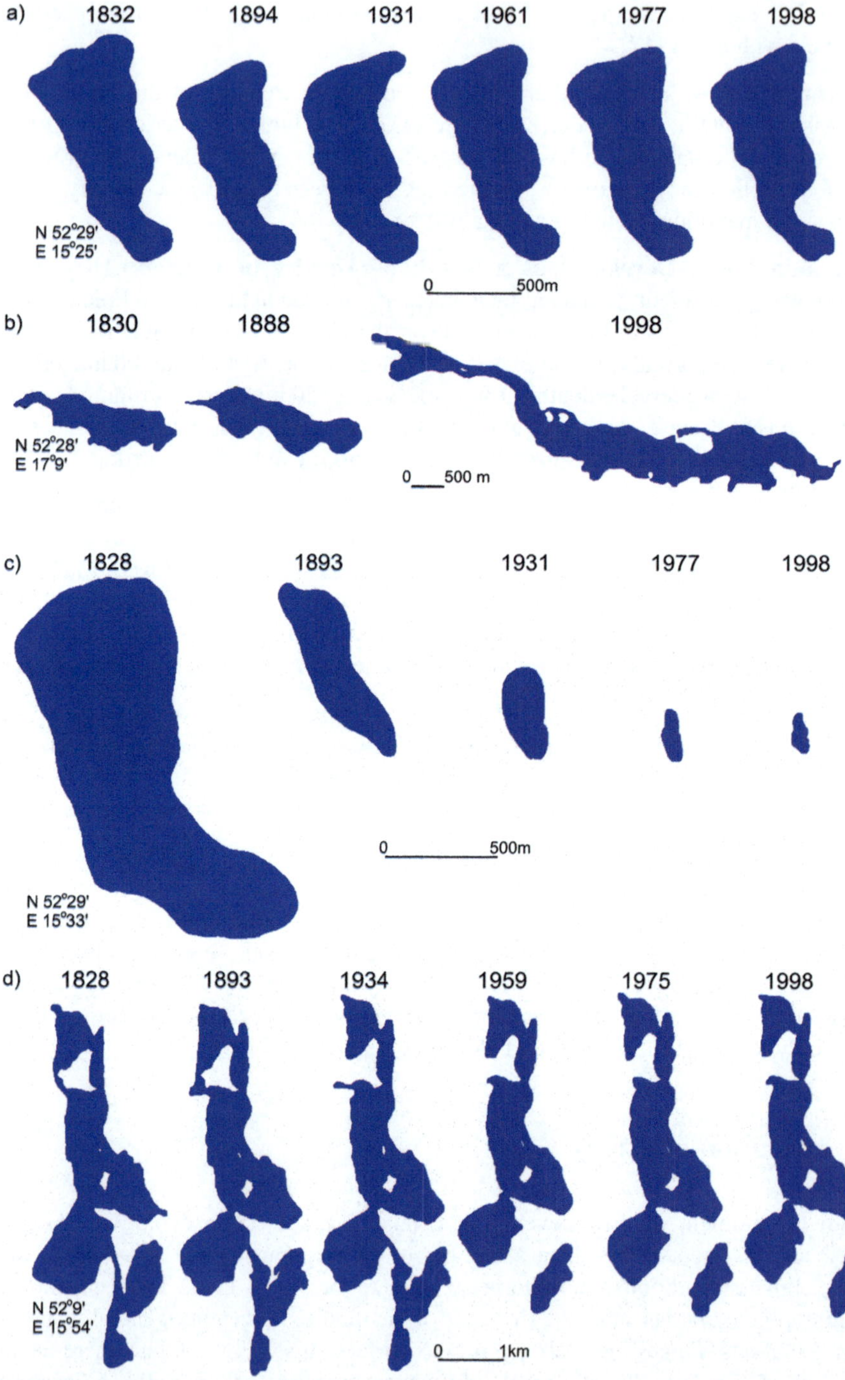

Fig. 3 Examples of surface areas of lakes: (**a**) Lake Cisie – stability, (**b**) Lake Kowalskie – an increase in the surface area as a result of damming, (**c**) Lake Grążyk – considerable decline,

decrease in the surface area of a lake and its division into several smaller water bodies (Fig. 3d). It should be emphasised that the majority of lakes as concave landforms are subject to fast decline.

From the geological point of view, the age of lakes is very short term. This particularly concerns postglacial lakes. Their time of existence is estimated for several thousand years. According to calculations performed by Kalinowska [28], lakes currently constitute only 40% of the original state (counted from the moment of retreat of the Scandinavian ice sheet) in the territory of Poland (Fig. 4).

Modern processes causing the decline of lakes are very dynamic and constitute a resultant of natural and anthropogenic factors. A considerable decrease in the number and surface area of lakes occurred from the end of the nineteenth century. According to Majdanowski [6], the number of lakes with an area of 1 ha and more amounted to 9,296, and their surface area was 316,614 ha. Lake density in Poland amounted to 1.02%. Over a period of less than half a century, the number of lakes decreased by 2,215, and their surface area by 35,550 ha, i.e. by 11.20 % [29]. This resulted in a decrease in lake density to 0.90%.

For the purpose of determination of how the passage of time affects the life span of lakes, the following analysis was performed. In areas located between the ranges of subsequent phases of the last glaciation, i.e. Leszczyn and Poznań phase, Poznań and Pomeranian phase, and north of the Pomeranian phase, the number of lakes was determined, their total water resources, and total surface area. Based on the known surface area of the designated areas, three indices were calculated for each of them: lake density, mean depth of lakes, and layer of resources of lake waters in mm. The analysis evidently shows that towards the south the lakes are “older”. This is confirmed by the values of all indices. Lake density consequently decreases towards the south. Mean depths of lakes and layer of water resources decrease in a similar way (Table 3, Fig. 5).

Two factors can be generally designated, usually co-occurring, responsible for the decline process. The first one is lake water level fluctuations, and the second is aggradation of sediments within the lake basin. The causes of the oscillation of the water level in lakes can be as follows: short- or long-term climate fluctuations causing variable alimentation of aquifers feeding the basin; deforestation of the catchment area (natural – plant pests and that related to human activity); effect of local factors, e.g. variable time of inclusion of the lake to the hydrographic network and variable erosion base [31]; and performance of different kinds of hydrotechnical works, both within the lake (e.g. outflow regulation) and in the catchment area (e.g. meliorations). The latter in the case of the discussed case have been particularly evident over the last several centuries. The turning point was the seventeenth century: the beginning of the “olender colonisation”. The period of expanding arable land began through the removal of excess water [32]. The activities substantially

Fig. 3 (continued) (**d**) Lakes Chobienickie-Wielkowiejskie-Kopanickie – an increase in the number of lakes as a result of division into several smaller water basin (following [27], changed)

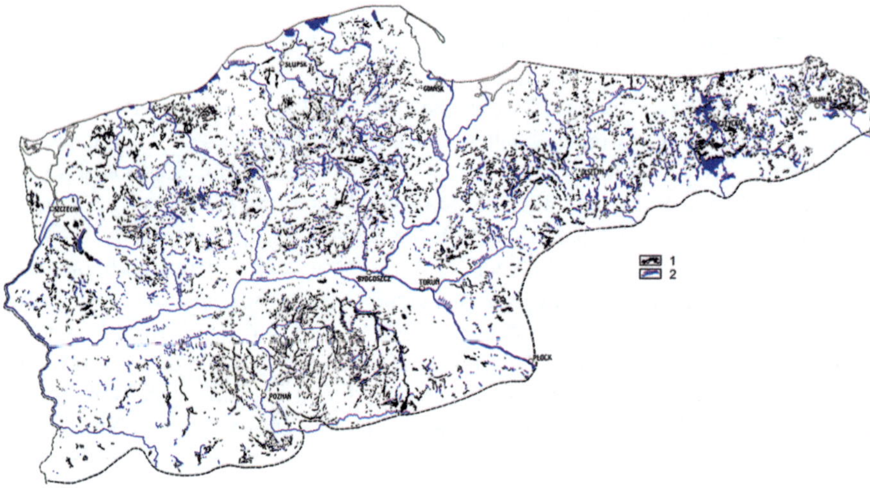

Fig. 4 Non-existent (1) and currently existing lakes (2) in Poland (according to [28])

Table 3 Data and indices concerning the surface area, number, and water resources in lakes within the designated phases of the ice sheet [30]

Area	Surface area in km^2	Number of lakes	Water resources in km^3	Surface area of lakes in km^2	Lake density in %	Mean depth in m	Water resources per surface area in mm
North of the Pomeranian phase (III)	42,800	2,443	8.694247	1,216.003	2.84	7.15	203
Between the Poznań and Pomeranian phase (II)	62,470	3,998	10.063343	1,446.983	2.32	6.95	161
Between the Leszczyn and Poznań phase (I)	11,470	352	0.498868	121.168	1.06	4.12	43

Notice: designations with Roman numbers in accordance with those in Fig. 5

translated into a decrease in the surface area of lakes, and further transformations (an increase in the intensity of agriculture, industrial revolution, etc.) intensified the process. Examples of a considerable decrease in the surface area of the lake as a result of melioration works and complete disappearance of the lake as a result of such activities are presented in Fig. 6. In the first case, the surface area of Lake Wielimie decreased by 2,800 ha, and in the second case, a lake with an area of 490 ha was completely meliorated.

A change in the water level affects an important process in the scope of morphological transformations of lakes, namely, growing over. Succession of vegetation is

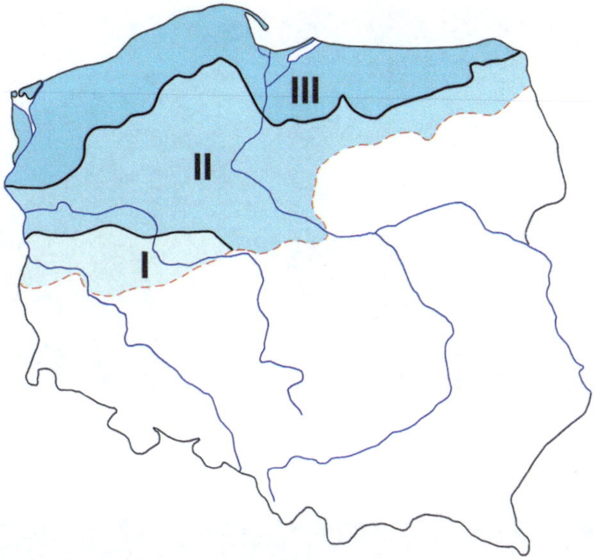

Fig. 5 Phases of the Scandinavian glaciation: I – Leszczyn phase, II – Poznań phase, III – Pomeranian phase, ([30], changed)

a natural element of evolution of lakes, and its acceleration can occur through the exposure of new littoral zones. Conditions favourable for an increase in the rate of overgrowing of lakes cause an increase in the supply of biogenic substances (nitrogen and phosphorus compounds), among others, as a result of permanent intensification of agriculture and increase in the use of artificial fertilisers. Data of the Central Statistical Office in Poland [35] shows an increase in the use of nitrogen fertilisers by 54 kg (per 1 ha) and by 6 kg in the case of phosphorus fertilisers (per 1 ha) over the period of the last 60 years. Selected examples of overgrowing of lakes in Poland were presented in Figs. 7 and 8.

The aforementioned processes, i.e. change in the water level and vegetation succession, are relatively evident, unlike processes occurring under the water surface, resulting in filling the lake basin with sediments. Their intensification results from the location of the lakes in the catchment, and the lake basins constitute recipients of matter transported by rivers, supplied during surface overflows or as a result of aeolian activity causing filling the basin and consequently its nivelation. The aforementioned processes overlap with the production of biomass within the lake itself. The aggradation of bottom sediments can be a result of aggradation of biogenic mass (rapid process in shallow lakes), precipitation of chemical compounds, sedimentation of detrital sediments supplied by rivers, and colluvial and deluvia deposited in the lake, and effect of aeolian processes causing burying [19]. One of the methods permitting the determination of changes occurring in the morphometry of lake bottoms is the analysis of bathymetric plans from at least two periods. An example of such a procedure is presented in Fig. 9.

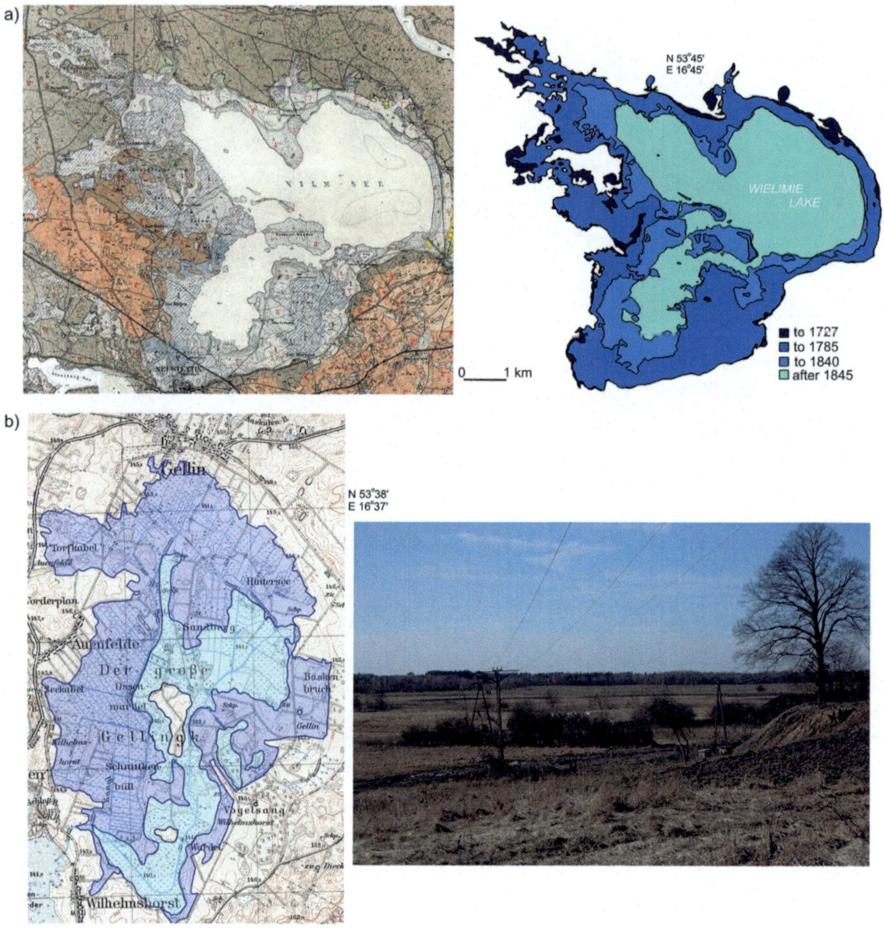

Fig. 6 Changes in the surface area of lakes: (**a**) a considerable decrease in the surface area of Lake Wielimie, (**b**) complete disappearance of Lake Jelenino [33, 34]

Both comparisons show considerable shifts of isobaths over the period of the last 50 years towards the middle of the lake. According to numerous studies [38–40], such a situation is commonly recorded in the case of many lakes in Poland. Importantly, the obtained results showed that the decline is faster vertically (shallowing) than horizontally (changes in the surface area). The three-dimensional approach to the process of decline of the lakes is very important due to the assessment of changes in water resources accumulated in their basins. The water resources of Poland belong to the lowest in Europe. Their averaged volume per one inhabitant amounts to approximately 1,600 m^3 $year^{-1}$ (it constitutes a ratio of mean annual river outflow to number of inhabitants). Due to this, it is important to possibly decrease water outflow from the catchment. The role is fulfilled by lakes. One of the

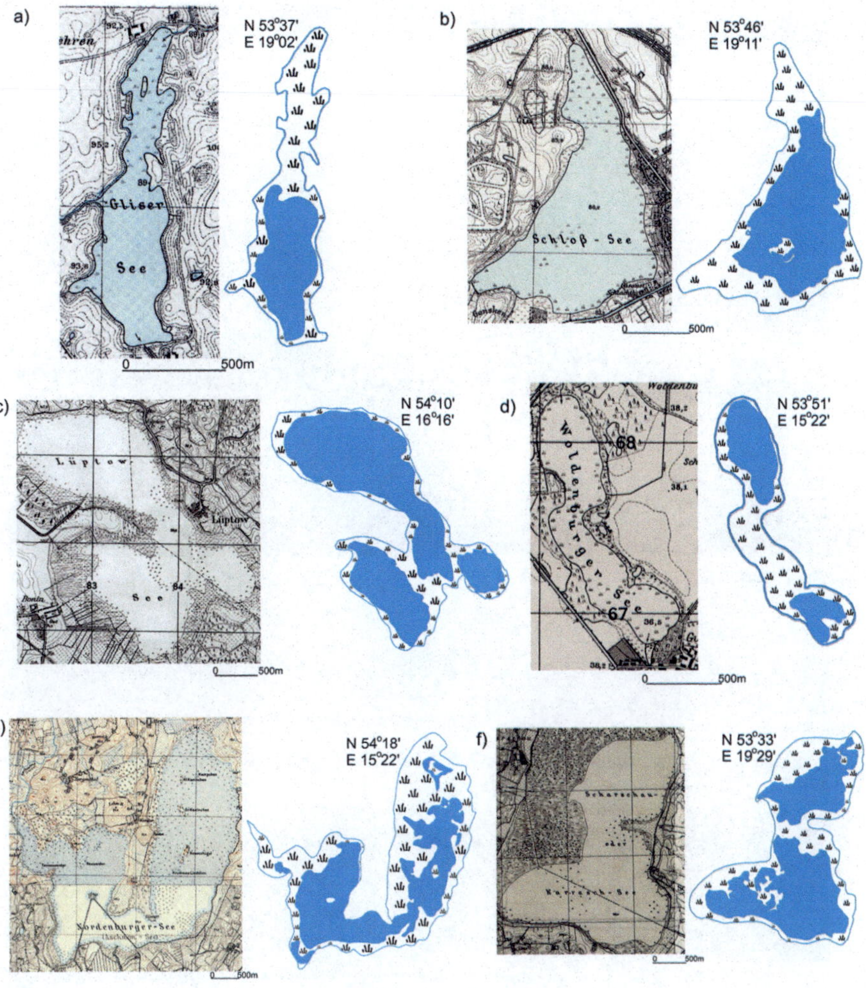

Fig. 7 Selected examples of overgrowing of lakes: (**a**) Czarne Dolne, (**b**) Liwieniec, (**c**) Lubiatowo, (**d**) Dąbie, (**e**) Oświn, (**f**) Karaś (after [36])

main features of lakes is the ability to retain water and therefore de facto slowing down of the transformation of precipitation into outflow.

The retention abilities of lakes contribute to the mitigation of hydrological phenomena with extreme character, i.e. both during violent floods (suppression of the flood wave) and in situations of draught ("sustenance" of the surface runoff). Due to the possibility to retain water, lakes are key element for the existence and development of many branches of the economy (energy engineering, agriculture, tourism, etc.). Due to the aforementioned conditions, lakes become an important factor shaping the course of natural processes as well as the economic development of a given region. Therefore, all efforts should be taken to eliminate or largely

a)

b)

Fig. 8 Progressing process of overgrowing – selected examples: (**a**) Łebsko, (**b**) Stępuchowskie (Phot. A. Choiński, M. Ptak)

limit processes accelerating their decline (meliorations, uncontrolled fertilisation in catchments, etc.). In the context of water deficits, increasing attention is paid to the possibilities of water retention in Poland. Artificial reservoirs are constructed, and natural lakes are dammed for that purpose.

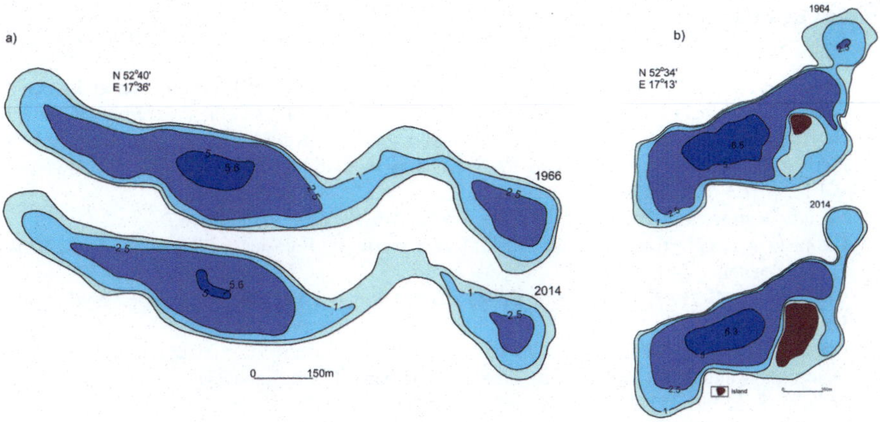

Fig. 9 Selected examples of changes in the bathymetry of lakes: (**a**) Mielno, (**b**) Turostowo [37]

5 Conclusions

More than 7,000 lakes exist in Poland. Their genesis is particularly related to the presence of the Scandinavian ice sheet. Lakes constitute a very important element of the natural environment, affecting water circulation, microclimate, biodiversity, economy, etc. However, benefits resulting from the occurrence of lakes can already in a relatively short time be lost as a result of the progressing process of their decline. Such a situation results from the natural process of evolution, accelerated by human activities related to the regulation of water relations and an increase in productivity of agriculture, and consequently an increase in the supply of nitrogen and phosphorus compounds to water. Due to this, all efforts should be taken to limit harmful aspects of human activity in lake catchments and in a further perspective to strive for the renaturation of degraded or non-existent lakes.

To sum up, the current rate of decline of lakes in Poland suggests that the perspective age of the majority of them can be estimated for several hundred to 2–3 thousand years. This, however, only concerns a decrease in the surface areas of lakes. It should be remembered that the process of decline also occurs through successive shallowing of lakes, and the process is often several times more intensive than the visible rate of decrease in the surface area. This suggests that in the conditions of the current tendency for shallowing, in exceptional situations, the perspective age of some lakes will vary from several tens to several hundred years.

References

1. Bajkiewicz-Grabowska E, Hillbricht-Ilkowska A, Kajak Z, Kufel L (1989) Metodyka oceny odporności i obciążenia jezior, stanu eutrofizacji i czystości ich wód oraz wpływu zlewni. In: Hillbricht-Ilkowska A (eds) Jeziora Mazurskiego Parku Krajobrazowego, Stan eutrofizacji, kierunki ochrony. Zeszyty Naukowe PAN, KN przy Prezydium PAN "Człowiek i Środowisko", vol 1, pp 163–190
2. Pociask-Karteczka J (eds) (2003) Zlewnia, właściwości i procesy. Kraków
3. Choiński A (1991) Katalog jezior Polski. Część pierwsza: Pojezierze Pomorskie. Wyd. Nauk. UAM, Poznań
4. Choiński A (1991) Katalog jezior Polski. Część druga: Pojezierze Mazurskie. Wyd. Nauk. UAM, Poznań
5. Choiński A (1992) Katalog jezior Polski. Część trzecia: Pojezierze Wielkopolsko-Kujawskie i jeziora na południe od linii zasięgu zlodowacenia bałtyckiego. Fundacja "Warta". Poznań
6. Majdanowski S (1954) Katalog jezior Polski, Warszawa
7. Bajerlein J (1929) Geneza jezior Sierakowskich. Bad Geogr nad Polską Płn-Zach z. 4–5, Poznań
8. Żynda S (1961) Morfologia południowo-zachodniej części Wysoczyzny Lubuskiej. (streszczenia referatów i przewodnik wycieczkowy Zjazdu Regionalnego PTG), Poznań
9. Krygowski B et al (1953) Mapa morfologiczna Niziny Wielkopolsko-Kujawskiej (maszynopis). IG UAM, Poznań
10. Kondracki J (1981) Geografia fizyczna Polski. PWN, Warszawa
11. Wilgat T (1953) Jeziora Łęczyńsko-Włodawskie. Annales UMCS, vol 8, Lublin
12. Kostrowicki J (1968) Środowisko geograficzne Polski. PWN, Warszawa
13. Borowiak D (1993) Pochodzenie i morfologia niecek jeziornych. In: Lange W (eds) Metody badań fizycznolomnologicznych. UG Gdańsk
14. Pietkiewicz S (1958) Wody kuli ziemskiej. PWN, Warszawa
15. Augustowski B (1972) Niziny nadmorskie (Pobrzeże Bałtyckie). In: Geomorfologia Polski, t. 2, PWN, Warszawa
16. Łomniewski K (1958) Zalew Wiślany. Prace Geogr., vol 15, Inst. Geogr., PAN, Warszawa
17. Bartkowski T (1965) Rzeźba, budowa geologiczna, geomorfologia i stosunki wodne. Województwo Koszalińskie, Monografia geograficzno-gospodarcza, Inst. Zach., Poznań
18. Choiński A, Skowron R (1999) Jezioro Tobellus przykładem nowego typu genetycznego misy jeziornej. Acta Univ Nicolai Copernici z. 103
19. Choiński A (2007) Limnologia fizyczna Polski. Wyd. Nauk. UAM, Poznań
20. Rutkowski J, Król K, Krzysztofiak L (2001) Jezioro Pietronajć w Wigierskim Parku Narodowym – geneza i stan obecny. Chrońmy Przyrodę Ojczystą R LVII(57):z. 2
21. Choiński A, Strzelczak A (2011) Bathymetric measurements of Morskie Oko Lake. Limnol Rev 11(2)
22. Plany Batymetryczne (1960–1988) Inst. Rybactwa Śródlądowego, Olsztyn
23. Sawicki L (1929) Atlas jezior tatrzańskich. PAU, Prace kom Geogr T. XXII
24. Choiński A, Ptak M (2016) Batymetria jezior meteorytowych w rezerwacie "Meteoryt Morasko". Acta Socetatis Metheoriticae Polnorum 25–31
25. Choiński A (2017) Changes in morphometrics of the coastal lakes. In: Obolewski K, Astel A, Kujawa R (eds) The Hydroecological Determinants of Functioning of Southern Baltic Coastal Lakes. PWN, Warszawa
26. Mrula A (1981) Charakterystyka i możliwości eksploatacyjne gminy Dziwnów. In: Gmina Dziwnów. Monogr. Geogr. Stud. Koło Naukowe Geogr. im. S. Pawłowskiego, UAM, Poznań
27. Choiński A (2009) Changes in the area of lakes from the Obra River drainage basin taking place from the beginning of the 19th century. Limnol Rev 9(4):159–164
28. Kalinowska K (1961) Disappearance of the Post-glaci.al Lakes on the Territory of Poland. Przegl Geogr 33(3):511–518
29. Choiński A (2006) Katalog jezior Polski. Wyd. Nauk. UAM, Poznań

30. Choiński A (2017) Geneza i rozmieszczenie jezior. In: Jokiel P, Marszelewski W, Pociask-Karteczka J (eds) Hydrologia Polski. PWN, Warszawa, pp 223–229
31. Marsz A (1971) Zmiany linii brzegowej jeziora Komorze. Bad. Fizjogr. nad Polską Zach., t. 24
32. Kaniecki A (1991) Zmiany stosunków wodnych w rejonie Konina, związane z działalnością kopalnictwa odkrywkowego. In: Stankowski W (eds) Przemiany środowiska geograficznego obszaru Konin-Turek. IBCz UAM, Poznań
33. Wieczorek J, Choiński A, Ptak M (2017) Effect of human pressure on changes in the range of Wielimie Lake (Pomeranian Lakeland). Z badań nad wpływem antropopresji na środowisko 18:48–55
34. Ptak M, Choiński A, Strzelczak A, Targosz A (2013) Disappearance of Lake Jelenino since the end of the XVIII century as an effect of anthropogenic transformations of the natural environment. Pol J Environ Stud 22(1):191–196
35. Rocznik Statystyczny (1959, 2016) GUS, Warszawa
36. Ptak M (2013) Historical medium-scale maps as a source of information on the overgrowing of lakes. Limnol Rev 13(3):155–162
37. Choiński A, Ławniczak A, Ptak M (2016) Changes in water resources of Polish lakes as influenced by natural and anthropogenic factors. Pol J Environ Stud 25(5):1883–1890
38. Choiński A, Ptak M (2009) Lake infill as the main factor leading to Lake's disappearance. Pol J Environ Stud 18(3):347–352
39. Ławniczak AE, Choiński A, Kurzyca I (2011) Dynamics of lake morphometry and bathymetry in various hydrological conditions. Pol J Environ Stud 20(4):931–940
40. Ptak M (2013) Lake evolution in the Żnin region in the years 1912-1960 (Central Poland). Quaestiones Geographicae 32(1):21–26

Soft-Water Lobelia Lakes in Poland

Dariusz Borowiak, Ryszard Piotrowicz, Kamil Nowiński, and Piotr Klimaszyk

Contents

Abstract The lobelia lakes, unique and rare in Poland, are located mainly in northern part of country in moraine plateaus of Bytów and Kashubian Lakelands and outwash plains of Tuchola Forest and Charzykowy. In the middle of the twentieth century, there were over 190 lobelia lakes in Poland. Lobelia lakes are classified as so-called soft-water lakes due to low concentrations of calcium and magnesium. Their waters are weakly buffered and usually have acid reaction and also small amount of biogenic elements. Lobelia lakes are overgrown by specific plant species – isoetids – adapted to the poor habitat: *Lobelia dortmanna*, *Isöetes lacustris*, *Littorella uniflora* and several other accompanying species. The lobelia lakes in Poland are prone to degradation and loss of their unique values. The most serious threats include acidification, humification deposition,

D. Borowiak
Department of Limnology, University of Gdańsk, Gdańsk, Poland

Limnological Station in Borucino, University of Gdańsk, Borucino, Poland

R. Piotrowicz · P. Klimaszyk (✉)
Department of Water Protection, A. Mickiewicz University, Poznań, Poland
e-mail: pklim@amu.edu.pl

K. Nowiński
Department of Limnology, University of Gdańsk, Gdańsk, Poland

E. Korzeniewska, M. Harnisz (eds.), *Polish River Basins and Lakes – Part I*,
The Handbook of Environmental Chemistry 86,
https://doi.org/10.1007/978-3-030-12123-5_5

alkalinization, eutrophication by agricultural activities or recreational use and lowering of water level induced by climate changes.

Keywords Eutrophication · Humification · *Isöetes lacustris* · Lakes · *Littorella uniflora* · *Lobelia dortmanna*

1 Introduction

The "lobelia lake" term originates from a plant species, a boreal-Atlantic relict *Lobelia dortmanna* – consisting of basal rosette of evergreen elongate leaves typically found in oligotrophic soft-water lakes [1–4]. It is mostly in Poland and Denmark that this term is used in scientific literature in reference to isoetid lake [5]. Apart from *L. dortmanna*, these lakes can be overgrown by other characteristic hydromacrophytes, namely, *Littorella uniflora* and *Isöetes lacustris* (Fig. 1). The lobelia lakes are most frequent in the boreal and temperate zones of northern hemisphere as well as at higher elevations of subtropic regions. These lakes are particularly unique within limnetic ecosystems in Poland and therefore deserve special protection and attention.

The investigations on Polish lobelia lakes (PLL) had been initiated in the middle of the twentieth century [1, 6, 7]. The primary research in this area had focused on physico-chemical description of water and sediments and analyses of coverage of characteristic plant species. At the turn of the twentieth and twenty-first century, the studies on PLL has flourished and focused on the biological and chemical basis of their functioning and identification of main threats, particularly from human pressure [8–20]. The conducted research identified about 190 lobelia lakes in Poland, most of which are water bodies with no outflow and small surface area. Elaboration of

Fig. 1 Characteristic plant species (**a**) *Lobelia dortmanna*, (**b**) *Littorella uniflora*, (**c**) *Isöetes lacustris*

physical, chemical and biological properties of PLL allowed Kraska and Piotrowicz [16] to distinguish four lake subtypes: (1) acidified (oligohumic and polyhumic), (2) balanced, (3) eutrophicated and (4) degraded.

The lobelia lakes in Poland are particularly prone to degradation and loss of their unique character due to poorly buffered status of water, specific catchment physiography and lake basin morphology. The greatest threats to PLL include acidification from atmospheric deposition, alkalinization, eutrophication by agricultural activities or recreational use, lowering of water level induced by climate changes, draining of peatbog humic water into the lakes, liming and fish stocking – all leading to critical changes in the physico-chemical properties of their waters and, consequently, to the disappearance of characteristic vegetation.

The publication discusses the physical and chemical aspects of the functioning of Polish lobelia lakes in the context of both natural changes associated with ageing of lakes and human pressure.

2 Distribution of Polish Lobelia Lakes

Soft-water lobelia lakes may be found in temperate and boreal zones of the northern hemisphere. They may also occur in subtropic or tropic regions although in such areas they are usually located at high elevations. In the Northern Europe and North America, they are situated on granite bedrock, while in Atlantic regions of Western and Central Europe, lobelia lakes occur within low calcareous, sandy soils or develop in proximity of high or transition bogs. They usually are small, non-throughflow water bodies which are fed exclusively by precipitation or surface flow from catchments which are poor in nutrients and calcium. Literature data indicate that in the middle of the twentieth century, over 190 lobelia lakes in Poland were present [1, 16, 21, 22]. In Poland, lobelia lakes are located in Northern part of the country within the terminal moraine and on outwash plains formed during the Wistulian Glaciation (North-Polish Glaciation, Weichselian) (Fig. 2). The majority of them are located in moraine plateaus of Bytów and Kashubian Lakelands and outwash plains of Tuchola Forest and Charzykowy [23]. Western range of PLL reach Drawsko and Ińsko Lakelands. In the north-east part of Poland (near Olsztyn city), there are three lakes in which occurrence of *Isöetes lacustris* was found. One of PLL – Lake Dołgie Wielkie – developed in the vicinity of Baltic Sea and initially was a sea lagoon. The only one of the PLL is located in mountain region of southern Poland (Karkonosze Mountains). It occupies the glacial cirque excavated in granite bedrock. The characteristic lobelian vegetation is represented here only by *Isöetes lacustris*.

Catchments of these lakes are characterized by soils which are poor in nutrients and calcium and which are primarily overgrown with acidophilous coniferous (pine) or deciduous (beech) forests. In case of several lakes, quaking bogs developed on its edges, while in some other peatbogs occur within the catchment areas.

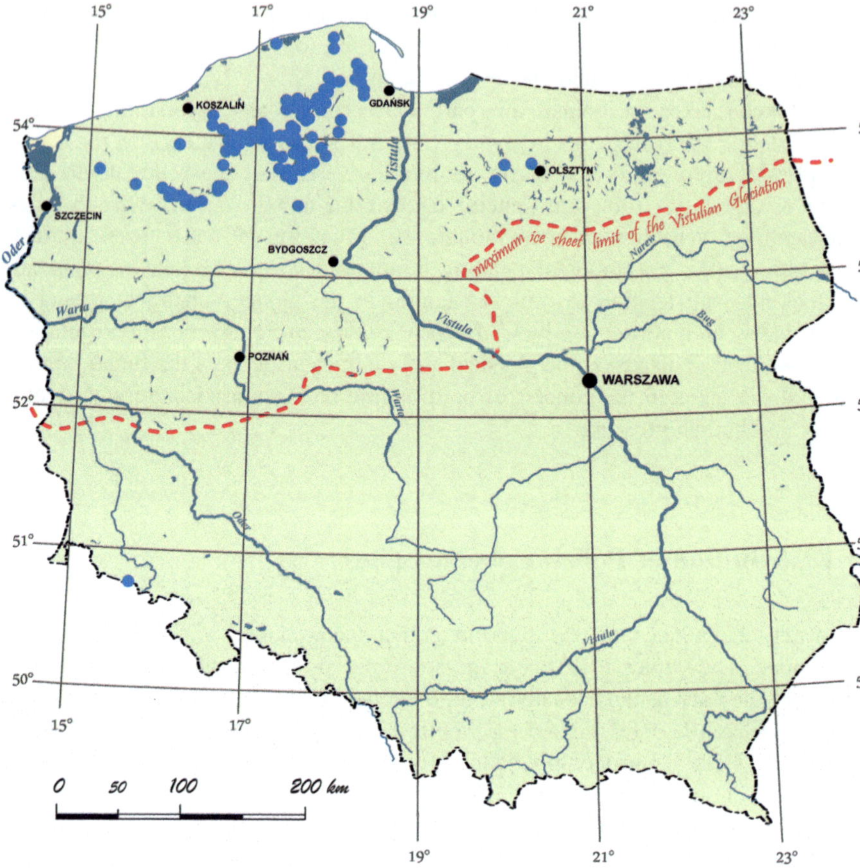

Fig. 2 Distribution of Polish lobelia lakes

The lobelia lakes are usually located near the main water divide zone separating the catchments of coastal rivers from the Vistula and Oder basins, with slightly more (57% of lakes) lying on its northern side. At a distance of up to 10 km from the Pomeranian Water Divide (PWD), there are 46%, while in the zone up to 15 km about 61% of the PLL.

In terms of the location of the former and current lobelia lakes against the hydrographic units, those on the north side of the PWD are within the basins of the following rivers: the Dziwna, Rega, Parsęta, Wieprza, Słupia, Łupawa, Łeba, Piaśnica, Reda and Radunia.

Lakes located on the south side of the PWD are located in the basins of the Wierzyca, Wda and Brda as well as the Noteć through the Gwda and Drawa basins. Two of the three lakes located farthest to the east lie in the Drwęca basin (Lake Długie near Łukta and Lake Czarne near Ostróda), while Lake Żbik is in the Łyna basin. Wielki Staw, located in the Karkonosze Mts., is drained via the Biały Potok and Łomnica to the Bóbr (the left tributary of the Oder River).

The number of PLL is expected to gradually decrease due to increase in human impact within their catchment areas.

3 Catchments of Polish Lobelia Lakes

The variability of geomorphological forms is accompanied by lithological heterogeneity of surface deposits of the lake catchments. It is manifested by the occurrence of loose sands and gravels around lakes located on outwash plains, the domination of tills, glacial sands and gravels on undulated moraine plateaus, as well as sands, gravels, boulders and tills with a predominance of clay fractions on terminal moraine hills.

About 50% of lobelia lakes are located on sandy and gravel formations of outwash plains. This creates very good conditions for infiltration, while reducing the importance of surface supply of lakes. A characteristic feature of lobelia lakes is the importance of the infiltration function, which is based on the significant underground water outflow. The underground outflow is also facilitated by the location of lakes near deeply cut subglacial channels and river valleys, which form the basis for groundwater drainage. Increased importance of surface and subsurface supply can be observed in the case of lakes which catchment is built, among others, of glacial tills. These types of sediments, apart from sands and gravels, dominate in the catchments built mainly of ground moraines (~37% of lakes) and terminal moraines (~6% of lakes) deposits.

In addition to the above-mentioned surface deposits, in the catchments of some lobelia lakes, there are kame sands and silts (~1.5% of lakes, e.g. lakes Boruja Mała, Wisełka, Świdno), aeolian sands locally on dunes (e.g. lakes Dołgie Wielkie, Nowoparszczenickie) as well as peats and fluvial silts, sands and gravels (e.g. lakes Orzechowo, Racze, Chełm, Dołgie Wielkie). The catchment of Wielki Staw, the only Polish lobelia lake in the mountains, is made of Carboniferous granites.

The young age of the postglacial relief and the location of the lobelia lakes in the vicinity of hills determining a course of the first-order water divide influence a significant differentiation of relative and absolute heights within the catchment. Despite mostly small areas of the catchments, the denivelations often exceed 50 m (e.g. lakes Boruja Mała, Choczewskie, Czarne near Salino), reaching values even above 90 m, e.g. in the case of the Płocica lake catchment. The absolute heights of the water surface elevation of the Polish lobelia lakes are within a wide range of 0.1 m a.s.l. (Lake Dołgie Wielkie) to 1,224.1 m a.s.l. (Lake Wielki Staw in the Karkonosze Mts.). Extreme values relate to exceptional lakes, the first of which is located on the Slovincian Coastland, at a distance of about 1 km from the sea, while Lake Wielki Staw is a high-mountain corrie lake. The remaining lakes are mostly located in the lake district areas, and most of them (90% of lakes) are at an altitude of over 100 m a.s.l. The elevation of the water surface of some of them (8% of lakes) is over 190 m, reaching a maximum of 208.4 m (Lake Święte) and 214.0 m a.s.l. (Lake Stacinko). The mean absolute altitude of the water level of the Polish lobelia lakes is 149.5 m a.s.l. The median is slightly higher reaching 149.9 m a.s.l.

The mean slope of the lobelia lake catchments is about 35‰, although this parameter shows a considerable variation. The lowest values are characteristic for catchments located on outwash plains (e.g. 6.8‰, Lake Duże Sitno; 11.3‰, Lake Sierzywko; 11.4‰, lakes Kiedrowickie and Cietrzewie). However, the largest slopes occur in geomorphologically diverse catchments, where, apart from outwash plains or ground moraine plateaus, there are areas occupied by terminal moraine (78.0‰, Lake Herta; 70.9‰, Lake Czarne near Borzytuchom; 68.7‰, Lake Płocica).

The location of lakes in close proximity to the main watershed along the highest peaks of Pomerania means that their catchment areas are usually small (mean of 3.5 km^2). The smallest catchments are only about 0.12 km^2 (e.g. lakes Czarne near Borzytuchom and Morskie Oko), but about 10% of lobelia lakes have more extensive supply areas, exceeding 20 km^2 (Lake Bobięcińskie Wielkie, 28.8 km^2; Lake Salińskie, 21.9 km^2).

The consequence of the small size of the total catchment is a frequent lack of surface outflow from lobelia lakes. In general, the hydrographic network of lake catchments is usually very poorly developed, so lakes are often its only elements. The catchments of individual lakes also show internal inhomogeneity of their hydrographic structure, which is indicated mainly by the large proportion of endorheic areas (approx. 40% on average). In the case of about 25% of the lakes, the endorheic areas occupy over 60% of the total catchment area, reaching maximum values of over 80% (e.g. Lake Sierzywko, 96%; Lake Cietrzewie, 92%; Lake Moczadło near Męcikał, 87%). In the discussed group, there are also lakes whose catchment area is completely devoid of endorheic areas (Lake Morskie Oko, Lake Wielki Staw in the Karkonosze Mts.) or their share does not exceed 5% (Lake Jelonek, Lake Osowskie, Lake Herta). Endorheic areas can be considered as passive parts of the catchment, whose share reduces the surface actively involved in the transport of matter to the lakes.

About 70% of lobelia lakes are closed lakes, and most of them do not have a tributary also (Table 1). Probably almost all of the lobelia lakes were originally endorheic links of the hydrographic network, which were artificially included into the surface outflow system through drainage ditches. In a large part of this type of lakes, surface outflow occurs only periodically or episodically. As a result of the former melioration works, about 50% of lobelia lakes have permanent or periodic inflows. This is a serious threat to the proper functioning of aquatic environments

Table 1 Surface inflows and outflows in former and current lobelia lakes (based on a topographical map in the scale of 1:10,000 and own field research)

Inflow	Outflow	Number of lakes	Percentage
–	–	87	44.8
+	–	45	23.2
–	+	10	5.2
+	+	52	26.8
	Total	194	100.0

due to the unnatural supply of humic compounds (yellow substances) from drained peat bogs, as well as increased supply in biogenic elements from areas used for agriculture.

Comparing the lake surface area with the size of its alimentation area, it is possible to predetermine its role in the surface water circulation. Lakes, whose ratio of the catchment area to the lake area (Ohle's coefficient) is relatively low, usually constitute endorheic components of the hydrographic network. The low values of this ratio are the distinguishing features of the most lobelia lakes, which largely justifies their closed hydrologic type or significantly longer water residence times (lower flushing rates). According to calculations made by Lange [24], the surface outflow from Pomeranian lakes is initiated only in conditions if the Ohle's coefficient is higher than 7.8.

The outflow from the lakes is also driven by other main factors such as permeability of the deposits that build the catchment, the dominant direction of groundwater exchange (drainage lakes/seepage lakes) and the share of endorheic areas (drainage blind depressions) in the catchments. Lobelia lakes have a large proportion of closed drainage areas, a significant elevation above the sea level and location on sediments of fairly good permeability. Therefore, despite sometimes extensive total catchments, their active parts are small, which means that even if the value of Ohle's coefficient exceeds 100 (Lake Cietrzewie, Lake Sierzywko), these lakes still maintain their closed nature. The majority of lobelia lakes show, however, much lower coefficients (mean ~20, median ~10). For instance, the minimum values are 1.3, Lake Morskie Oko; 2.0, Lake Piasek; and 2.3, Lake Smołowe.

The measure of the impact of alimentation on the lake is the Schindler's coefficient, which determines the relationship between the catchment area and the lake's volume. It decides the lakes' resistance to degradation. The mean and median of the coefficient calculated for the lobelia lakes are, respectively, ~5.5 and ~3.5. About 18% of the lakes show the values below 1. Among them are, for example, lakes Morskie Oko, Piasek, Łąkie and Jeleń, in which the Schindler's coefficient varies from 0.3 to 0.4. This parameter exceeds the value of 10 in only about 10% of lakes. High values concern, among others, the lakes Sierzywko (69.6), Cietrzewie (46.6), Żabionek (33.4) and Stary Staw (31.1).

Lakes characterized by zero surface outflow, which are the largest group, exchange water mainly with the atmosphere through precipitation and evaporation. Alimentation from the catchment is marginal in these conditions. It is usually limited to inefficient and unorganized surface and subsurface runoff and possibly small groundwater drainage. In the lakes associated with open surface drainage systems, in addition to vertical exchange, surface exchange is also recorded. However, the riverine outflow is usually an insignificant element of the water balance. The mean value of the flushing rate in lobelia lakes with surface outflow is 0.30 $year^{-1}$ (median 0.16 $year^{-1}$). Majority of lobelia lakes are hydrologically passive types [25]. The smallest water exchange is recorded in the lakes in which the outflow functions only periodically. An example can be Lake Smołowe with an exchange rate of around 1% per year. Few of the drainage lobelia lakes (less than 10% of total number) have flushing rates of over 1.0, representing the medium hydrological type. Such values were noted, for example, in the lakes Wysokie, 1.4; Brzeżonko, 1.31; and Otalżyno, 1.04 $year^{-1}$ [25].

The geological, morphological and hydrological conditions of the catchment determine that the lobelia lakes are not connected to the drainage network or are connected to it via watercourses being only at initial stages of organization within the hierarchically developed systems of riverine outflow (Table 2). So, majority of lobelia lakes are hydrologically isolated closed-basin lakes of order –3 (68%). Few of them are connected with the drainage systems by wetlands, constituting of order –2 lakes (1.5%). Lakes of order –1, which are connected to the surface drainage network by temporary streams or streams of very low flow, account for 16% of former and contemporary lobelia lakes. Among lakes drained by permanent streams, the largest group consists of headwater lakes with no inlets and assigned as lakes of order 0 (4.6%). The number of the first-order lakes is slightly lower (4.1%), and the smallest number of lakes is drained by streams higher than second order (2.6%).

The dominant category of the land use structure of most lobelia lake catchments are forests. The mean proportion of total forest from total active parts of the lake catchments is about 75%, with the share of agricultural and built-up areas at the level of, respectively, ~20% and ~4% (median ~12% and ~1%). In the total catchments, the most important types of land use are, on average, forests ~60, agricultural areas ~30 and built-up areas ~1%.

In the case of around 25% of the total number of lakes, the active parts of their catchments are totally covered with forests (e.g. lakes Moczadło near Męcikał, Sosnówek, Żabionko, Sporacz, Czarne near Borzytuchom, Duże Zmarłe and Nierybno). However, such favourable conditions do not apply to all lakes. Less

Table 2 Number of lobelia lakes in particular hydrological types and their corresponding order defining the hierarchical location of the lakes in the surface outflow system [26] (based on a topographical map in the scale of 1:10,000 and own field research)

Hydrological characteristic of lakes	Order	Number of lakes	%
Seepage lakes (closed and hydrologically isolated lakes) without hydrological connection to the surface drainage network	–3	132	68.0
Seepage lakes (closed lakes) connected to the surface drainage network through wetlands where the channelized flow does not occur	–2	3	1.5
Seepage lakes (periodically closed lakes) connected to the surface drainage network by temporary streams or streams of very low flow	–1	31	16.0
Headwater lakes (outflow lakes) drained by a permanent streams and with no permanent inlets	0	9	4.6
Flow-through lakes drained by a first-order stream [stream order determined by the Strahler method (1964)]	1	8	4.1
Flow-through lakes drained by a second-order stream [stream order determined by the Strahler method (1964)]	2	6	3.1
Flow-through lakes drained by streams of the order higher than second [stream order determined by the Strahler method (1964)]	>2	5	2.6

than half of the catchment overgrown by forest is found in about 20% of lakes. The lowest forest cover, below 10%, is recorded in the catchments of such lakes as Osowskie, Wielkie Świeniebudy, Ząbinowskie and Księże. In these lakes the predominant type of land use is mainly agricultural land. In the case of about 13% of lakes, agricultural land occupies over half of the catchment area. The maximum share of this type of land use in the active parts of the catchment of several lakes reaches over 70% (e.g. lakes Ząbinowskie, Wielkie Świeniebudy, Glinno, Gubisz and Czarne Jezioro). Built-up areas are very rarely the dominant form of land use of the lobelia lake catchments. A significant (over 30%) share of buildings in the active parts of the catchment occurs, for example, in the lakes Księże, Sitno, Osowskie and Kamień near Szemud.

4 Morphometric Diversity of Polish Lobelia Lakes

The influence of the size and shape of lake basins on the intensity and nature of physical, biological and chemical processes in lakes is widely recognized as significant. Thus, it allows to treat parameters and morphometric indices of lakes as important limnological classification criteria [27–30] and independent variables used in productivity, eutrophication and sedimentation models [31–34]. The morphometric characteristics are often also closely related to the morphogenesis of lake basins and their geological history, assuming specific values for particular types of lake origin [35].

A detailed morphometric analysis of over 190 former and contemporary lobelia lakes of north-western Poland indicates that it is a very diverse group of water bodies. In terms of surface area, the range among lobelia lakes is huge and amounts to nearly three orders of magnitude. The smallest of the lakes have an area of less than 1 ha (Lake Moczydło near Lubnia, Lake Diabelskie), and only 10 of them have an area exceeding 100 ha. Poland's largest lobelia lake is Lake Bobięcińskie Wielkie, whose water surface is 524.6 ha. Almost 43% of the lobelia lakes have areas smaller than 10 ha, 67% smaller than 20 ha, and as many as 85% of them are water bodies whose area does not exceed 50 ha. The "average" lobelia lake, defined by the mean value, has an area of 27.6 ha, while taking as a criterion the median value – only 11.2 ha. For comparison, the "average" lake of the lakeland belt of northern Poland has a mean surface area 1.5 times larger (42.9 ha) but a very similar median value (about 9.9 ha). Although lobelia lakes are mostly small-sized water bodies, and such are dominant among Polish lakes, the size structure in this group of lakes is clearly different. Among the lobelia lakes, the percentage of very small lakes (<5 ha) and very large lakes (>50 ha) is relatively small (in each size class <15%). Small lakes (5–10 ha) and medium-size lakes (10–20 ha) dominate in number, with a combined share exceeding 52% (Fig. 3). The size range of the surface area, in which 50% of all lobelia lakes are contained, takes the values of 6.6 and 27.9 ha.

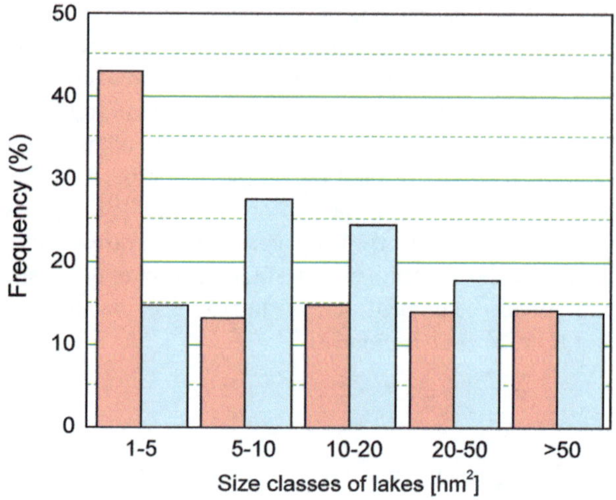

Fig. 3 Frequency distribution of the surface area for all Polish lakes (red bars, $n = 6{,}793$) and contemporary and former lobelia lakes (blue bars, $n = 194$)

The total area of lakes currently considered as lobelia (174 such lakes in Poland) is 45.7 km^2, which is the equivalent of 1.57% of the total area of all lakes in Poland. Almost all lobelia lakes occur in the north-west of Poland (Pomerania), and their percentage share in the area of this region is 0.1%.

The discussed lakes are equally much varied in terms of depth. With reference to the Bogoslovsky's depth typology (1960), which has the greatest practical application to Polish lakes, lobelia lakes have their representatives in all four distinguished typological classes. The largest groups are medium-deep (z_{max} of 5–10 m, z_{mean} of 2–5 m) and deep (z_{max} of 10–20 m, z_{mean} <10 m) lakes, whose share is, respectively, 41.2 and 33.0%. The occurrence frequency of shallow lakes (z_{max} <5 m, z_{mean} <3 m) does not exceed 17.5%, and of very deep lakes (z_{max} >20 m, z_{mean} >9 m) 8.3%. The structure of the lobelia lakes outlined in this way is also directly reflected in the values of central measures (mean and median) calculated for the maximum and mean depth. Based on the criterion of the arithmetic mean for the whole set of lakes ($z_{max} = 11.2$ m, $z_{mean} = 4.5$ m), these are deep lakes, and on the basis of the median value ($z_{max} = 9.1$ m, $z_{mean} = 3.7$ m), these are medium-deep lakes. Extreme values of the maximum depth measured in Lake Bobięcińskie Małe and Lake Bobięcińskie Wielkie are, respectively, 2.0 and 48.0 m. The range of variation in the mean depth ranges from 0.8 (Lake Chełm) to 12.6 m (Lake Krzemno).

The ratio of the mean depth to the maximum depth (depth ratio) allows an approximate assessment of the lake basin form. Among the lobelia lakes, the most common are those whose basins have a hyperboloid shape – typical for outwash plain kettle lakes (depth ratio between 1/3 and 1/2). Lakes with this form of a basin account for nearly 75% of all studied lakes. The mean and median values calculated for this index are identical (0.43), and the interquartile range is determined by the values of 0.38 and 0.48. Less numerous are the lakes with the ellipsoidal form of the basin (depth ratio between 1/2 and 2/3) – 17%, and the lakes whose basins have the shape of a trumpet (depth ratio <1/3) – 11%. The remaining lakes have exactly a conical (3%) and paraboloid (2%) shape of their basin. The extreme values of depth ratio are 0.19 (lakes Zawiat, Bobięcińskie Wielkie, Jelenie Wielkie, Kamień) and 0.62 (lakes Kocioł and Nawionek).

The consequence of the considerable diversity of surface area and depth of lake basins are very large disproportions in lake volume. As the value of the lake volume is more influenced by the surface of the lake, the largest volumes of water are stored by the largest lakes: Bobięcińskie Wielkie (48,985.2 dam^3), Krzemno (17,511.2 dam^3) and Ciemino (14,394.1 dam^3). The total volume of the last two lakes is only 30–35% of the volume of Lake Bobięcińskie Wielkie. The lowest values are recorded in Lake Moczydło near Lubnia (11.2 dam^3), as well as in lakes Diabelskie and Kczewo (25.2 dam^3). The median value calculated from the volume of the discussed lakes is 483.7 dam^3, and the mean value is 1,564.3 dam^3. More than three times smaller median indicates a clear dominance of lakes with very small volumes. Thus, nearly 77% of lakes store less water than the mean volume calculated for the whole group. A generalized picture of the relations binding surface-depth-volume parameters of the lobelia lakes is shown in Fig. 4.

Surface-depth relations also describe the values of the exposure index which is the ratio of the lake surface (in hectares) to its mean depth (in metres). This ratio expresses the individual susceptibility of lakes to the influence of external factors

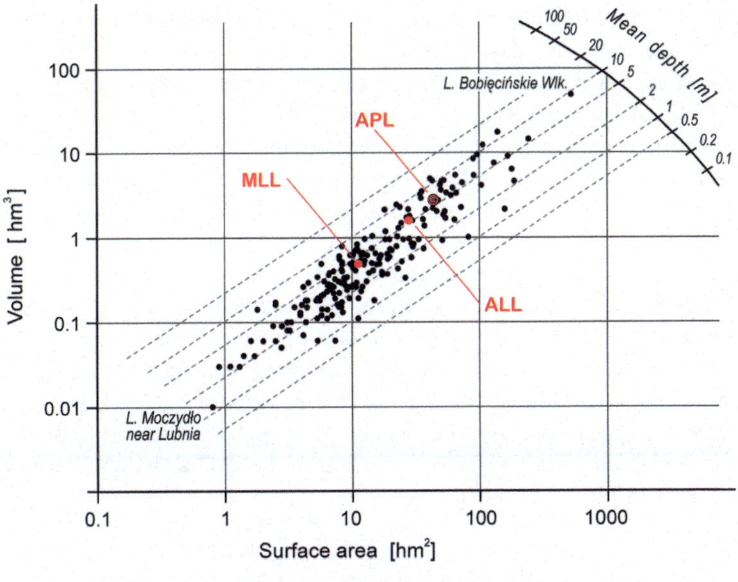

Fig. 4 Differentiation in surface area, volume and mean depth of the contemporary and former lobelia lakes ($n = 194$). Explanation: *MLL* "average" lobelia lake defined by median values, *ALL* "average" lobelia lake defined by mean values, *APL* "average" Polish lake defined by mean values

(mainly wind action) shaping the intensity of hydrodynamic processes and vertical zonation of water temperature. The value of the exposure index in lobelia lakes varies from less than 1.0 in small and deep lakes which are the least exposed to the wind effects (Lake Czarne near Borzytuchom, 0.2; Lake Morskie Oko, 0.7; Lake Herta, 0.8) and medium-deep lakes (Lake Ostronek, 0.5; Lake Cietrzewie, 0.8) to over 60.0 in very shallow and extensive reservoirs (Lake Dołgie Wielkie, 111.7; Lake Wielatowo, 74.6; Lake Otalżyno, 67.3). In deep and very deep lakes with low exposure index values, the morphometric factor promotes the formation of shallow and sharply defined thermoclines and epilimnion in the summer with a thickness of no more than 2–3 m. This is a consequence of the delay and shortening of the spring circulation phase – bradymixis. In comparison with the lakes of northern Poland, lobelia lakes (in the size class of 10–50 ha) show exposure index values lower by approx. 26% (median) and 40% (mean). Increased resistance of lobelia lakes to wind mixing is also demonstrated by the values of relative depth (mean 2.7%, median 2.5%), which in these lakes have higher values than those found in comparable other Polish lakes. In terms of the median value, the differences reach 38%, while in the case of the mean value – 29%. Therefore, the probability of bradymictic circulation occurrence in lobelia lakes is theoretically much higher. This observation is confirmed by the studies of a group of 33 lobelia lakes in Pomerania, of which as much as 42% had a summer thermal stratification typical or similar to bradymixis [19].

Another, often referenced in ecological studies, morphometric index which is shoreline development does not differentiate lobelia lakes from other Polish lakes. In both groups of lakes, the mean and median values are in the range of 1.4 to 1.6. Accepting the common interpretation of this index, lobelia lakes do not show a significantly different potential for the littoral communities development in comparison with other Polish lakes. The range of the shoreline development index values within lobelia lakes determine the extreme values of 1.01 (Lake Oczko Małe) and 3.60 (Lake Bobięcińskie Wielkie). Most often, this index takes the values of 1.21–1.68. The influence of littoral on the lobelia lake ecosystems can also be expressed by the index of lake basin permanence (IBP) [36]. Characteristic values of the IBP obtained for lobelia lakes (average: 0.42, median: 0.32) are 14–15% higher than those calculated for similarly sized lakes of northern Poland. The IBP values below 0.1, reflecting the significant advancement of lake ageing processes, are found in only 14 lakes (7% of the total number of those studied). The most extreme case among them is Lake Moczydło near Lubnia, whose IBP is 0.03. In 14 lakes, the IBP is higher than 1, and its maximum value (1.78) is found in Lake Krzemno. The largest lobelia lake in Poland – Lake Bobięcińskie Wielkie – has an index value of 1.68. The most frequent value of the IBP in the lobelia lakes ranges from 0.16 to 0.46 (50% of lakes).

5 Physical and Chemical Features Water of Polish Lobelia Lakes

The PLL mainly develop on poor calcium, manganese and nutrients sandy deposits; thus their waters are usually poor in these elements, weakly buffered with low concentrations of CO_2 and bicarbonates and characterized by low conductivity. Due to varying morphology of PLL and their catchment areas, geological features and differences in human impact, the physico-chemical properties of water can display significant inter-lake differences. According to physical and chemical features, Kraska and Piotrowicz [16] divided PLL into four general types: acidified, balanced, eutrophicated and degraded.

Acidified type has a 30% share in the PLL. The lakes of this type are mostly small (mean lake area does not exceed 15 ha) and at the same time relatively deep water bodies (average maximum depth over 8 m). Acidified lobelia lakes are usually located within land hollows with significant inclination of slopes. Catchment areas of these lakes are typically overgrown by forests (mostly coniferous with pine dominance). The characteristic feature of water of that group of lakes is acidic reaction (pH 3.8–5.9). Such low water pH is a result of insignificant amount of calcium and manganese (Table 3) leading consequently to low buffer capacity [16]. Acidification may result from increased atmospheric deposition of sulphur and nitrogen oxides [4, 5] and/or inflows of large amounts of humic acids from the catchment area [37–39]. The acidified lobelia lakes may also reveal significant variation of other chemical features. According to Kraska and Piotrowicz [16], this group should also be divided into two further subgroups: acidified oligohumic and acidified polyhumic lakes (Table 3).

Water of acidified oligohumic lakes is characterized by lower concentration of humic substances (dissolved organic carbon (DOC) <10 mg C L^{-1}) because humic compounds form organometallic complexes and undergo sedimentation [40]. Prevalence of fulvic acids, which are less coloured but chemically active [41], results in slight water colour (even below 1 mg Pt L^{-1}, mean 9 mg Pt L^{-1}), acidic reaction (as low as pH 3.8 in Wygoda or Szare lakes) and high water transparency (average Secchi depth (SD) visibility 5.8 m) frequently reaching the bottom of lake. Penetration of light to the bottom causes heating of overbottom water layer and convection mixing of lakes. Acidified oligohumic lakes are poor in nutrients. The mean concentration of nitrogen is below 1 mg N L^{-1} and varies from 0.3 to over 2 mg N L^{-1}. In all of these lakes, organic form predominates and consists of about 75% of total nitrogen content. Phosphorus occurs in very low concentration (Table 3), and most of its load is incorporated within plankton assemblages. Soluble P in many oligohumic lobelia lakes is below detection limit. Lakes classified into this group are also very poor in soluble mineral salts. The mean conductivity of water is below 45 μS cm^{-1} (Table 3), and in selected lakes it does not even exceed 10 μS cm^{-1}. Due to low concentrations of calcium and magnesium, water of acidified oligohumic lakes is characterized by the lowest hardness among all groups of PLL.

Table 3 Selected physical and chemical features of water (epilimnia) of Polish lobelia lakes (after Kraska and Piotrowicz [16] supplemented with unpublished data)

	Colour	Ca	Mg	TH	EC	Nmin	Norg	TP	SRP	Chl
pH	mg Pt L^{-1}	mg Ca L^{-1}	mg Mg L^{-1}	°n	μS cm^{-1}	mg N L^{-1}	mg N L^{-1}	mg P L^{-1}	mg P L^{-1}	μg L^{-1}
Acidified oligohumic $n = 30$										
3.7–5.9	9.0 (4.02)	3.1 (1.45)	0.6 (0.47)	0.5 (0.23)	44 (12.5)	0.16 (0.188)	0.72 (0.293)	0.025 (0.011)	0.003 (0.005)	5.9 (9.2)
Acidified polyhumic $n = 21$										
4.1–5.9	72.1 (41.16)	3.0 (1.23)	0.8 (0.54)	0.6 (0.22)	50 (13.7)	0.62 (0.401)	1.29 (0.46)	0.085 (0.053)	0.008 (0.005)	10.5 (8.6)
Balanced $n = 54$										
6.0–7.6	20.1 (8.17)	7.9 (3.4)	0.7 (0.56)	0.9 (0.52)	61 (21.3)	0.39 (0.209)	0.995 (0.423)	0.048 (0.031)	0.002 (0.003)	7.3 (8.7)
Eutrophicated $n = 38$										
6.3–9.8	27.8 (11.19)	15.8 (9.21)	1.4 (0.95)	2.5 (1.37)	128 (54.1)	0.56 (0.439)	1.42 (0.505)	0.053 (0.039)	0.006 (0.005)	8.4 (9.1)
Degraded (former lobelia lakes) $n = 27$										
6.5–9.8	33.2 (17.8)	20.6 (16.07)	2.4 (1.6)	3.5 (2.44)	168 (88.2)	0.65 (0.479)	1.52 (0.723)	0.085 (0.042)	0.002 (0.003)	18.6 (19.5)

Mean values and standard deviations (in parenthesis) and range for pH were given

TH total hardness, *EC* electrical conductivity, *Nmin* mineral nitrogen, *Norg* organic nitrogen, *TP* total phosphorus, *RSP* soluble reactive phosphorus, *Chl* chlorophyll-*a*

Water of acidified polyhumic lobelia lakes is characterized by high content of humic substances. The presence of humic and fulvic acids results in the water colour encompassing intense yellow to reddish-brown colour, with its intensity ranging from 40 to over 200 mg Pt L^{-1}. Water of that group of lakes is also characterized by low electrical conductivity (at level comparable to oligohumic lakes) as well as high concentrations of DOC [40], the main component of all humic substances. In contrary to oligohumic, polyhumic lobelia lakes are rich in nitrogen and phosphorus (Table 3). Compared to acidified oligohumic PLL, the polyhumic lakes are characterized by threefold and twofold higher mean concentrations of mineral and organic nitrogen, respectively (Table 3). Moreover, the mean level of phosphorus in their waters is significantly higher. Despite high loads of nutrients, nitrogen and especially phosphorus are complexed by humic acids and biologically unavailable for autotrophic organisms. Photodegradation of humic acids supplies the significant concentrations of polyhumic lakes [42], and their decomposition causes oxygen depletion.

Balanced lobelia lakes constituted approximately 30% of all currently existing PLL. This group is diversified according to their morphology and comprises the largest and deepest PLL. The characteristic feature of balanced lobelia lakes is the neutral pH of their waters and low trophic state. The pH of epilimnetic water (during vegetation season) range from 6.0 to 7.6. Compared to acidified PLL, the water of the balanced lobelia lakes contain twofold higher calcium (Table 3) and bicarbonates (0.21 vs 0.1 meq HCO_3 L^{-1}) concentrations. Altogether, this results in much higher buffer capacity and further decreases impacts of inflows of organic (humic) and mineral acids from the catchment areas. The mean electrolytic conductivity of these water bodies is slightly higher than of acidified PLL, while the concentration of dissolved organic salts is insignificant. Content of nitrogen and phosphorus in balanced lobelia lakes is still very low and comparable to that usually observed in acidified PLL (Table 3). The organic nitrogen consists about 30% of total N pool. The mineral N forms are dominated by nitrates in the well-oxygenated zone, while ammonium nitrogen predominates in deep anoxic layers. During vegetation season, concentration of orthophosphates in epilimnetic water is very low – an undoubtedly key factor that limits intensive proliferation of algae and consequently preserving high water transparency. Within balanced PLL, some diversity of physical and chemical properties of waters can be found. A few of balanced PLL display increased availability of nutrients (especially phosphorus) and mineral salts and, simultaneously, higher algae density (as manifested by increased chlorophyll content in surface layer and decreased transparency of water). These changes are observed in those of PLL which are under increasing human impact and can be considered as a transition to a next group of water bodies: eutrophicated PLL.

The eutrophicated PLL are also, similarly to balanced PLL, very diversified as far as lake basin morphology is concerned although their catchment areas are usually under significant human impact. Compared to acidified and balanced PLL, eutrophicated type is characterized by increased pH of epilimnion water. The pH during vegetation period ranges from 6.3 to over 9.0 (Table 3). The alkalic water reaction results from depletion of carbon dioxide from water column by developed algae assemblages. Significant amount of algae suspended in the water decrease water transparency. Eutrophicated PLL are characterized by higher concentrations of biogenic elements, calcium, magnesium (Table 3) and organic matter [40]. The content of total nitrogen is twofold higher than in balanced PLL and even more higher compared to acidified PLL. The concentration of total phosphorus is higher comparing to balanced and acidified oligohumic PLL but lower than in polyhumic lobelia lakes. The conductivity values are also much higher compared to other PLL (Table 3). High concentration of nutrients and other chemical elements are the result of their intensive inflow from the catchment, which are transformed by human, and mostly constitute of arable land, pastures and urbanized areas.

Degraded lobelia lakes constitute about 14% of all PLL which were described in literature since the beginning of twentieth century. This group includes lakes which are very diversified according to their surface area and depth. Characteristic features of these water bodies include considerable human impact and catchment transformation leading to their significant eutrophication. Changes of physical and chemical features of water led to the disappearance of characteristic vegetation. This group of lakes can be described as former or historical lobelia lakes. The water of degraded lobelia lakes contain significantly high concentrations of nutrients, calcium, magnesium and dissolved mineral salts (Table 3). Increased nutrients availability stimulates proliferation of different groups of algae, including cyanobacteria. The mean chlorophyll concentration in degraded PLL is twofold as high as in eutrophicated PLL and almost four times higher than in acidified oligohumic. Maximal observed values of chlorophyll concentration (over 70 μg L^{-1}) are typical for lakes in hypertrophic state [16]. Algal blooms decrease light availability and SD, resulting in rapid thermal and oxygen stratification. During vegetation period deeper water layers are deoxygenated. Some of the degraded lobelia lakes are converted into midfield fish breeding ponds (e.g. lakes Baroczno, Chojnackie) or serve as sewage receivers from animal farms (e.g. Lake Drzywko).

6 Optical Properties of Water of Polish Lobelia Lakes

One of the distinguishing features of the abiotic environment of this type of ecosystem are the favourable conditions for light transmission in the water column. A widely used, albeit approximate, parameter for the assessment of the optical properties of lake waters is water transparency measured with a Secchi disc (Secchi depth). Against the background of the entire set of the Pomeranian lakes, lobelia lakes generally show a much greater transparency of water. The range of its variability among lakes is also greater (Fig. 5). In the last decade of the twentieth century, in the midsummer, the "average" lobelia lake showed a transparency of 2.8 (median) and 3.3 m (mean). At the same time, in Pomeranian lakes these values were over 40% lower and amounted to 1.6 and 1.9 m, respectively. The range of variability of water transparency in lobelia lakes ($n = 148$), which was observed

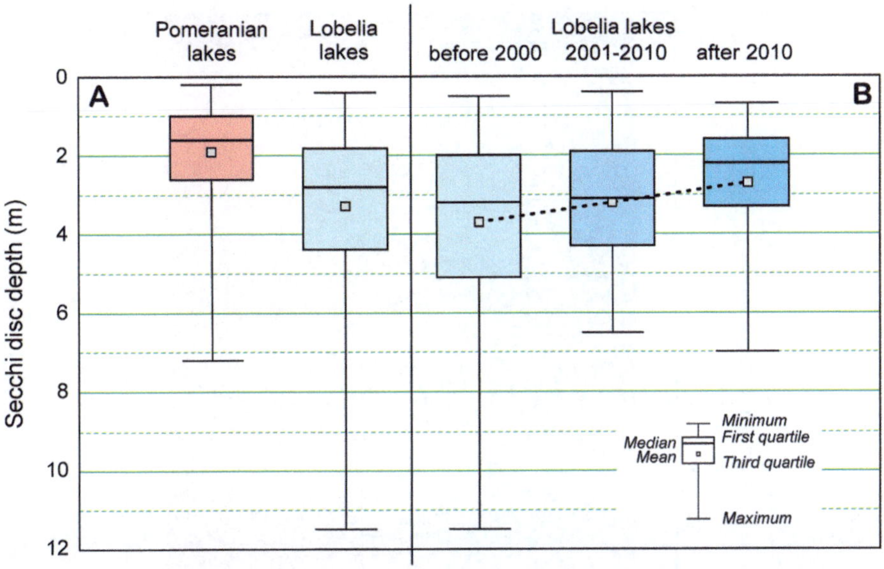

Fig. 5 Box-and-whisker plots showing differences in water transparency (Secchi depth) between Pomeranian ($n = 249$) and lobelia lakes ($n = 148$) (**a**) and within group of 63 lobelia lakes during three decennary monitoring periods (**b**)

then, was determined by the extreme values of 0.4 (Lake Dołgie Wielkie) and 11.5 m (Lake Modre). In the first two decades of the twenty-first century, the cleanliness of the lobelia lakes deteriorated significantly, which was also reflected in water transparency.

Analysing the homogeneous suit of 63 lobelia lakes of Pomerania, monitored in 1991–2017, one can observe a clear fall in the transparency of their waters. Assuming the mean value to be a distinguishing factor of the changes taking place, water transparency of the lobelia lakes decreased in this period by 27%. With respect to the median value, the decrease in transparency was slightly higher and amounted to over 30%.

As a result of the intensification of the eutrophication and humification processes, the number of the clearest lakes drastically dropped, most of all of those with the transparency higher than 4 m (oligotrophy). In the years 1991–2000, there were 27 such lakes, and after 2010 only 11. The extreme example of changes in the water transparency of the lobelia lakes caused by cultural eutrophication is Lake Modre. The introduction of alien fish species, and thus the change in the species structure of the native ichthyofauna, led to a reduction of water transparency to 4–5 m. At the turn of the 1980s and 1990s, the water transparency in this lake was 11–12 m [15].

Spectral analysis of underwater light field carried out in 54 lobelia lakes in Pomerania shows that these lakes represent a wide variety of optical types of water, evaluated based on the spectral shape of diffuse attenuation coefficient for downwelling irradiance (Fig. 6), which is the quantity very sensitive to changes in

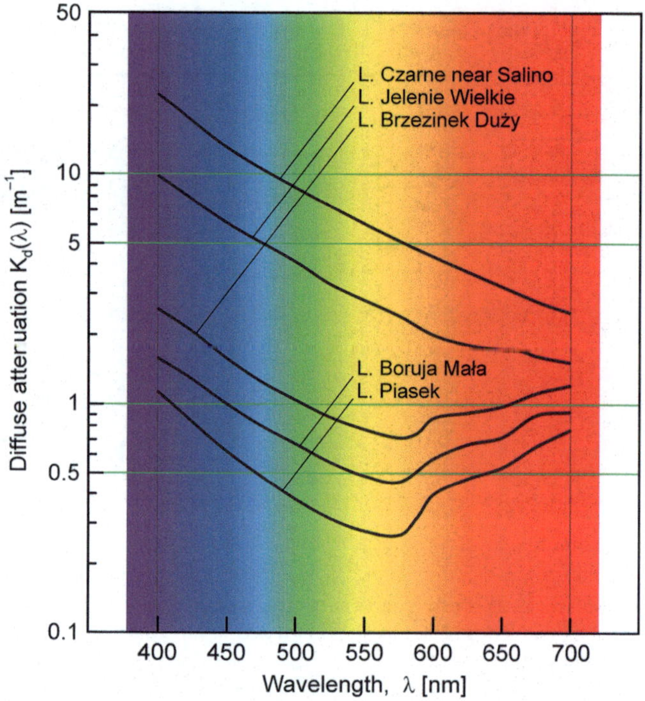

Fig. 6 Spectra of the diffuse attenuation coefficient of downwelling irradiance for selected Pomeranian lobelia lakes

the concentrations of optically significant constituents of water [43, 44]. Averaged over the entire euphotic zone, values of the diffuse attenuation coefficients for irradiance of PAR (spectral range 400–700 nm), $K_{d,PAR}$, in the lobelia lakes are from 0.412 (Lake Rekowskie) to 5.895 m^{-1} (Lake Czarne near Salin). The corresponding transmittances (T_{PAR}) are, respectively, 66.2 and 0.3%. In the polyhumic Lake Czarne near Salino ($K_{d,PAR}$ = 5.895 m^{-1}), solar radiation penetrating the water column is almost completely absorbed in the water layer of only 1 metre thickness. In the clearest lakes, the highest transmittance is observed in the spectral band of 550–570 nm (green-yellow light). As the concentration of the coloured dissolved organic matter (CDOM) increases, the maximum spectral transmittance moves towards the longer wavelengths. In lakes with a true colour of water of more than 30 mg Pt L^{-1}, the maximum transmittance falls on the red light region (670–700 nm).

The lobelia lakes show high values of the product of the Secchi depth (SD) and the diffuse attenuation coefficient for irradiance of PAR ($K_{d,PAR}$), $\kappa = SD \cdot K_{d,PAR}$. The extreme κ-values determine its variability range from 1.48 (Lake Płocica) to 4.29 (Lake Oblica). In the light of recent research, however, the presence of indicator species (*L. dortmanna*, *I. lacustris* and *L. uniflora*) has not been confirmed in the latter lake [45]. As many as 85% of studied lobelia lakes have κ-values higher than

2.0, indicating that they are clean or coloured lakes [44, 46]. Since the κ product links the inversely proportional correlation with the ratio of scattering to light absorption (b/a) [47], the effect of absorption on the total light attenuation in the lobelia lakes is slightly greater. Lobelia lakes are generally less turbid. Concentrations of suspended solids (dry mass of seston) in the surface water of lobelia lakes are on average 4.7 mg L^{-1} (range: 1.6–10.1 mg L^{-1}) and are almost twice lower than those observed in other Pomeranian lakes. The main water admixtures determining the spectral shape of the total absorption coefficients, as well as diffuse attenuation coefficients for irradiance of PAR in the lobelia lakes, are the coloured dissolved organic matter and phytoplankton suspensoids, which totally dominate the optical properties of their waters. The statistical relationships between the diffuse attenuation coefficient for downward irradiance of PAR and the true colour of water and chlorophyll-*a* concentrations found in the discussed group of lakes are shown in Table 4.

The lower values of the scattering-to-absorption ratio found in the lobelia lakes are also expressed in the fact that the amount of the surface incident light penetrating to the Secchi depth is also lower in these lakes (in average conditions ~9%) than in other Pomeranian lakes (~15%). As an effect, the relations that bind the euphotic zone and Secchi depths are substantially different in both groups of lakes. For lobelia lakes the equation describing the euphotic zone depth (EZD) as a function of the Secchi depth (SD) is as follows: $EZD = 1.79 \cdot SD$ ($R^2 = 0.79$, $n = 54$). The slope value of the regression line, most often adopted for non-lobelia Pomeranian lakes, is significantly greater and ranges from 2.4 to 2.5 [44, 48]. The most frequent euphotic zones observed in lobelia lakes range from 3.3 to 6.4 m (average 5.2 m), and only in a few exceed 10 m (lakes Piasek, Rekowskie, Kwisno Duże).

The availability of light for photosynthesis is one of the most important factors regulating both the abundance and distribution of autotrophic aquatic plants. Strong and selective attenuation of light in the water column causes its quantity and spectral composition to change significantly with depth. The intensity of these changes in combination with the individual light requirements of particular macrophyte species determines the maximum depths of their colonization. In reference to indicator species of lobelia lakes (*L. dortmanna*, *I. lacustris*, *I. echinospora*, L.), a clear regularity, manifested by an increase in their light needs along with an increase in water transparency, is observed [49, 50]. In other words, these species colonizing deeper areas of the littoral zone need more and more light. In the case of

Table 4 Empirical relationships between diffuse attenuation coefficient of downward irradiance ($K_{d,PAR}$; m^{-1}) and true colour of water (TC) and chlorophyll-*a* concentration (CHL) in Pomeranian lobelia lakes ($n = 55$)

Water quality parameter	Equation	r^2
TC [mg Pt L^{-1}]	$K_{d,PAR} = 0.036$ TC + 0.569	0.89
TC [mg Pt L^{-1}]	$K_{d,PAR} = 0.364\ TC^{0.436}$	0.66
CHL [μg L^{-1}]	$K_{d,PAR} = 0.039$ CHL + 0.468	0.28
CHL [μg L^{-1}]	$K_{d,PAR} = 0.216\ CHL^{0.569}$	0.49
TC [mg Pt L^{-1}] and CHL [μg L^{-1}]	$K_{d,PAR} = 0.033$ TC + 0.012 CHL + 0.378	0.91

L. dortmanna, the light demand of individuals growing at a depth of 2.0 m is 47% of the surface irradiance of PAR. These requirements can only be met if the water transparency is about 7–8 m. The light needs of individuals growing at a depth of 1.0 m are about 40%, and at a depth of 0.5 m, only 34% of the surface irradiance [50]. Limitations in access to light mean that even in the cleanest Pomeranian lakes, the maximum depth of colonization of the water lobelia is not greater than 2.0–2.2 m. The light requirements of the *Isoëtes lacustris* species are twice as low. Individuals growing in the littoral zone at depths of 1–5 m need only 20–25% of the surface irradiance of PAR.

7 Vegetation of Polish Lobelia Lakes

In Poland, the term "soft-water lobelia lake" is associated with the natural habitat of Natura 2000, code 3110 – "Oligotrophic waters containing very few minerals of sandy plains (Littorelletalia uniflorae)". This habitat is designated as "oligotrophic water bodies with a small amount of mineral and alkaline compounds, with aquatic or bi-habitat perennial vegetation belonging to the order Littorelletalia uniflorae", occurring on the oligotrophic soils of lake and pond shores (sometimes on peat soils). The vegetation forms one or more clearly defined zones, dominated by *Littorella uniflora*, *Lobelia dortmanna* or *Isöetes lacustris* (rarely *Isöetes echinospora*), although not all zones must be present in one water body [51]. These small perennial plants, commonly called isoetids, are often accompanied by other plant species, both vascular, e.g. *Myriophyllum alterniflorum*, *Juncus bulbosus* and *Luronium natans* (priority species in the Natura 2000 network), and bryophytes from the genera *Drepanocladus*, *Warnstorfia*, *Sphagnum* or *Fontinalis* and algae from the family Characeae, especially from the genus *Nitella*.

Vegetation conditions for plants in the specific habitats of soft-water lakes are not favourable, due to low availability of carbon dioxide (CO_2) in the water and in the bottom sediments, as well as nutrients – nitrogen and phosphorus. CO_2 is a key factor determining the occurrence of isoetid communities. This is because carbon dioxide is the only source of carbon necessary for them to conduct photosynthesis. Most of the remaining species of aquatic plants can, thanks to the appropriate enzymes (carbonic anhydrase), take C from bicarbonates. The presence of free CO_2 depends on the pH of the water. In neutral and alkaline water environments, there are mainly carbonates and bicarbonates, and the amounts of free CO_2 are minimal. In addition, dissolved carbon dioxide is very quickly assimilated by autotrophs, primarily phytoplankton. In such ecosystems, permanent CO_2 deficits occur. This restriction is an effective barrier that prevents the growth of isoetids. When the water pH is about 6.0, the equivalent amounts of free CO_2 and bicarbonates are present in the environment. Further lowering of the pH results in an increase in the amount of free carbon dioxide, with the simultaneous disappearance of bicarbonates. At pH of around 4.0, only mineral carbon in the form of free CO_2 is present. Regardless of the water pH, the total amount of mineral carbon remains constant, and only its chemical form changes.

The lack of carbonic anhydrase in isoetids means that these plants can only vegetate in lakes with low levels of water reactivity. An additional factor favouring the occurrence of these species is the low trophy of waters (low concentrations of nitrogen and phosphorus) which limits the development of other primary producers (plankton and higher plants), competitors with respect to CO_2 [52]. In addition, isoetids have evolved some morphological and biochemical adaptations enabling their vegetation within limited carbon availability. They have developed mechanisms that allow for the uptake of CO_2 from the substrate where the concentration of this gas is usually 10–100 times higher compared to the water layer. Well-developed root systems serve this purpose. In the case of isoetids, shoot-to-root ratio is as low as 0.3 enabling not only the uptake of carbon dioxide but also an increased nutrient supply [2, 3]. In addition, isoetids are able to recover some of the CO_2 from respiration by storing it in an extensive air cell system [3, 53–55].

Another adaptation of isoetids to the low availability of mineral carbon forms are additional physiological mechanisms supporting carbon binding. Isoetids growing in an oligotrophic environment – like many terrestrial plants, especially those vegetating in the unfavourable conditions of the hot climate of deserts and semi-deserts – belong to the C4 cycle plants. In this cycle, as a result of CO_2 incorporation, a four-carbon compound oxaloacetate occurs. It undergoes further transformations and is transported to specialized leaf bundle sheath cells. Consequently, the C4 pathway leads to a concentration of C in bundle cells. It is estimated that the concentration of carbon in the cytosol of these cells exceeds 10–20 times the concentration in mesophyll cells.

Littorella and *Isöetes* are also mentioned in the group of plants that carry out metabolic processes involving crassulacean acid (CAM – crassulacean acid metabolism). This metabolic cycle is in some respects similar to C4 metabolism. The differences relate to the location and the time-dependent processes of initial carboxylation and decarboxylation. In C4 plants, these processes occur simultaneously but in different types of cells. They are therefore spatially separated. In CAM plants, both processes run in the same cells but are separated in time. This is because carboxylation occurs at night, when the concentration of free CO_2 (respiration, limitation of CO_2 capture by plankton) is greatest, and decarboxylation takes place during the day. The adaptive mechanisms described above occur simultaneously in the same isoetid species, enhancing their effect. According to Farmer and Spence [56], intensive CAM metabolism and CO_2 uptake from sediments were recorded in *Isöetes lacustris*, while in *Lobelia dortmanna* CO_2 uptake from sediments and the efficient functioning of the CO_2 capture mechanism from respiration, the so-called photorespirational trap, have been described. In contrast, *Littorella uniflora* efficiently uses all three of the above mechanisms. Diversified adaptation mechanisms make *Littorella* the most flexible species among all species characteristic for lobelia lakes, and this is why this species occurs in Polish lobelia lakes, both acidified and typical lakes, as well as in lakes with increased trophic state, where it shares the zone of littoral with elodeids or even representatives of stoneworts.

The first studies [1, 7] to attempt an inventory examination and description of the functioning of Polish lobelia lakes revealed the significant diversity of these water bodies in terms of vegetation structure. The reasons for such diversity have been investigated and determined by many scientists dealing with the aquatic environment [6, 8, 12].

The occurrence of bryophytes in lobelia lakes may indicate a threat to the isoetid habitat resulting from the acidification of waters in these lakes, and the emergence of elodeids and stoneworts can indicate threats resulting from eutrophication – often originating from human impact [14]. The spatial and qualitative structure of plant communities depends on the trophic character of waters.

In lobelia lakes with acidic waters and at the same time with high water transparency, the zone of littoral and shallow eulittoral (up to about 1.5 m) is overgrown by *Lobelia dortmanna* and *Littorella uniflora*. They are often accompanied by *Juncus bulbosus*. In lakes with water characterized by extremely low pH (pH <4.5), characteristic species occur with much smaller density or even disappear. Below a depth of 1.5 m, *Isöetes lacustris* begins to dominate, and the density of the two remaining species decreases rapidly with depth. Below a depth of 3–5 m (depending on the water transparency), *I. lacustris* disappears quite rapidly, and Sphagnum mosses occur with the dominating of *Sphagnum denticulatum* [37].

In acidified polyhumic lakes, due to the intense colour of the water resulting small water transparency, the phytolittoral zone is narrow. Rapid reduction of light penetration strongly limits the occurrence of *L. dortmanna* and *L. uniflora*. Among the characteristic species, *Isöetes lacustris* dominates, but its coverage rarely exceeds 40%, and it occurs at much smaller depths than in acidic oligohumic lakes. In some polyhumic lakes, conditions of limited availability of light in the water depth restricting submerged vegetation will lead to the prevalence of plants with floating leaves. *Nuphar pumila* and *Sparganium affine* often occur here. The leaves floating on the water surface will additionally shade the bottom and limit the possibility of isoetid occurrence [37].

In typical balanced lobelia lakes, characteristic plant species develop lush and create specific spatial zones. In the shore zone, the shallowest part of the lake, *Lobelia dortmanna* and *Littorella uniflora* grow very abundantly – creating dense beds overgrowing the sandy or sandy-stony type of surface. Suitable habitat conditions usually allow the emergence of a two-layered vegetation structure. Small isoetids co-occur along with *Eleocharsis palustris* or *Carex rostrata* and *C. lasiocarpa* (in lakes with a slightly more acidic pH). In deeper parts, species such as *Myriophyllum alterniflorum* occur, and in lower densities, *Elodea canadensis* and *Ceratophyllum demersum*, along with isoetids. With increasing depth, *L. dortmanna* and later *L. uniflora* gradually disappear, and *Isöetes lacustris* begins to dominate. Such species zonation can be observed up to a depth of 5 m. At greater depths *I. lacustris* quickly disappears, and algae of the genus *Nitella* and *Chara delicatula* dominate, covering the bottom with a thick coverage. The maximum range of plants in this type of lobelia lake is 12 m, although due to the large diversity of physical and chemical properties of waters it may be much lower.

The lobelia lakes that are subjected to accelerated eutrophication are characterized by a simplified floristic composition, and the composition of characteristic species is always limited. Littoral surfaces overgrown by characteristic species are also limited, and they also occur in less compact patches. Much larger phytolittoral surfaces are overgrown by *Myriophyllum alterniflorum* and by other species characteristic of waters with higher trophic conditions. In extreme cases, *Myriophyllum spicatum* develops massively; it is a symptom of strong eutrophication [16].

Many lakes have already ceased to be lobelia lakes due to rapidly progressing eutrophication, and characteristic species have been replaced by typically eutrophic ones. These water bodies are now surrounded by a wide reed belt. Due to the turbidity of the water, the submerged macrophytes are poor, and there are species with floating leaves.

8 Threats and Protection of Polish Lobelia Lakes

Most of the PLL are covered by different forms of protection. Nine of PLL are located in National Parks, Tuchola Forest National Park (7 lakes), Slovincian National Park (Lake Dołgie Wielkie) and Karkonosze National Park (Lake Wielki Staw), 40 lakes are located in landscape parks and 26 lakes are nature reserves. The PLL fully correspond to the definition of EU habitat 3110 Natura 2000 programme described as "Oligotrophic waters containing very few minerals of sandy plains (Littorelletalia uniflorae)" consisting of "oligotrophic waters with few minerals and base poor, with an aquatic to amphibious low perennial vegetation belonging to the Littorelletalia uniflorae order, on oligotrophic soils of lake and pond banks (sometimes on peaty soils). This vegetation consists of one or more zones, dominated by *Littorella*, *Lobelia dortmanna* or *Isöetes*, although not all zones may be found at a given site" [51]. Therefore, lobelia lakes were the basis for designating many Natura 2000 areas, and now 119 of PLL are protected within them [57]. Almost 25% of PLL (about 40 sites) were assigned for the monitoring of the 3110 habitat conservation status and since 2009 monitoring is carried out [20, 58, 59].

Soft-water lobelia lakes are vulnerable to acidification, humification, eutrophication and alkalinization. As a result of these pressures, a number of these lakes have undergone substantial deterioration in European countries. It is estimated that 14% of PLL had significantly changed their physical and chemical water properties since the beginning of twentieth century, resulting in loss of all plant species characteristic for lobelia lakes. In various other PLL, some characteristic species disappeared entirely, while those which remained had significantly decreased in their coverage. Kraska and Piotrowicz [16] found that within the group of acidified PLL (both oligo- and polyhumic), 32% of ecosystems lost at least one characteristic plant species and about 4% lost two plant species. Within the group of eutrophicated PLL, almost 40% ecosystems are inhabited only by two of three characteristic species and 15% only by one characteristic plant species.

Acidification of lobelia lakes and deterioration of characteristic plant species is phenomenon well documented in Europe [4, 5]. Increased atmospheric deposition of nitrogen and sulphur compounds leads not only to the drop of water pH but also to increase of C and N availability. According to Arts et al. [60], *Lobelia dortmanna* is the last tolerant species for acidification and then *I. lacustris* and *L. uniflora* as the species with the highest tolerance. This is not confirmed by the observations carried out in PLL. In the most acidified of PLL (lakes Szare, pH 3.9; Regnice, pH 4.4; Wygoda, pH 3.8), *L. uniflora* is absent, and among other characteristic species, *I. lacustris* occur with highest abundance [16]. Decrease of water pH may be also a result of intensive inflows of humic substances from catchment areas. Overland flow from coniferous forests and waters outflowing from sphagnum or high bogs is characterized by a very acidic reaction – even below pH 4.0 – and contains significant levels of humic substances [39]. Humic substances not only acidify the lake water but also significantly change underwater light conditions, decrease euphotic zone and change hydrochemistry of water [41]. In natural catchments (undisturbed by humans), delivery of humic substances is usually balanced with their photodegradation and sedimentation. Additionally, natural and well-developed ecotonal zones disperse and modify the transport of organic matter from catchments to lobelia lakes [37]. The increase in the total precipitation may intensify the transport of humic substances to the lakes. Particularly event of heavy rain can enhance inflow of humic acids and further trigger acidification and dystrophication processes [38]. In many cases improper forest management is the key factor inducing the acidification and humification of lobelia lakes. Clear logging of tree stands (especially in proximity of lakes) and plowing of raw humus layer to prepare new plantations cause the flush of forest litter directly to the lake. Such methods of logging were in the past carried out within catchment area of majority of PLL. The other activity that can accelerate lobelia lake dystrophication is draining of the wetlands located within catchment areas. Examples from the past show that humus-rich inflows from drained peat bogs changed the optical properties and lake hydrochemistry causing the deterioration of characteristic plant species (e.g. lakes Czarnówek, Skąpe). Currently, forest management takes into account conservation needs of PLL because majority of these ecosystems are located in areas of legal protection [61]. In many cases, according to protective recommendations, artificial inflows that drained the wetlands were cut off.

Eutrophication of PLL is the result of many different processes although it is almost always associated with intense human impact on lakes and catchments. Deforestation and transformation of catchment areas into arable lands, pastures and human settlements lead to changes in water cycle. Decrease of water retention in the catchment accelerates water outflow and increases the loads of nitrogen, phosphorus, carbon, calcium and other chemical elements reaching the lakes. It promotes the development of "hard water" plant species (what in first stages of eutrophication lead to the increase of biodiversity) and algae growth. The loss of isoetid species is usually gradual and results from several overlapping factors. One of the most important is light limitation and competitive exclusion of isoetid species by plants well adapted to turbid water (nympheids, elodeids). The reaction of characteristic plant species on eutrophication is different, and undoubtedly the *Littorella uniflora* is the species with the highest tolerance and plasticity [62, 63]. In eutrophicated PLL *L. uniflora* is often the only species among vegetation characteristic for these water bodies. Because of its amphibious character, *L. uniflora* survive out of water (Fig. 7), and therefore it is resistant to periodic drying of the littoral zone. In eutrophicated lobelia lakes, even small changes in chemical elements balance, caused by natural or human-driven factors, can cause degradation and vanishing of characteristic vegetation. Klimaszyk et al. [64] described the increase of water trophy, decrease of isoetids bottom coverage and development of elodeids and nympheids triggered by nutrient inflow from the abundant cormorants (*Phalacrocorax carbo*) roost established on the shore of the Lake Dołgie Wielkie. Between 1998 and 2009, the mean bottom coverage by *L. uniflora* in the lake decreased from over 20% to about 5%, and the maximum depth of occurrence has decreased of about 50% [64]. Studies conducted in 2016 (made within the monitoring of habitats 3110 Natura 2000) showed complete disappearance of *Littorella* from the lake [20].

Fishery, even on a small scale, constitutes a threat to the number of lobelia lakes. A poor food base results in the impoverishment of the quantitative and qualitative structure of fish fauna [65, 66]. Another factor reducing the occurrence of fish in

Fig. 7 Patches of *Littorella uniflora* on the dried edge of Lake Krzemno

lobelia lakes is the acidic reaction of the water [67, 68]. The significant thread to PLL are both fishery and recreational fishing. Usually in acidified lobelia lakes, the quantity and quality structure of fish assemblages is scarce. Significant threat to the lobelia lakes may be fish stocking. Case study of the Modre Lake [15] shows that attempts of fishery on acidified lobelia lakes led to long-lasting changes in water hydrochemistry and isoetids structure. Liming and fish stocking of this lake led to rapid changes in physico-chemical water parameters. Within several years water transparency decreased from 12 to 3 m. Summer thermal and oxygen stratification developed, and total oxygen depletion occurred in the water layer over bottom sediments. Changes in the physical and chemical parameters of the lake water led to changes in the species composition of vegetation patches (significant decrease of bottom coverage by *L. dortmanna* and *I. lacustris* and decrease in the area of the phytolittoral) [15]. Intensive commercial and recreational fishery may thread the functioning of lobelia lakes by the changes in the species structure of fish assemblage. Overfishing of predator species (which is a widespread phenomenon) leads to an increase in the number of small cyprinids. This in turn, due to complex processes, increases the number of phytoplankton and decreases the transparency of water. Recreational fishery thread the lobelia lakes also by other impacts: supplementation of water in nutrients and other chemical elements (with baits) and mechanical destruction of plant habitats during wadding [15]. According to Kapusta and Czarkowski [69], many of Polish lobelia lakes were stocked with alien and invasive fish species. The introduction of grass carp (*Ctenopharyngodon idella*) or silver/bighead carps (*Hypophthalmichthys* sp.) observed in some PLL [69] seems to be particularly in danger. These fish species of Asian origin are known to exert very strong impact on the food chains in newly colonized ecosystems.

Due to their natural values, most of the PLL are subjected to intense recreational pressure. Clear-transparent water, sandy bottom, and forested shorelines attract high numbers of tourists for bathing, resting and water sports. The major consequences of the human pressure are:

- Patches trampled within catchment area what decrease water retention and accelerate transport of chemical elements to the lake
- Destruction of significant fragments of ecotonal zones by sunbathers
- Havoc of hydromacrophytes (isoetids) within bathing areas
- Littering of lake and catchment

These effects are particularly pronounced in PLL which are located close to human aggregations (urban areas, holiday resorts). The most profound example of the impact of recreational pressure is Lake Jeleń, one of the biggest PLL (88 ha) located within the borders of City of Bytów. The significant part of the lake catchment is used for recreational purposes. In the immediate proximity of the shoreline, one can found camp site, holiday houses, a restaurant, playgrounds and different infrastructure – roads and parking lot. Such impact and transformation of lake shore promote the overland flow and increase loading of chemical elements. Hundreds of tourists resting annually at the lake shore cause significant damage in the littoral zone. On the considerable length of the shoreline, the isoetids and other aquatic plants disappeared due to the mechanical destruction of the habitat by people wading in the water. This pressure together with increasing transformations of catchment into residential building areas led to the changes in vegetation structure but also increased trophy of the lake. Since the 90th of the twentieth century, the concentrations of nutrients in lake water increased over twofold, while water transparency decreased by approximately 50%. The Carlson Trophic Index changed from values characteristic for mesotrophy in 1991 to eutrophy in 2013. Within the last three decades, the abundance of isoetids decreased by 50%, and plants characteristic for eutrophic lakes (e.g. *Elodea canadensis*) started to expand their coverage.

The lobelia lakes are rare and endangered ecosystems in Poland. Due to the low buffer capacity of water, they are particularly prone to various natural and human-driven impacts. These threats are of different types, namely, urbanization, agriculture, improper forest management, excessive recreation and global climate changes. In the last decades, significant part of PLL lost their ecological values. It is our duty to broaden knowledge about their functioning and to implement comprehensive methods of its protection and restoration.

References

1. Szmal Z (1959) Badania hydrochemiczne jezior lobeliowych Pomorza Zachodniego. (Hydrochemical studies of lobelia lakes of Western Pomerania.) Pozn Tow Przyj Nauk Prace Kom Biol 19:1–106. (in Polish)
2. Szmeja J (1994) Dynamics of the abundance and spatial organization of isoetid populations in an oligotrophic lake. Aquat Bot 49:19–32
3. Pedersen O, Sand-Jensen K, Revsbech NP (1995) Diel pulses of O2 and CO2 in sandy lake sediments inhabited by *Lobelia dortmanna*. Ecology 76:1533–1545
4. Smolders AJP, Lucassen ECHET, Roelofs JGM (2002) The isoetid environment: biogeochemistry and threads. Aquat Bot 73:325–350
5. Arts GHP (2002) Deterioration of atlantic soft water macrophyte communities by acidification, eutrophication and alkalinisation. Aquat Biol 73:373–393
6. Dąmbska I (1965) Vegetation of littoral of lobelia lakes of Kartuzy Lakeland. PTPN Prace Kom Biol 30(3):55
7. Szmal Z, Szmal B (1965) Badania hydrochemiczne jezior lobeliowych województwa gdańskiego i koszalińskiego. (Hydrochemical studies of lobelia lakes of Gdańsk and Koszalin Voivodeships). Pozn Tow Przyj Nauk Prace Kom Biol 30(1):1–56. (in Polish)
8. Szmeja J (1992) Struktura, organizacja przestrzenna i demografia populacji isoetydów. (Structure, spatial organization and demography of the population of isoetids).Wyd. Uniw. Gdańskiego, Gdańsk, 137 p. (in Polish, English summary)
9. Szmeja J (1996) Rejestr polskich jezior lobeliowych (Register of Polish lobelia lakes). Fragm Flor Geobot 3:347–367. (in Polish, English summary)
10. Szmeja J (1997) Evolution and conservation of lobelia lakes in Poland. Fragm Flor Geobot 42(1):89–94
11. Kraska M (1994) Jeziora lobeliowe. Charakterystyka, funkcjonowanie i ochrona. Część I (The lobelia lakes. Characteristics, functioning and protection. Part I). Idee Ekologiczne 6(4):9–177. (in Polish, English summary)
12. Kraska M (1994) Jeziora lobeliowe. Charakterystyka, funkcjonowanie i ochrona. Część II (The lobelia lakes. Characteristics, functioning and protection. Part II). Idee Ekol 7(5):9–105. (in Polish, English summary)

13. Kraska M (2004) 3110 – Jeziora lobeliowe. W: J. Herbich (red.), Siedliska przyrodnicze. Poradniki ochrony siedlisk i gatunków Natura 2000 – podręcznik metodyczny. Tom 2. Wody słodkie i torfowiska. [3110 – Lobelia Lakes. In: Herbich J (ed) Natural habitats. Guides the protection of habitats and species Natura 2000 – methodological manual, vol 2. Freshwater and bogs]. Ministerstwo Środowiska, Warszawa
14. Kraska M, Piotrowicz R, Klimaszyk P (1996) Jeziora lobeliowe w Polsce (Lobelia lakes in Poland). Chrońmy Przyr Ojcz 52(3):5–25. (in Polish, English summary)
15. Kraska M, Klimaszyk P, Piotrowicz R (2013) Anthropogenic changes in properties of the water and spatial structure of the vegetation of the lobelia lake Lake Modre in the Bytów Lakeland. Oceanol Hydrobiol Stud 42(3):3012–3313
16. Kraska M, Piotrowicz R (2000) Lobelia Lakes: specificity, trophy, vegetation and protection. In: Malinowski B (ed) Protection of lakes and wetlands of Pomerania region. vol 3, pp 48–52. (in Polish)
17. Bociąg K (2003) The impact of the acidic organic matter on the diversity of underwater vegetation in soft water lakes. Acta Soc Bot Pol 72(3):221–219
18. Szańkowski M, Kłosowski S (2006) Habitat variability of the Littorelletea uniflorae plant communities in Polish Lobelia lakes. In: Caffrey JM, Dutartre A, Haury J, Murphy KJ, Wade PM (eds) Macrophytes in aquatic ecosystems: from biology to management. Developments in hydrobiology, vol 190. Springer, Dordrecht, pp 117–126
19. Borowiak D, Nowiński K, Borowiak M, Grabowska K (2016) Środowisko abiotyczne ekosystemów jezior lobeliowych (Abiotic environment of lobelia lake ecosystems). In: Bociąg K, Borowiak D (eds) Jeziora lobeliowe w drugiej dekadzie XXI wieku (Lobelia lakes in the second decade of the 21st century), Wydaw. FRUG, Gdańsk, pp 20–51. (in Polish, English summary)
20. Kolada A, Piotrowicz R, Wilk-Woźniak E, Dynowski P, Klimaszyk P (2017) Conservation status of the Natura 2000 habitat 3110 in Poland: monitoring, classification and trends. Limnol Rev 17(4):215–222
21. Fijałkowski D (1973) Occurrence of *Isöetes lacustris* in Lubelskie voivodeship. Fragm Flor Geobot 19(1):5–7
22. Gos K, Bociąg K (2003) Uzupełnienie do listy jezior lobeliowych w Polsce (Supplement to the list of lobelia lakes in Poland). Bad Fizjogr Pol Zach B 52:151–158. (in Polish, English summary)
23. Kondracki J (1994) Geografia Polski, Mezoregiony Fizyczno-Geograficzne, (Geography of Poland, Physico-Geographic Mezoregions) Wyd Nauk PWN, Warszawa, p 340. (in Polish)
24. Lange W (1986) Fizyczno-limnologiczne uwarunkowania tolerancji systemów jeziornych (Physical and limnological conditions of tolerance of lake systems). Zesz Nauk UG Ser Rozpr Monogr 79:1–177. (in Polish)
25. Pasławski Z (1975) Typologia hydrologiczna jezior Pojezierza Wielkopolskiego (Hydrological typology of the lakes of the Wielkopolska Lake District). Prz Geofiz 28(3):271–280. (in Polish, English summary)
26. Riera JL, Magnuson JJ, Kratz TK, Webster KE (2000) A geometric template for the analysis of lake districts applied to the Northern Highland Lake District, Wisconsin, USA. Freshwater Biol 43(3):301–318
27. Bogoslovsky B (1960) Skhema gidrologicheskoj klassifikacji ozer i rajonirowania ozer SSSR (Scheme of hydrological classification of lakes and territorial division of lakes of the USSR). Vestn Mosk Univ Ser 5 Geogr 2:18–36. (in Russian)
28. Gołębiewski R, Lange W (1975) Stosowalność niektórych typologii limnologicznych na przykładzie jezior Pojezierza Kaszubskiego (The use of selected limnological typologies on the example of, lakes located on Kaszuby lakeland). Zesz Nauk Wydz BiNoZ 5:25–56. (in Polish, English summary)
29. [EC] European Commission (2000) Directive 2000/60/EC of the European Parliament and of the Council of 23 October 2000 establishing a framework for community action in the field of water policy. Off J Eur Union L 327:1–72

30. Touchart L (2000) Les lacs. Origine et morphologie. Edit de L'Harmattan, Paris, 208 p
31. Hondzo M, Stefan HG (1996) Dependence of water quality and fish habitat on lake morphometry and meteorology. J Water Res Plan Man 122(5):364–373
32. Prepas EE, Planas D, Gibson JJ, Vitt DH, Prowse TD, Dinsmore WP, Halsey LA, McEachern PM, Paquet S, Scrimgeour GJ, Tonn WM, Paszkowski CA, Wolfstein K (2001) Landscape variables influencing nutrients and phytoplankton communities in Boreal Plain lakes of northern Alberta: a comparison of wetland- and upland-dominated catchments. Can J Fish Aquat Sci 58:1286–1299
33. Håkanson L (2005) The importance of lake morphometry for the structure and function of lakes. Int Rev Hydrobiol 90(4):433–461
34. Knoll LB, Hagenbuch EJ, Stevens MH, Vanni MJ, Renwick WH, Denlinger JC, Hale RS, González MJ (2015) Predicting eutrophication status in reservoirs at large spatial scales using landscape and morphometric variables. Inland Waters 5:203–214
35. Timms B (2010) Geomorphology of lake basins. In: Likens GE (ed) Lake ecosystem ecology. Academic, San Diego, pp 203–210
36. Kerekes J (1977) The index of lake basin permanence. Int Revue ges. Hydrobiol 62
37. Kraska M, Piotrowicz R, Radziszewska R (1999) Dystrophication as a chief factor of changes in the physico-chemical properties of water and vegetation of lobelian lakes of Bory Tucholskie National Park. Acta Hydrobiol 41(6):127–135
38. Klimaszyk P, Rzymski P (2011) Surface runoff as a factor determining trophic state of Midforest Lake. Pol J Environ Stud 20(5):1203–1210
39. Klimaszyk P, Rzymski P (2013) Catchment vegetation can trigger lake dystrophy through changes in runoff water quality. Ann Limnol Int J Limnol 49:191–197
40. Barałkiewicz D, Kraska M, Siepak J (1996) The content of DOC, POC and TOC in Lobelian Lakes. Polish J Environ Stud 5(6):17–22
41. Kortelainen P (1999) Source of aquatic organic carbon. In: Keskitalo J, Eloranta P (eds) Limnology of humic waters, Backhuys, Leiden, pp 43–57
42. Hongve D (1999) Production of dissolved organic carbon in forested catchments. J Hydrol 224:91–99
43. Jerlov NG (1976) Marine optics. Elsevier, Amsterdam, 230 p
44. Borowiak D (2011) Właściwości optyczne wód jeziornych Pomorza (Optical properties of water ecosystems of the Pomeranian lake waters). Wydaw. UG, Gdańsk. 275 p. (in Polish, English summary)
45. Bociąg K, Rudowska A (2016) Zasoby i stan populacji isoetydów w jeziorach lobeliowych Pojezierza Pomorskiego (Isoetids resources and population numbers in lobelia lakes in the Pomorskie Lake District). In: Bociąg K, Borowiak D (eds) Jeziora lobeliowe w drugiej dekadzie XXI wieku (Lobelia lakes in the second decade of the 21st century), Wydaw. FRUG, Gdańsk, pp 52–79. (in Polish, English summary)
46. Koenings JP, Edmundson JA (1991) Secchi disk and photometer estimates of light regime in Alaskan lakes: effect of yellow color and turbidity. Limnol Oceanogr 36(1):91–105
47. Davies-Colley RJ, Vant WN (1988) Estimation of optical properties of water from Secchi disk depth. Water Res Bull 24(6):1329–1335
48. Faraś-Ostrowska B, Lange W (1998) Przezroczystość wody jako miara nasilenia eutrofizacji jezior (Water transparency as a measure of the intensity of lake eutrophication). In: Lange W, Borowiak D (eds) Zagrożenia degradacyjne a ochrona jezior (Degradation threats and lakes protection), Bad. Limnol. vol 1, KLUG, Gdańsk, pp 181–191. (in Polish)
49. Middelboe AL, Markager S (1997) Depth limits and minimum light requirements of freshwater macrophytes. Freshwater Biol 37:553–568
50. Borowiak D, Bociąg K, Nowiński K, Borowiak M (2017) Light requirements of water lobelia (*Lobelia dortmanna* L.). Limnol Rev 17(4):171–182
51. [EC] European Commission (2013) Interpretation manual of European Union Habitats – EUR28, European Commission, DG Environment, Nature ENV B.3, Brussel, 144 pp
52. Moss BR (2018) Ecology of freshwaters: Earth's bloodstream, V edn, Willey, New York, p 554

53. Wium-Andersen S, Andersen JM (1972) The influence of vegetation on the redox profile of the sediment of Grane Langsø, a Danish Lobelia lake. Limnol Oceanogr 17:948–952
54. Sand-Jensen K, Søndergaard M (1979) Distribution and quantitative development of aquatic macrophytes in relation to sediment characteristics in oligotrophic Lake Kalgaard, Denmark. Freshwater Biol 9:1–11
55. Roelofs JGM, Schuurkes JAAR, Smits AJM (1984) Impact of acidification and eutrophication on macrophyte communities in soft waters. Part II. Experimental studies. Aquat Bot 18:389–411
56. Farmer AM, Spence DHN (1985) Studies of diurnal acid fluctuations in British isoetid-type submerged aquatic macrophytes. Ann Bot 56:347–350
57. Szmidt K, Bociąg K (2016) Wprowadzenie. Zakres i cele projektu (Introduction. Range and aims of the project). In: Bociąg K, Borowiak D (eds) Jeziora lobeliowe w drugiej dekadzie XXI wieku (Lobelia lakes in the second decade of the 21st century), Wydaw. FRUG, Gdańsk, pp 8–13. (in Polish)
58. Wilk-Woźniak E, Kraska M, Piotrowicz R, Klimaszyk P (2012) 3110 Jeziora lobeliowe (3110 Lobelia lakes), In Mróz W (ed) Monitoring siedlisk przyrodniczych. Przewodnik metodyczny. Część druga (Monitoring of natural habitats. Methodological guide. Part two), GIOŚ, Warszawa, pp 114–129. (in Polish)
59. Wilk-Woźniak E, Kraska M, Piotrowicz R, Klimaszyk P (2013) 3110. Lobelia lakes – Oligotrophic waters containing very few minerals of sandy plains (Littorelletalia uni-florae). In: Mróz W (ed) Monitoring of natural habitats. Methodological guide for 5 natural habitats: 3110, 6210, 6520, 7230 and 9180, GIOŚ, Warszawa, pp 30–44
60. Arts GHP, de Haan AJ, Siebum MB, Verheggen GM (1989) Extent and historical development of the decline of Dutch soft waters. Proc Kon Ned Akad Wetensch C92:281–295
61. Bociąg K, Szymański P, Gos K (2016) Ochrona jezior lobeliowych a gospodarka leśna (Lobelia lakes protection and forest management). In: Bociąg K, Borowiak D (eds) Jeziora lobeliowe w drugiej dekadzie XXI wieku (Lobelia lakes in the second decade of the 21st century), Wydaw. FRUG, Gdańsk, pp 142–160. (in Polish, English summary)
62. Andersen T, Pedersen O, Andersen FO (2005) Nutrient concentrations in a *Littorella uniflora* community at higher CO2 concentrations and reduced light intensities. Freshwater Biol 50:1178–1189
63. Pedersen O, Andersen T, Ekejima K, Hossain MZ, Andersen FØ (2006) A multidisciplinary approach to understanding the recent and historical occurrence of the freshwater plant *Littorella uniflora*. Freshwater Biol 51:865–873
64. Klimaszyk P, Piotrowicz R, Rzymski P (2015) Changes in physico-chemical conditions and macrophyte abundance in a shallow soft-water lake mediated by a Great Cormorant roosting colony. J Limnol 74(1):114–122
65. Hesse T (2000) Ichtyofauna of selected lobelia lakes (north Poland). Folia Univ Agric Stetin Piscaria 27:83–92
66. Morzuch J, Kapusta A (2010) Management of fish populations in lobelia lakes in the vicinity of Bytów (northern Poland). Arch Pol Fish 18:101–113
67. Rahel FJ (1984) Factors structuring fish assemblages along a bog lake successional gradient. Ecology 65(4):1276–1289
68. Baker JP, Schofield CL (1982) Aluminum toxicity to fish in acidic waters. Water Air Soil Pollut 18:289–309
69. Kapusta A, Czarkowski TK (2016) Gospodarka rybacka w jeziorach lobeliowych (Fisheries management in lobelia lakes). In: Bociąg K, Borowiak D (eds) Jeziora lobeliowe w drugiej dekadzie XXI wieku (Lobelia lakes in the second decade of the 21st century), Wydaw. FRUG, Gdańsk, pp 162–170. (in Polish, English summary)

Environmental Conditions in Polish Lakes with Different Types of Catchments

Renata Tandyrak, Jolanta Grochowska, Katarzyna Parszuto, Renata Augustyniak, and Michał Łopata

Contents

Abstract The research covered three groups of lakes: urban lakes in Kartuzy in the East Pomeranian Lake District, a group of Barczewo lakes in the Olsztyn Lakeland, and a group of Leginy lakes in the Mrągowo Lakeland. The water bodies were evaluated in terms of water dynamics, oxygen saturation, mineral pollution, and trophic status. The lake complexes differed fundamentally in land-use types in the surrounding areas. The lakes from each group were different. Electrolytic conductivity was highest in urban lakes, especially in Lake Karczemne which has been a sewage receptacle for many years. In the group of urban water bodies, the trophic state index was highest in Lake Karczemne (TSI (SD) and (TP), within the limits of hypertrophy, $R^2 = 0.886$). In this group of lakes, organic matter limited visibility ($R^2 = 0.943$). In the lakes surrounded mainly by forests, TSI was lower and indicative of oligo/mesotrophy in Lakes Kierźlińskie and Dłużek. Lake Dłużek was characterized by extremely low electrolytic conductivity (43–87 μS cm^{-1}). TSI (TP) exceeded the reference standards in all examined lakes. In eight water

R. Tandyrak (✉) · J. Grochowska · K. Parszuto · R. Augustyniak · M. Łopata
Department of Water Protection Engineering, Faculty of Environmental Sciences, University of Warmia and Mazury in Olsztyn, Olsztyn, Poland
e-mail: renatat@uwm.edu.pl

E. Korzeniewska, M. Harnisz (eds.), *Polish River Basins and Lakes – Part I*, The Handbook of Environmental Chemistry 86,
https://doi.org/10.1007/978-3-030-12123-5_6

bodies, primary production was limited by nitrogen (TSI (Chl) – TSI (TN) > 0 and TSI (TN) – TSI (TP) < 0, TN/TP < 10), which promoted cyanobacterial blooms.

Keywords Catchment · Lakes · TSI · Urban · Water dynamics

1 Introduction

Lakes are relatively unstable elements of the landscape, and they are characteristic of young postglacial landscapes. Lakes occupy depressions and naturally accumulate the substances that flow from the basin, both in dissolved and slurry form. These factors are responsible for eutrophication which determines the quality of lakes [1]. According to Borowiak [2], clear and blue lake waters contribute to the aesthetic appeal of young glacial landscapes and confirm their good ecological status. The green color of lake water is indicative of significant changes not only in the lake but also in other reservoirs that feed into lakes. Lakes are a part of an ecosystem that extends beyond the shoreline to the catchment area. These ecosystems are highly sensitive to anthropogenic pressure [3].

The functioning of water bodies is strictly dependent on the supply of water from the basin [4]. Land-use types in the catchment have a decisive influence on the quantity and type of pollutants supplied to surface water and groundwater. Anthropogenic changes in catchment areas, such as intensification of agriculture, drainage, sewage and rainwater disposal [5–7] as well as excessive recreational use, contribute to the disappearance of lake ecosystems around the world [8, 9]. In practice, this is manifested by intensified primary production, deteriorating aesthetics and sanitary conditions, increasing complexity and cost of water treatment, and, in the long run, closure of lakes for recreational use and fishing.

Urban lakes are particularly exposed to degradation. These water bodies are often small and shallow. Changes in the catchment area and hydrotechnical treatments further contribute to the unnatural and intensified degradation of urban lakes [10]. The lakes in Kartuzy, which have been used as sewage receptacles for many years, belong to this category. Lakes in agricultural catchments undergo eutrophication due to artificial fertilizers which are often applied in excessive doses or in the wrong period. Fertilizer ingredients are largely washed out from the soil surface. In periods of extremely adverse weather and high rainfall, up to 40% of fertilizer ingredients can be washed out from the soil [11]. This is the case in the group of lakes near Reszel in the Mrągowo Lake District. Forest catchments are least conducive to the eutrophication of lakes.

The aim of this work is to compare environmental conditions in lakes located in different areas (urban, agricultural, and forest).

2 Materials and Methods

This study did not set out to investigate the relationship between the analyzed parameters and land-use type in sub-catchments. The main focus was on the location of lakes in the spatial sense. Three groups of lakes were examined in the study:

(a) A group of four urban lakes in Kartuzy (Mielenko, Karczemne, Klasztorne Małe, and Klasztorne Duże) (Fig. 1) in the Eastern Pomeranian Lakeland
(b) A group of 11 lakes situated near Barczewo (Kiermas, Kierzlińskie, Dłużek, Tumiany, Pisz, Podąbek, Umląg, Orzyc, Orzycek, Kiełdynek, and Świętajno) in the Olsztyn Lakeland (Fig. 2)
(c) A group of six lakes (Klawój, Pasterzewo, Legińskie, Mutek, Widryńskie, and Trzcinno) in the Mrągowo Lake District (Fig. 3)

These lakes form compact groups, and they are exposed to anthropogenic pressure to a varied extent. Their basic morphometric parameters are presented in Table 1.

The study was carried out on an annual cycle, during all seasons of the year. Water samples were collected at a depth of 1 m (surface) and 0.5 m above the bottom (bottom) in one measuring station located above the deepest part of the lake.

Fig. 1 The location of urban lakes in Kartuzy with a marked boundary of the catchment area (Source: based on [6])

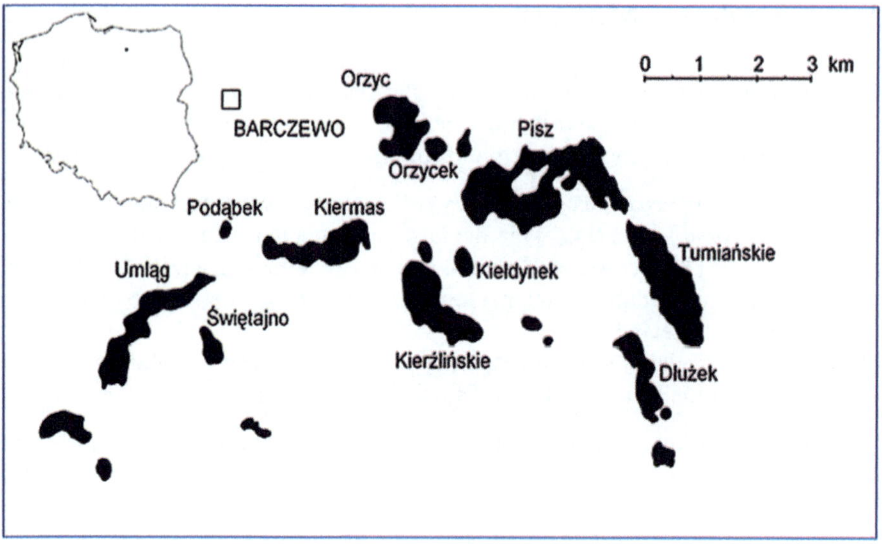

Fig. 2 The location of the lakes in Barczewo (Source: based on [9])

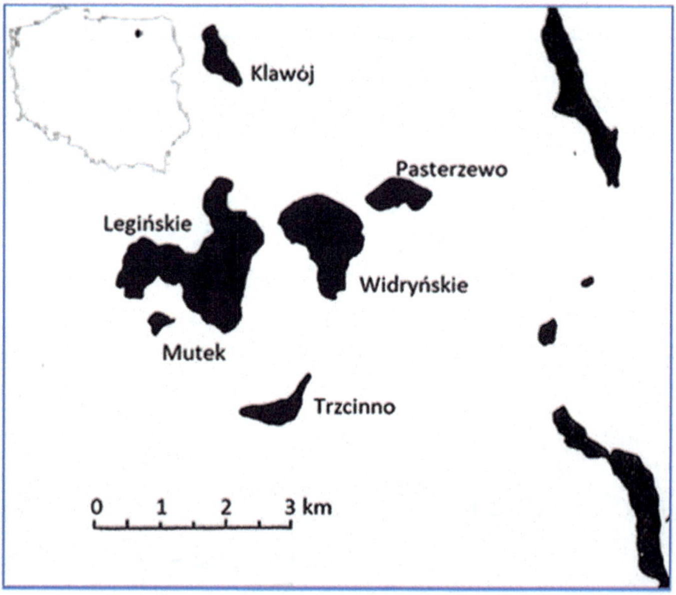

Fig. 3 The location of the lakes in Leginy (Source: based on [13])

Secchi disc (SD) visibility and the thermal-oxygen profile were determined during each measurement. Oxygen content and temperature were determined with the ProOdo optical oxygen probe (YSI), and electrolytic conductivity was measured

Table 1 Basic morphometric parameters of the investigated lakes

The group of lakes	Location a.s.l (m)	Water surface (ha)	Volume (m^3)	Max. depth (m)	Aver. depth (m)
Kartuzy					
Mielenko[a]	204.0	7.8	102,900	1.9	1.3
Karczemne[a]	203.7	40.4	798,300	3.2	2.0
Klasztorne Małe[a]	203.0	13.7	1,106,000	20.0	8.1
Klasztorne Duże[a]	202.3	57.5	2,780,000	8.5	4.8
Barczewo					
Kiermas	109.0	70.0	2,150,000	9.3	3.1
Kierźlińskie	113.7	92.8	10,861,100	44.5	11.7
Dłużek	143.2	43.3	3,078,000	29.8	7.1
Tumiańskie	112.9	120.6	8,110,900	17.0	6.7
Pisz	112.8	205.0	11,570,800	25.2	5.1
Podąbek[b]	108.2	3.3	228,222	17.5	5.3
Umląg	109.2	85.0	2,335,700	5.0	1.9
Orzyc	67.5	57.6	2,666,800	11.5	4.6
Orzycek[b]	112.8	7.1	No data	2.2	No data
Kiełdynek	113.4	9.0	186,200	4.0	2.1
Świętajno	109.1	18.0	494,100	6.6	2.7
Leginy					
Klawój	106.9	26.0	2,090,900	17.3	8.0
Pasterzewo	104.8	31.0	2,271,100	21.5	7.3
Legińskie	103.2	220.0	27,813,300	37.2	12.6
Widryńskie	114.0	123.9	10,557,400	27.0	8.5
Trzcinno	107.7	27.5	3,383,700	32.2	12.3
Mutek[b]	107.1	7.0	250,000	15.3	3.6

Source: based on: [a][6, 12, 13], [b]lake area was measured with the tools available at geoportal pl. [14]

with the MultiLine probe (WTW). Colorimetric measurements of total phosphorus (TP) were conducted after digestion with sulfuric acid and potassium persulfate with ammonium molybdate and $SnCl_2$ using the Macherey-Nagel Nanocolor spectrophotometer. Ammonium nitrogen (NH_4) concentration was determined with the Merck SQ 118 spectrophotometer. The concentrations of nonvolatile total organic carbon (TOC) and total nitrogen (TN) were determined with the Hach IL 550 TOC-TN analyzer. Chlorophyll a was determined by pheopigment-corrected spectrophotometry [15].

The trophic state of the studied lakes was assessed on the basis of Carlson's trophic state index [16].

Partial and detailed indexes were calculated from the following transformed equations:

$$\mathrm{TSI\ (SD)} = 60 - 14.43\ \ln\ (\mathrm{SD})$$

$$\mathrm{TSI\ (Chl)} = 9.81\ \ln\ (\mathrm{Chl}) + 30.6$$

$$\mathrm{TSI\ (TP)} = 14.42\ \ln\ (\mathrm{TP}) + 4.15$$

In addition, TSI (TN) was calculated in simplified terms according to Kratzer and Brezonik [17]:

$$\mathrm{TSI\ (TN)} = 14.43\ \ln\ (\mathrm{TN}) + 54.45$$

and TSI (TOC) was calculated according to Dunalska [18]:

$$\mathrm{TSI\ (TOC)} = 15.71\ \ln\ (\mathrm{TOC}) + 20.59.$$

TSI < 40 is characteristic of oligotrophic lakes, $40 <$ TSI < 60 – of mesotrophic lakes, $60 <$ TSI <80 – of eutrophic lakes, and TSI > 80 – of hypertrophic lakes.

The data did not have normal distribution; therefore, R^2 was used. The results were regarded as statistically significant at a probability level of <0.05.

3 Characteristics of the Research Area

3.1 Lakes in Kartuzy

The catchment area of the group of lakes in Kartuzy covers the area of 12.25 km^2 [6] and is characterized by a varied glacial landscape that was formed as a result of the Nordic Glacier's stop. The catchment is situated on a rolling patch of corrugated bottom moraine, strongly eroded and cut by numerous melt basins with no outflow. Lakes Karczemne, Klasztorne Małe, and Duże are situated in a clear cavity and form a line with the width of several 100 m in the direction of NNE-SSW. A large melt depression is partially occupied by Lake Mielenko and adheres to its southwestern part. Klasztorna Struga is the main stream which delimits a distinct hydrographic axis in the area [19].

The areas surrounding this group of lakes are characterized by two extremely different land-use types (Fig. 4): forests and urbanized areas.

Direct catchments are difficult to identify in urban areas. Fragments of the topographic basin are excluded from the catchment area because they are drained by storm water drainage systems. Therefore, built-up land was not taken into account in the direct catchment of Lakes Mielenko, Karczemne (Fig. 5), and Klasztorne Małe. However, this does not mean that the lakes are not supplied with storm water from the catchment. Storm water is carried by the drainage system and is discharged into the water body. The direct catchment of Lakes Klasztorne Małe, and Klasztorne Duże is occupied by forests in 100%.

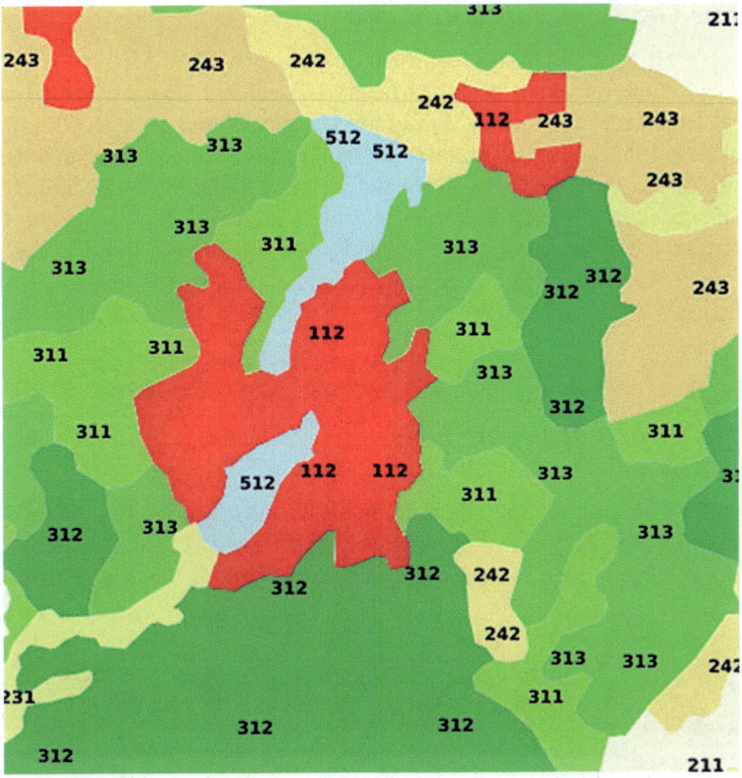

Fig. 4 Land cover and land-use type in the catchment of the group of lakes in Kartuzy (according to CLC 2012 [20]): 112, urban area, low-density development; 242, cropped land; 243, cropped land with a high share of natural vegetation; 311, mixed forest; 312, coniferous forest; 313, deciduous forest; 512, water body

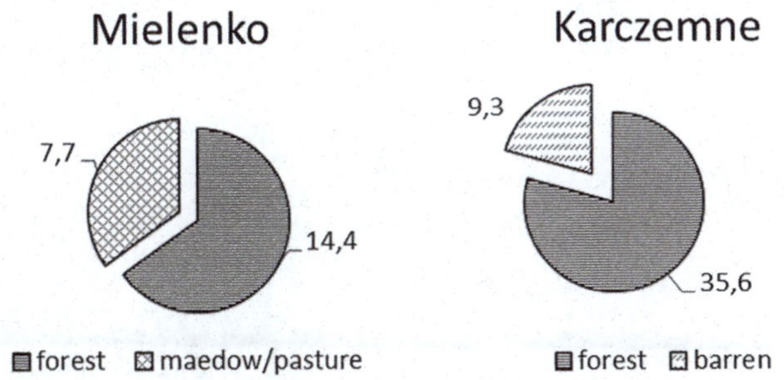

Fig. 5 Land-use types in the direct catchment of Lakes Mielenko and Karczemne

3.2 *The Group of Barczewo Lakes*

It is situated in the western part of the Mazury Lake District in the Olsztyn Lakeland. The local landscape was formed during the last glaciation period. Its decay phases are marked by the arches of moraine embankments. Moraine height is less than 200 m above sea level. Glacial valleys and former lake basins are occupied by peat bogs and meadows [19]. The location and the features of this part of the Olsztyn Lake District are characteristic of glaciated areas in northern Poland. The area is largely covered by mixed and deciduous forests, with a small share of agricultural land (Fig. 6).

3.3 *The Group of Leginy Lakes*

It constitutes a separate complex of water bodies in the Masurian Lake District, in the northern part of the Mrągowo Lake District, near Reszel. These lakes were formed during the sixth and seventh phase of the last glacial period. These water bodies were

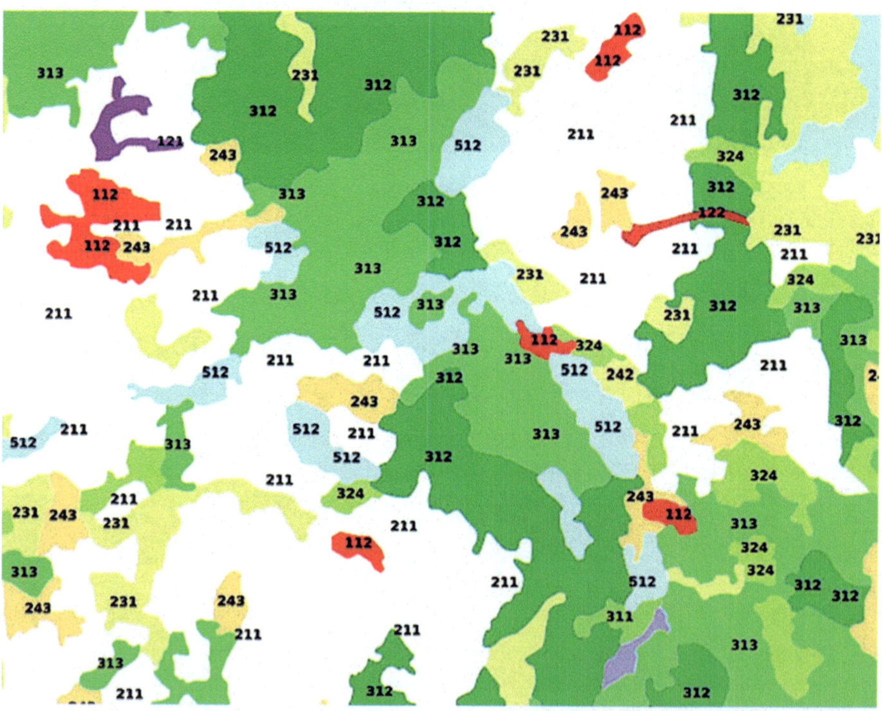

Fig. 6 Land cover and land-use types in the catchment of the group of Barczewo lakes (according to CLC [20]): 112, urban area, low-density development; 211, arable land without drainage systems; 243, cropped land with a high share of natural vegetation; 312, coniferous forest; 313, deciduous forest; 512, water body

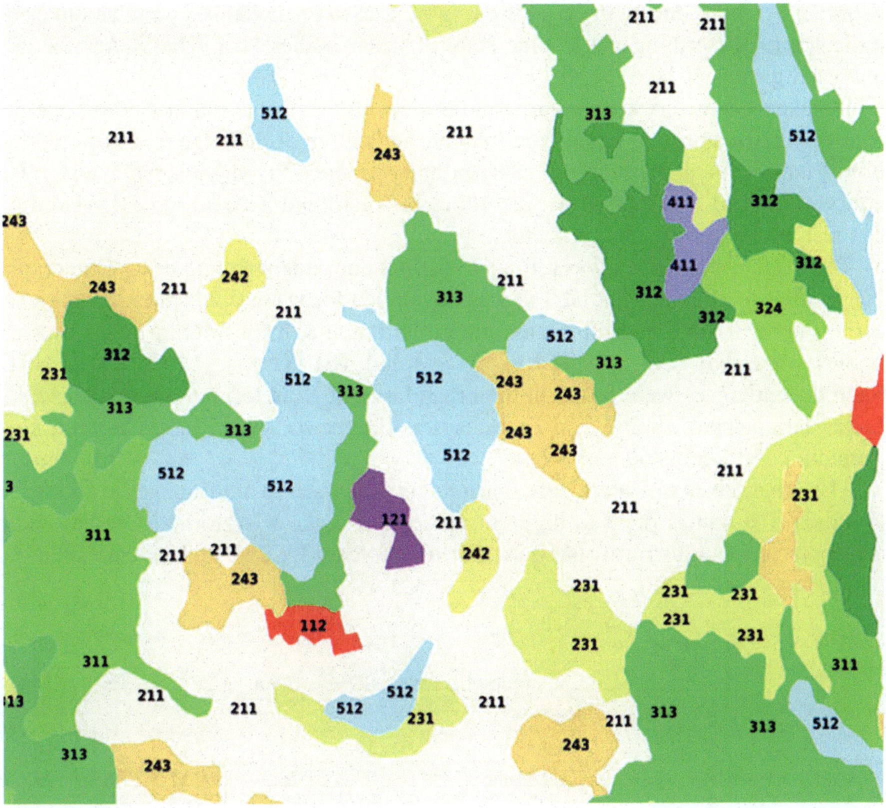

Fig. 7 Land cover and land-use types in the catchment of the group of Leginy lakes (according to CLC 2012 [20]): 112, urban area, low-density development; 121, industrial or commercial zone; 211, arable land without drainage systems; 231, meadow; 242, cropped land; 243, cropped land with a high share of natural vegetation; 311, mixed forest; 313, deciduous forest; 512, water body

grouped based on a shared catchment, the same glaciation phase, and compact location. The catchment is drained by the upstream section of river Sajna and the Łyna–Pregoła river basin [21, 22]. The meridional fringe orientation of glacial basins and the latitudinal course of seven moraine chains are the characteristic features of this area [19]. The catchment is covered mainly by arable land, with a high share of meadows (Fig. 7).

4 Results and Discussion

4.1 Trophic State

The trophic state of lakes was assessed by analyzing causal factors (phosphorus, nitrogen) and the resulting biological conditions (Secchi disc visibility, chlorophyll a, organic carbon) during the growing season. Carlson's model is a popular and easy indicator for describing the trophic state of water bodies [16]. The values for the analyzed groups of lakes are listed in Table 2.

In lakes characterized by harmonious development and minor anthropogenic changes, TSI values are equal, and they are indicative of the degree of eutrophication. Visibility is most often limited by primary production, which is reflected by the concentration of chlorophyll a. According to Kuehl and Troelstrup [23], visibility can also be limited by abiotic factors.

The trophic state index was refined to further our understanding of lake functioning and the food chain. Partial TSI was developed for water color [24, 25], organic carbon content [18], and nitrogen content. Nitrogen plays a limiting role in water bodies when TSI (Chl a) − TSI (TN) $>$ 0 and TSI (TN) − TSI (TP) $<$ 0 [17]. Such dependencies were found in the urban lakes of Klasztorne Małe and Duże; in Lakes Tumiańskie, Umląg, and Kiermas near Barczewo; and in Lakes Widryńskie, Trzcinno, and Legińskie near Reszel.

The urban lakes in Kartuzy are considerably exposed to anthropogenic pressure, and their TSI values point to high eutrophication. Lake Klasztorne Małe has been a receptacle of industrial wastewater for many years [5, 26], and its catchment is

Table 2 TSI values in the investigated lakes

	TSI				
The group of lakes	SD	Chl a	TP	TN	TOC
Kartuzy					
Mielenko	73.2	62.2	78.5	74.6	66.9
Karczemne	83.2	76.5	102.5	79.5	74.6
Klasztorne Małe	75.1	70.9	85.4	70.2	63.9
Klasztorne Duże	70.0	67.6	77.0	67.4	61.4
Barczewo					
Kiermas	60.7	67.1	84.1	64.1	61.4
Kierźlińskie	38.4	41.6	70.3	55.5	47.5
Dłużek	37.2	35.3	67.2	47.5	47.1
Tumiańskie	57.4	60.1	76.9	59.9	64.3
Pisz	56.6	52.6	73.5	56.3	62.8
Podąbek	56.5	56.4	70.5	62.6	60.4
Umląg	58.6	66.6	81.7	61.2	61.9
Orzyc	61.5	50.1	74.5	63.3	62.3
Orzycek	50.7	56.2	76.8	57.7	60.9
Kiełdynek	83.2	70.5	86.3	71.2	69.5
Świętajno	60.0	53.6	74.2	65.9	59.1
Leginy					
Klawój	49.3	52.3	69.8	58.6	55.6
Pasterzewo	39.7	37.3	69.0	60.1	54.7
Legińskie	48.0	47.4	65.4	55.9	56.1
Widryńskie	46.5	54.8	70.1	57.8	57.8
Trzcinno	41.5	42.7	69.0	56.4	74.8
Mutek	66.2	66.0	77.3	61.5	62.9

still disorganized [27]. The above is confirmed by the extreme values of total nitrogen (up to 22.38 mg N dm^{-3}) and total phosphorus (up to 20.6 mg P dm^{-3}, including 8.95 mg P dm^{-3} in mineral form), especially in the bottom layer. In this group of water bodies, TSI (TP) was always highest. In Lakes Karczemne and Klasztorne Małe, TSI (TP) was indicative of hypertrophy, and it approached the upper limit of eutrophication in other water bodies (Table 2). Significant differences in TSI (TP) and TSI (SD) values were observed in the range of 10.0–19.3. Similar results were noted in Lake Starodworskie in Olsztyn [7], where this difference reached up to 33. Similar or somewhat greater differences in the values of TSI (TP) and TSI (SD) were also observed by Kubiak [1] in the lakes of Western Pomerania, by Grochowska [28] in the lakes of the Upper Pasłęka River, and by Bajkiewicz-Grabowska [29] in the lakes of the Kashubian Landscape Park. The differences reported by the cited authors were generally higher. According to the above authors, phosphorus is the main cause of eutrophication.

The relationship between the main trophic state indicators in the group of lakes in Kartuzy was constant: TSI (TP) > TSI (SD) > TSI (Chl a). The noted values were characteristic of eutrophication (60–80). In Lake Karczemne, the values of TSI (TP) and TSI (SD) were indicative of hypertrophy.

The values of TSI (SD) and TSI (TOC) were bound by a significant relationship which was described by the following equation:

$$y = 1.333 \text{ TSI (TOC)} - 1.4538, R^2 = 0.943$$

and the relationship between TSI (SD) and TSI (TP) by:

$$y = 0.5393 \text{ TSI (TP)} = 27.85, R^2 = 0.886$$

The lowest value of $R^2 = 0.4654$ was calculated for the relationship between TSI (SD) and (Chl a).

TSI (SD) > TSI (Chl a) points to a predominance of small algae which limit visibility [18]. Low water transparency can be caused by high concentrations of organic matter, turbidity, or water color [30]. In the discussed group of lakes, SD (Fig. 8) was characteristic of highly eutrophic water bodies [2, 31]. The significant relationship between TSI (SD) and TSI (TOC) is confirmed by the limiting effect of organic matter at $R^2 = 0.943$.

The amount of nitrogen and phosphorus is as important as the nitrogen-to-phosphorus ratio in the primary production process. A high TN/TP ratio points to low light penetration caused by algae [31], whereas a low TN/TP ratio points to the limiting effects of nitrogen and an increase in undesirable species of cyanobacteria. According to Kosten et al. [32], TP and TN are better indicators of blue-green algal dominance than TN/TP in a warming climate. In the group of lakes in Kartuzy, TP and TN values were low, especially in Lake Karczemne ($2 < N/P < 6$) and Lake Klasztorne Małe ($3 > N/P < 7$), which created a favorable environment for the development of blue-green algae. The above was confirmed by a phytoplankton

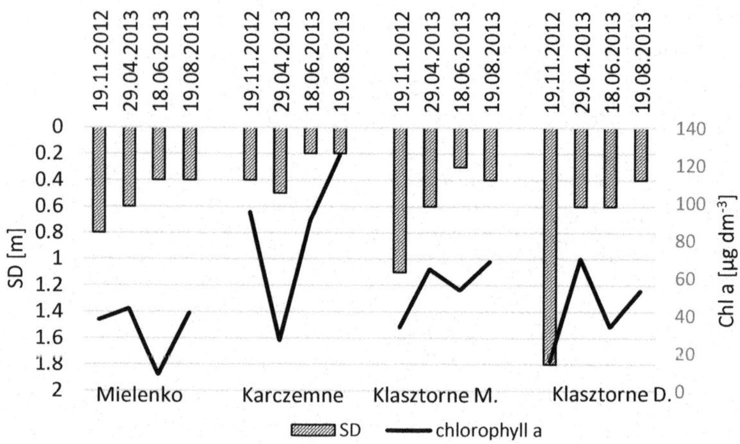

Fig. 8 Water transparency (SD) and chlorophyll a (Chl a) concentration in the group of lakes in Kartuzy

study [26] which revealed a predominance of cyanobacteria. In Lake Karczemne, blue-green algae accounted for 38% of phytoplankton in June and 93% in August, and the dominant species were *Microcystis wesenbergii* (56%), *Microcystis aeruginosa* (20%), and *Microcystis viridis* (10%). In the phytoplankton of Lake Klasztorne Małe, blue-green algae accounted for 53% of phytoplankton in June and 94% in August, with a predominance of *Microcystis aeruginosa* (78%) and *Microcystis wesenbergii* (11%). Cyanobacterial blooms were not observed in Lake Mielenko (12 < N/P < 15). In June, green algae (*Pediastrum boryanum*) were predominant, and the share of cyanobacteria was below 10%, whereas in August, cyanobacteria accounted for 36% of phytoplankton, with a predominance of *Microcystis aeruginosa* (18%). In Lake Klasztorne Duże, TN/TP was determined at 11, and the limiting effect of nitrogen was the main contributor to TSI in June [14] when blue-green algae accounted for 47% of phytoplankton (mainly *Cuspidothrix issatschenkoi*), whereas in August, TN/TP was determined at 9 [20] and blue-green algal blooms were not noted.

In the group of lakes near Barczewo, most of which are situated in a forest, TSI values differed significantly between individual lakes. The greatest variations of up to 32 were noted in Lakes Kierźlińskie and Dłużek. In most cases, TSI values were indicative of eutrophication. Similarly to urban lakes, the lakes near Barczewo were generally characterized by the highest values of TSI (TP). In Lakes Kiermas, Umląg, and Kiełdynek, TSI (TP) exceeded the limit for hypertrophy, but the above was not confirmed by other TSI values.

In the deepest lakes, Kierźlińskie and Dłużek, TSI (SD) and TSI (Chl a) were indicative of oligotrophy. In these lakes, TSI (SD) = TSI (Chl a) < TSI (TP), which suggests that visibility was determined by primary production ($R^2 = 0.668$) (Fig. 9) and could be modified by other factors, such as zooplankton feeding or toxic compounds [30]. In Lakes Kiermas, Tumiańskie, Umląg, and Orzycek,

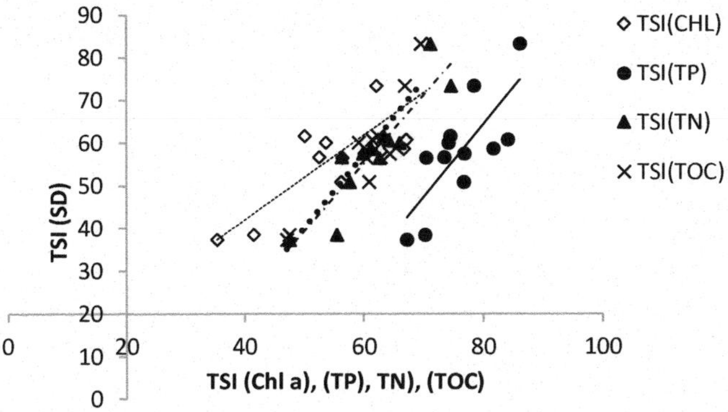

Fig. 9 The relationships between TSI (SD) vs. TSI (Chl a), (TP), (TN), and (TOC) in the group of Barczewo lakes

TSI (SD) < TSI (Chl a), which indicates that the phytoplankton was dominated by cyanobacteria [28]. The above was confirmed by low visibility (0.7–1.5 m) and the limiting effects of nitrogen on primary production [17]. In the above water bodies, the values of TN and TP (7 < N/P < 8) were lower than in the remaining lakes from this group.

In other water bodies, TSI (SD) > TSI (Chl), which points to the predominance of small algae which limited SD visibility [18]. The values of TSI (SD) and TSI (TOC) were bound by a significant relationship.

In this group of water bodies, the following relationships were observed between TSI(SD) and other partial indexes:

$$y = 0.9955\ \mathrm{TSI\ (Chl\ a)} + 2.0755;\ R^2 = 0.668,$$

$$y = 1.6905\ \mathrm{TSI(TP)} - 70.958;\ R^2 = 0.5899,$$

$$y = 1.5865\ \mathrm{TSI\ (TN)} - 39.961;\ R^2 = 0.819,$$

$$y = 1.719\ \mathrm{TSI(TOC)} - 45.895;\ R^2 = 0.834.$$

The group of Leginy lakes is situated in an area characterized by moderate agricultural production, absence of rehydration, and presence of natural vegetation. Most TSI values were within the limit of mesotrophy (40 < TSI < 60). The values of TSI (TP) were generally determined within the upper limit of eutrophication.

The most frequently observed relationship was TSI (SD) = TSI (Chl a) > TSI (TP). The above implies that light attenuation in water was significantly influenced by the phytoplankton. Algal biomass could also be limited by other factors, including extermination of zooplankton, toxic compounds, or the limiting effects of nitrogen. Based on visibility and other factors, the relationships between these indicators were described by the following equations:

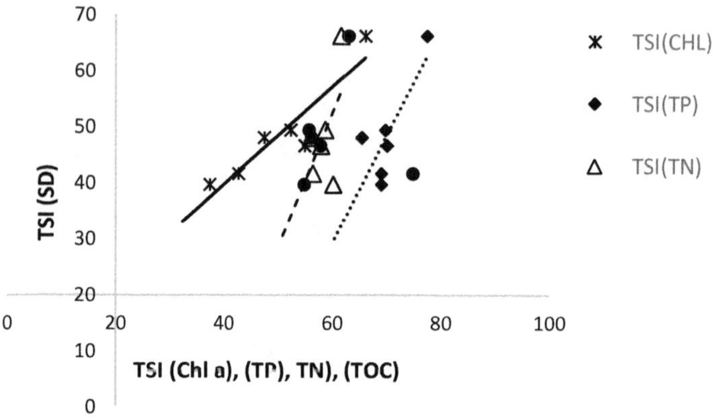

Fig. 10 The relationships between TSI (SD) vs. TSI(Chl a), (TP), (TN), and (TOC) in the group of Leginy lakes

$$y = 0.8696\ \mathrm{TSI(Chl\ a)} + 4.9688;\ R^2 = 0.859,$$

$$y = 1.9178\ \mathrm{TSI\ (TP)} - 85.948;\ R^2 = 0.6336,$$

$$y = 2.4166\ \mathrm{TSI\ (TN)} - 92.586;\ R^2 = 0.3119.$$

In this group of water bodies, organic matter (TOC) did not limit Secchi disc visibility (negligible R^2). The only exception was Lake Legińskie where TSI (TP) < TSI (TOC), which implies that such a relationship could theoretically exist, but was not confirmed in practice.

The limiting effects of nitrogen on primary production, determined based on the dependence described by Kratzer and Brezonik [17], were observed in Lakes Widryńskie, Trzcinno, and Legińskie. In these water bodies, TN/TP ranged from 8 to 19 and was higher than in urban lakes where cyanobacteria decreased Secchi disc visibility. In these lakes, visibility levels were dependent on chlorophyll a concentration (Fig. 10) and characteristic of eutrophication [2, 31] (Fig. 11). The differences between partial TSI values were significant, especially in lakes where nitrogen was a limiting factor. The highest values of TSI (TP) could be attributed to the fact that phosphorus was not fully utilized in the primary production process. According to Kubiak [1] and Marszelewski [11], TSI (TP) values in other Polish lakes could be higher than the actual trophic state of the lake environment.

4.2 Water Circulation

The trophic state of a water body is directly influenced by the availability of nutrients supplied from the catchment or released from bottom sediments and the trophogenic layer during the growing season [7]. The circulation of matter and energy in a water

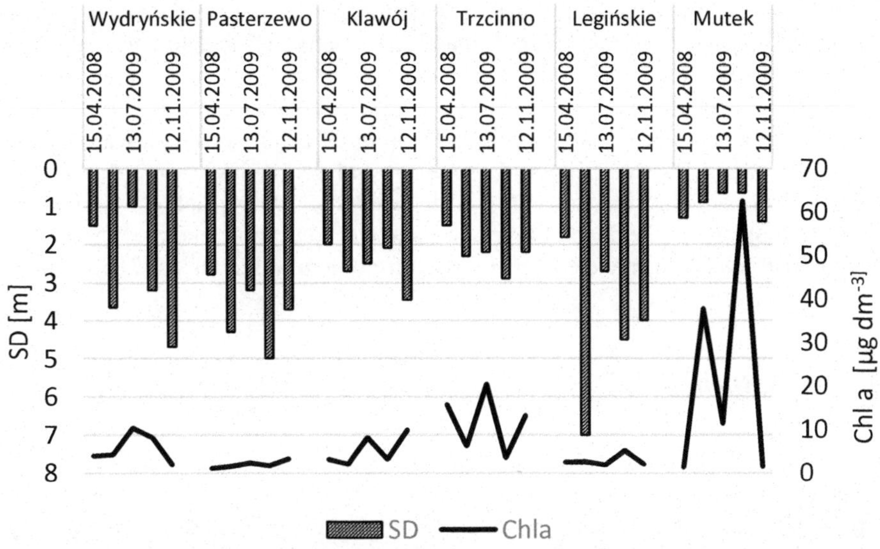

Fig. 11 Water transparency (SD) and chlorophyll a (Chl a) concentration in the group of Leginy lakes

body is driven by water circulation. A lake's mixing regime and oxygen saturation in the bottom layer play an important role in these processes.

The discussed water bodies were morphometrically differentiated (Table 1). Each group of lakes included water bodies situated in open areas as well as lakes that were shielded from wind by forests. These factors contributed to variations in the water dynamics of the analyzed lakes. The intensity and extent of wind-induced mixing, which are determined by the morphometric characteristics of each lake, primary production, and the distribution of organic matter influence oxygen levels in a water body.

In the group of urban water bodies, Lakes Karczemne and Mielenko were polymictic water bodies, and summer stratification in Lake Klasztorne Duże was incomplete. Lake Klasztorne Małe is heavily stratified, and it is one of the few meromictic water bodies in Poland [5, 7]. Due to considerable sewage pollution and limited water dynamics, including in periods of circulation, deeper lake strata were deficient in oxygen [25, 27]. Oxygen levels were low in all water bodies in this group.

In the group of Barczewo lakes, Lakes Świętajno and Kiełdynek are shallow water bodies with a complex circulation pattern. Lakes Pisz, Dłużek, Kierźlińskie, and Podąbek are bradymictic water bodies with low water dynamics. Lake Podąbek periodically exhibits the characteristic features of meromixis [7]. Lakes Orzyc, Kiermas, and Tumiańskie are characterized by moderate water dynamics and can be classified as eumictic water bodies. Lakes Orzycek and Umląg are polymictic water bodies where different thermal layers were not observed at the end of summer [9].

The group of Barczewo lakes was characterized by a clinograde oxygen profile. During the growing season, oxygen saturation reached up to 140% in the upper layers of Lakes Orzyc and Świętajno, which points to intensive production processes. In contrast, deeper layers, in particular the bottom layer, were highly deficient or completely deprived of oxygen [9]. In the group of Leginy lakes, all water bodies were fully thermally stratified in the summer, and deeper lakes were deficient in oxygen. In Lake Trzcinno, oxygen was depleted even during autumn circulation. Oxygen depletion in the deepest layers leads to eutrophication, and it contributes to the circulation of biogenic elements released from bottom sediments [7, 33, 34].

4.3 *Conductivity*

Electrolytic conductivity is a very important indicator of mineral pollution [11]. High conductivity can be attributed to strong anthropogenic pressure, mainly long-term sewage discharges (municipal and industrial) and agricultural pollution [3, 7].

The highest average electrolytic conductivity was noted in the group of lakes in Kartuzy (Fig. 12). In urban lakes, this parameter was highest in the deepest layers of Lake Klasztorne Małe (640–685 μS cm^{-1}), which provides additional evidence that Lake Klasztorne Małe was the most polluted water body in the study. Only minor variations in electrolytic conductivity were observed in the group of Barczewo lakes (343–480 μS cm^{-1} on the surface; 372–720 μS cm^{-1} at the bottom). The only exception was Lake Dłużek where electrolytic conductivity was extremely low in the

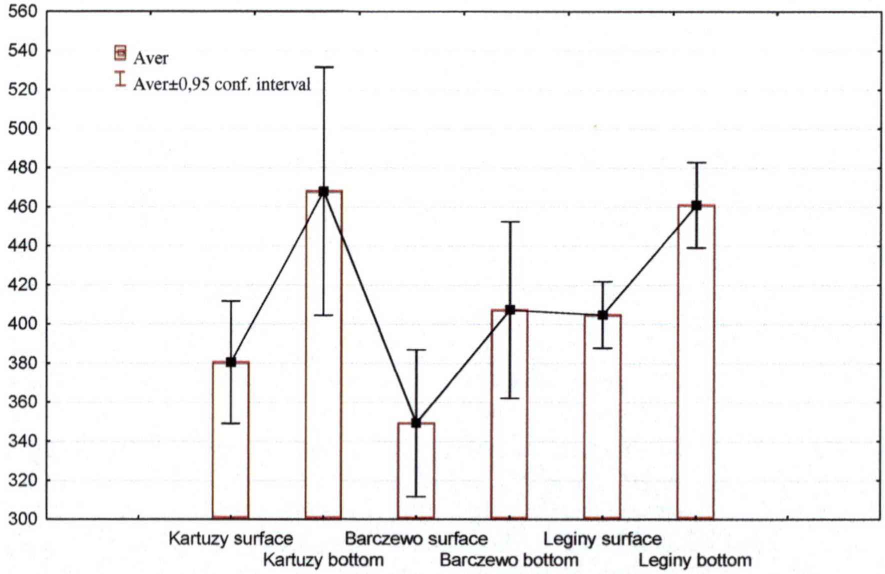

Fig. 12 Average values of electrolytic conductivity (μS cm^{-1}) in the studied lake groups

range of 43–48 μS cm^{-1} on the surface to 56–87 μS cm^{-1} at the bottom. This group of lakes was characterized by the lowest electrolytic conductivity (both on the surface and at the bottom) (Fig. 12). In the group of Leginy lakes, the analyzed parameter was highest in Lake Mutek, in particular in the bottom layer (580 μS cm^{-1}). Electrolytic conductivity was lower and somewhat variable in the remaining lakes from this group.

5 Conclusion

The trophic state of lakes is influenced by location as well as anthropogenic pressure associated with the transformation of catchment areas for agricultural use and urban development. The studied water bodies differed in morphometric parameters which contributed to variations in lake mixing types – from polymictic to bradymictic. All of the analyzed lakes were eutrophic water bodies in terms of oxygen saturation (which was also confirmed by Carlson's TSI). In periods of stagnation, excess organic matter was responsible for oxygen deficits. In eight lakes from the studied groups, nitrogen had a limiting effect on primary production. In these lakes, TSI (Chl a) – TSI (TN) > 0, TSI (TN) – TSI (TP) < 0, and TN/TP < 10. The above contributed to the development of blue-green algae, which was confirmed in the group of lakes in Kartuzy. In the group of urban lakes in Kartuzy, the role of organic matter was confirmed by the relationship between TSI (SD) vs. TSI (TP) and (TOC). A relationship was noted between the values of TSI (Chl a) and (TP) in the group of Leginy lakes (moderately intensive agricultural production) and between the values of TSI (SD) and TSI (TN) in the group of Barczewo lakes (forests). Urban lakes were contaminated mostly with mineral matter, whereas the lowest levels of pollution were noted in lakes with forest catchments.

Acknowledgments The authors would like to thank Marzena Karpienia, Agnieszka Grzegorczyk, and Bogumiła Kardaś for help in laboratory analyses. This study was supported by grant No 18.610.007-300 from the Ministry of Science and Higher Education (Poland).

References

1. Kubiak J (2003) Największe dymiktyczne jeziora Pomorza Zachodniego. Poziom trofii, podatność na degradację oraz warunki siedliskowe ichtiofauny (The largest dimictic lakes of Western Pomerania. Trophy level, susceptibility to degradation and habitat conditions of ichthyofauna). Dissertation, West Pomeranian University of Technology, Szczecin, Poland
2. Borowiak D (2007) Podwodny reżim światła [W]: Borowiak D (red.) Jeziora Kaszubskiego Parku Krajobrazowego (Underwater light regime [In]: Borowiak D (Ed.) Lakes of the Kashubian Landscape Park) Bad Limnol 5 Gdańsk

3. Sobolewski W, Borowiak D, Borowiak M, Skowron R (2014) Baza danych jezior Polski i jej wykorzystanie w badaniach limnologicznych (Database of Polish lakes and its use in limnological research). UMCS, Lublin
4. Bajkiewicz – Grabowska E, Mikulski Z (2010) Hydrologia ogólna (General hydrology). PWN, Warszawa
5. Januszkiewicz T (1969) Badania hydrochemiczne Jeziora Klasztornego jako odbiornika ścieków (Hydrochemical studies of the Klasztorne Lake as a sewage receiver). Prace Instytutu Gospodarki Wodnej 5(3):43–81
6. Dunalska J, Grochowska J, Wiśniewski G (2013) Program ochrony jezior kartuskich wraz z projektem rekultywacji (Protection program of the Kartuzy lakes with a restoration project), Typescript
7. Tandyrak R (2017) Ewolucja miksji i stan troficzny Jeziora Starodworskiego na tle warunków klimatycznych i postępującej urbanizacji (Evolution of mictic and trophic state of lake Starodworskie on the background of climate and urbanization). Dissertation, University of Warmia and Mazury, Olsztyn, Poland
8. Søndergaard M, Jeppsen E et al (2000) Lake restoration in Denmark. Lake Reservoirs Res Manage 5:151–159
9. Łopata M, Grochowska J, Augustyniak R, Tandyrak R, Parszuto K, Karpienia M (2017) Różnorodność warunków termicznych i tlenowych w ekosystemach limnicznych poddanych antropopresji – przykład grupy jezior barczewskich (Pojezierze Olsztyńskie) (Variety of thermal and oxygen conditions in limestone ecosystems subjected to anthropopression - an example of a group of Barczewskie lakes (Olsztyn Lakeland)), Toruń, 29–30 June 2017
10. Łopata M, Czerniawski R, Augustyniak R, Sługocki Ł, Goździk I, Krepski T, Kompowska A, Kowalska-Góralska M, Grochowska J, Parszuto K (2017) Uwarunkowania ochrony i rekultywacji jeziora usytuowanego w przestrzeni miejskiej na przykładzie zbiornika Rakowo Duże w Złocieńcu [W]: Ochrona i rekultywacja jezior. Wiśniewski R. (red.), Polskie Zrzeszenie Inżynierów i Techników Sanitarnych Oddział Toruń. (Conditions of protection and restoration of a lake situated in urban space on the example of Rakowo Duże reservoir in Złocieniec. In: Lake protection and restoration. Wiśniewski R (ed) Polish Association of Sanitary Engineers and Technicians Branch in Toruń, pp 39–54
11. Marszelewski W (2005) Zmiany warunków abiotycznych w jeziorach Polski północno – wschodniej (Changes of abiotic conditions in the lakes of north-eastern Poland). Dissertation, UMK, Toruń, Poland
12. Grochowska J, Łopata M, Parszuto K, Tandyrak R, Wiśniewski G (2011) Preliminary hydrochemical characteristic of barczewskie lakes. In: Marszelewski W (ed) Anthrophogenic and natural transformations of lakes, vol 5. UMK Toruń, pp 45–55
13. Choiński A (2006) Katalog jezior Polski (Catalog of Polish lakes). UAM, Poznań
14. Geoportal.gov.pl http://mapy.geoportal.gov.pl/imap/
15. Standard Methods for the Examination of Water and Wastewater (1985) American Public Health Association, Washington
16. Carlson RE (1977) A trophic state index for lakes. Limnol Oceanogr 22(2):361–369
17. Kratzer CR, Brezonik PL (1981) A Carlson – type trophic state index for nitrogen in Florida Lakes. Water Res Bull 17:713–715
18. Dunalska JA (2011) Total organic carbon as a new index for monitoring trophic states in lakes. Oceanol Hydrobiol Stud 40(2):112–115
19. Kondracki J (1998) Geografia regionalna Polski (Regional geography of Poland). PWN, Warszawa
20. Corine Land Cover-clc.gios.gov.pl
21. Olszewski P, Tadajewski A (1965) Morfometria jezior Legińskich (Morphometry of Leginskie lakes). Zesz Nauk WSR w Olsztynie 20:442 261–277
22. Kozłowski J, Poczyczyński P, Kozłowski K, Wziątek B, Teodorowicz M (2006) Operat Rybackiego Zagospodarowania Jeziora Legińskiego w zlewni rzeki Łyny nr 58. Olsztyn (A frame of fishing use of Legińskie lake in the catchment of Łyna, 58) Olsztyn

23. Kuehl L, Troelstrup Jr NH (2011) Relationships between net primary production, water transparency, chlorophyll a, and total phosphorus in Oak Lake, Brooking County, South Dakota. Proc S D Acad Sci 92:67–78
24. Eloranta P (1999) Humus and water physics. In: Keskitalo J, Eloranta P (eds) Limnology of humic waters. Backhuys Publisher, Leiden, pp 61–74
25. Håkanson L, Boulion VV (2001) Regularities in primary production, Secchi depth and fish yield and a new system to define trophic and humic state indices for lake ecosystems. Int Rev Hydrobiol 86:23–62
26. Grochowska J, Brzozowska R. Grzybowski M. Napiórkowska – Krzebietke A. Bigaj I., Szymański D. Zieliński R (2013) Raport końcowy. Badania fizyko – chemiczne wody i osadów dennych oraz pomiary hydrologiczne 4 jezior: Mielenko, Karczemne, Klasztorne Duże i Małe i ich zlewni pod kątem możliwości ochrony zbiorników przed degradacją. Maszynopis. (Final report. Physicochemical properties of water and sediments as well as hydrological measurements of 4 lakes: Mielenko, Karczemne, Klasztorne Duże and Małe and their catchment in terms of the possibility of their protecting from degradation) Typescript
27. Januszkiewicz T, Jakubowska Ł (1964) Dynamika zanieczyszczeń Jeziora Klasztornego (The pollution dynamic of the Klasztorne Lake). Prace Instytutu Gospodarki Wodnej 3(1):55–77
28. Grochowska J (2015) Obieg wybranych makropierwiastków w systemie rzeczno – jeziornym na przykładzie górnej Pasłęki (Circulation of selected macroelements in the river-lake system on the example of the Upper Pasłęka). Dissertation, PAN Lublin, Poland
29. Bajkiewicz – Grabowska E (2007) Zróżnicowanie troficzne jezior – stan obecny, miejsce w klasyfikacji troficznej [W]: Borowiak D (red.) Jeziora Kaszubskiego Parku Krajobrazowego (Trophic diversity of lakes present state, the place in trophic classification. In: Borowiak D (ed) Lakes of the Kashubian Landscape Park) Bad. Limol. 5, Gdańsk
30. Carlson RE, Simpson J (1996) A coordinator's guide to volunteer Lake monitoring methods. North American Lake Management Society, Madison
31. Faraś – Ostrowska B, Lange W (1998) Przezroczystość wody jako miara nasilenia eutrofizacji jezior (Transparency of water as a measure of the severity of lake eutrophication). Bad Limnol 1:181–191
32. Kosten S, Huszar VL, Bécares E, Costa LS, Donk E, Hansson LA, Meester L (2012) Warmer climates boost cyanobacterial dominance in shallow lakes. Glob Chang Biol 18(1):118–126
33. Brzozowska R, Wiśniewski G, Dunalska J, Filipkowska Z (2013) Sediment phosphorus fractions in an urban lake and its usability for predicting of the internal loading phenomenon. Int J Environ Health Res 6(4):340–349
34. Haggard BE, Moore PA, DeLaune PB (2005) Phosphorus flux from the bottom sediments in Lake Eucha, Oklahoma. J Environ Qual 34(2):724–728

Characteristics of Bottom Sediments in Polish Lakes with Different Trophic Status

Renata Augustyniak, Jolanta Grochowska, Michał Łopata, Katarzyna Parszuto, and Renata Tandyrak

Contents

Abstract An analysis of the chemical parameters of bottom sediments in selected Polish lakes was made, based on historical and recent data demonstrating the chemical composition of profundal sediments. The analysed lakes belong to a broad environmental spectrum, being different in the hydrological regime (seepage, flow-through), trophic state (mesotrophic, eutrophic, dystrophic), mixing type (polymictic, holomictic, meromictic), stratification (stratified, non-stratified) and sewage input presence (in the past or now). Moreover, some have been restored using technical methods (artificial mixing with and without destratification, phosphorus inactivation).

The results revealed that the main chemical components of the analysed sediments were silica, organic matter and calcium together with carbonates. Other sediment components occurred in low amounts, i.e. less than a few percent of d.w. Cluster analysis showed that the main factor which differentiated the sediments of the analysed Polish lakes seems to be their hydrological regime. The trophic state and stratification type of the lakes were less important factors affecting the character of bottom sediments.

Keywords Chemical characteristic · Lake · Sediment

R. Augustyniak (✉) · J. Grochowska · M. Łopata · K. Parszuto · R. Tandyrak
Department of Water Protection Engineering, Faculty of Environmental Sciences, University of Warmia and Mazury in Olsztyn, Olsztyn, Poland
e-mail: rbrzoza@uwm.edu.pl

E. Korzeniewska, M. Harnisz (eds.), *Polish River Basins and Lakes – Part I*,
The Handbook of Environmental Chemistry 86,
https://doi.org/10.1007/978-3-030-12123-5_7

1 Introduction

Bottom sediment in a lake is a product of deposition of autochthonous and allochthonous matter. This matter is subjected to various physical, chemical and biological transformation processes, commonly known as diagenetic ones, which occur during sediment formation [1, 2].

In the past, numerous studies of the chemical composition of bottom sediments attempted to determine dependencies between different variables (e.g. the trophic state of lakes, use of a catchment) and chemical properties of bottom sediments, especially organic matter or phosphorus content [3–9]. However, the results presented in the literature thus far are ambiguous. The content of organic matter content in sediment seems to be independent from the trophic state of lakes [7]. Calculating particular ratios between selected sediment components can implicate the origin of organic matter in sediment [3–5, 8] or the sediment's ability to absorb phosphorus [3].

A detailed study into phosphorus in Polish lakes and relationships between sediment components (including heavy metals) and phosphorus amounts was published by Bojakowska [10]. What makes this study particularly is a large number of analysed lakes (352 objects). The author emphasised the role of sewage inflow to lakes and recreational use of lakes as the main causes of high phosphorus content in sediments of urban lakes.

Certain observations of the role of internal loading in a lake-river system were made by Grochowska [11]. The author concluded that internal loading in flow-through lakes is one of the main factors affecting the retention of calcium, magnesium, iron, manganese, nitrogen and phosphorus in the profundal sediment of lakes in the Pasłęka River system.

The aim of this study has been to analyse the chemical composition of bottom sediments in selected Polish lakes in relation to the hydrological regime, mixing type, stratification, sewage inflow and a restoration method (if a given lake has undergone a technical treatment).

2 Material and Methods

The lakes taken for analysis belong to a broad environmental spectrum, being different in the hydrological regime (seepage, flow-through), trophic state (mesotrophic, eutrophic, dystrophic), mixing type (polymictic, holomictic, and one lake was meromictic) and, consequently, in stratification (stratified, non-stratified) and sewage input presence (in the past or now). Moreover, some have been restored using technical methods (artificial mixing with and without destratification, phosphorus inactivation), implemented by the Department of Water Protection Engineering. The localisation of the lakes is shown in Fig. 1.

Fig. 1 Localisation of lakelands, where the analysed Polish lakes are situated (source – [https://d-maps.com/carte.php?num_car=18765&lang=en], modified). 1 Kashubian Lakeland, 2 Western Pomeranian Lakeland, 3 Wielkopolskie Lakeland, 4 Olsztyńskie Lakeland, 5 Iławskie Lakeland, 6 Masurian Lakeland, 7 Drawskie Lakeland, 8 Mrągowskie Lakeland

Three of the lakes were described by Januszkiewicz [12, 13] as highly valuable lobelia lakes (Głębokie, Sitno and Karlikowo, located in Kashubian Lakeland), with the plant cover of *Lobelia dortmanna* L. Another lake, Tyrsko in Olsztyn, was the habitat of *Isoëtes lacustris* L., a very precious relict plant, but this species has vanished from the lake due to progressing eutrophication [14].

All sediment samples were analysed using methods described by Januszkiewicz [15] (including the results from this author's publications before 1978). The sediment was taken from the deepest point of every lake (profundal zone).

The sediment samples were mineralised with mixture of concentrated acids (H_2SO_4, $HClO_4$ and HNO_3 in 1 + 2 + 3 ratio). After mineralisation the samples were filtered through ash-free filter No 390. The residue on the filter was treated as silica and burned at 900°C using the electrical furnace Thermolyne®. And the filtrate was the material for chemical analysis of following components:

- Iron – was measured spectrophotometrically using Merck SQ 118
- Aluminium – was measured spectrophotometrically using Merck SQ 118
- Manganese – was measured spectrophotometrically using Merck SQ 118
- Calcium – was measured by titration with EDTA and murexide as an indicator
- Magnesium – was calculated as the difference between total hardness titration with EDTA (with eriochrome black T as an indicator) and calcium titration results with EDTA
- Total phosphorus – by molybdenum blue method with $SnCl_2$ as reducing agent (Nanocolor spectrophotometer by Macherey-Nagel)

Organic matter was measured as the loss of ignition at 550°C after carbonate regeneration using the distilled water saturated with CO_2. And carbonates were analysed by burning the sediment samples (after the regeneration with CO_2) at 1,000°C. Total nitrogen was analysed using Kjeldahl method for mineralisation and distillation in BÜCHI B-24.

Statistical analysis of the chemical composition of profundal sediments, including the main components (organic matter; silica, as SiO_2; iron, as Fe_2O_3; aluminium, as Al_2O_3; calcium, as CaO; magnesium, as MgO; manganese, as MnO; carbonates, as CO_2), was made based on data obtained from 47 Polish lakes (Table 1). The data consisted of historical literature data obtained from publications by Januszkiewicz and results of other sediment analyses which have been made at our department for the purposes of many environmental reports, mentioned in Table 1.

The sediment chemistry data published by Polish scientists in many other publications did not include all the variables taken into consideration in the current study. Hence, our decision was to take into consideration only the data obtained at our department so as to avoid data deficiency. Moreover, the data set is more stable owing to the same methods (sometimes with minor modifications) used for chemical analyses. All chemical analyses of sediments were made in the laboratory of the Department of Water Protection Engineering.

The statistical analysis was supported by a STATISTICA 12.5 Software package [38].

The organic carbon amounts in sediments were assessed according to [5].

3 Results and Discussion

The chemical composition of the profundal bottom sediment in the analysed lakes was highly varied. Silica, organic matter and calcium were the main components of deposits. The maximum silica content was noted in meromictic Starodworskie Lake in Olsztyn (79.1% SiO_2 in d.w.), whilst the minimum value of this element (10.9% SiO_2 in d.w.) was observed in Klasztorne Górne Lake in Strzelce Krajeńskie

Table 1 Basic data of the analysed Polish lakes and sources of bottom sediment chemistry data (ns non-stratified, s stratified, eu eutrophic, meso mesotrophic, hyper hypertrophic)

Lake	Area [ha]	Maximum depth [m]	Stratification	Trophic type	Sewage inflow	Restoration	Hydrological regime	Lakeland	Source
Mielenko	7.8	1.9	ns	eu	No	No	Flow-through	Kashubian	[16]
Karczemne	40.4	3.2	ns	hyper	Yes	No	Flow-through	Kashubian	[16]
Klasztorne Małe	13.7	20.0	s	hyper	Yes	Yes	Flow-through	Kashubian	[16]
Klasztorne Duże	57.5	8.5	s	hyper	Yes	No	Flow-through	Kashubian	[16]
Przywidz	15.7	15.1	s	eu	No	No	Flow-through	Kashubian	[17]
Wierzchołek	17.3	3.1	ns	eu	No	No	Flow-through	Kashubian	[17]
Średnik	19.3	12.5	s	eu	No	No	Flow-through	Kashubian	[17]
Gatno	72.6	25.2	s	eu	No	No	Flow-through	Kashubian	[17]
Zagnanie	143.0	19.5	s	eu	No	No	Flow-through	Kashubian	[18]
Smołdzino	2.2	8.0	s	eu	No	No	Seepage	Kashubian	[12, 13]
Techlinka	8.0	3.0	ns	eu	No	No	Seepage	Kashubian	[12, 13]
Głębokie	56.0	21.0	s	meso	No	No	Seepage	Kashubian	[12, 13]
Sitno	66.0	9.0	ns	eu	No	No	Flow-through	Kashubian	[12, 13]
Karlikowo	31.0	5.0	ns	dys	No	No	Seepage	Kashubian	[12, 13]
Borowo	3.0	1.5	ns	dys	No	No	Flow-through	Kashubian	[12, 13]
Dzierżążno	21.0	7.0	ns	eu	No	No	Seepage	Kashubian	[12, 13]
Mezowskie	41.4	10.0	s	eu	No	No	Seepage	Kashubian	[12, 13]
Okunkowo	9.0	13	s	dys	No	No	Seepage	Kashubian	[12, 13]
Szczyczno	7.0	4.0	ns	dys	No	No	Seepage	Kashubian	[12, 13]
Kiełpino Małe	7.7	5.0	ns	dys	No	No	Seepage	Kashubian	[12, 13]
Kiełpino Duże	20.0	20.0	s	eu	No	No	Seepage	Kashubian	[12, 13]
Grabowskie	140.7	28.1	s	eu	No	No	Flow-through	Kashubian	[19]
Wierzysko	61.5	6.0	ns	hyper	Yes	No	Flow-through	Kashubian	[20]

(continued)

Table 1 (continued)

Lake	Area [ha]	Maximum depth [m]	Stratification	Trophic type	Sewage inflow	Restoration	Hydrological regime	Lakeland	Source
Klasztorne Górne (S)	18.9	6.6	ns	hyper	Yes	Yes	Flow-through	Western Pomeranian	[21]
Wolsztyńskie	124.0	4.2	ns	eu	Yes	Yes	Flow-through	Wielkopolskie	[22]
Podkówka	6.92	6.0	ns	eu	No	No	Seepage	Olsztyńskie	[3]
Długie	26.8	17.3	s	eu	Yes	Yes	Seepage	Olsztyńskie	[23]
Redykajny	26.8	20.6	s	meso/eu	No	No	Seepage	Olsztyńskie	[3]
Track	52.8	3.8	ns	eu	Yes	No	Seepage	Olsztyńskie	[3]
Tyrsko	18.6	30.4	s	meso/eu	No	No	Seepage	Olsztyńskie	[14]
Sukiel	20.8	25	s	eu	No	No	Seepage	Olsztyńskie	[3]
Starodworskie	6.0	23.0	s(meromixis)	eu	No	Yes	Seepage	Olsztyńskie	[24]
Kortowskie	89.7	17.2	s	eu	Yes	Yes	Flow-through	Olsztyńskie	[25]
Wadąg	494.5	35.5	s	eu	No	No	Flow-through	Olsztyńskie	[26]
Skanda	51.1	12.0	s	eu	No	No	Flow-through	Olsztyńskie	[23]
Gim	175.9	25.8	s	eu	No	No	Flow-through	Olsztyńskie	[27]
Suskie	62.7	5.3	ns	eu	Yes	No	Flow-through	Iławskie	[28]
Dejguny	762.5	45.0	s	meso/eu	No	No	Flow-through	Masurian	[29]
Starokiejkuckie	33.1	30.2	s	meso/eu	No	No	Flow-through	Mrągowskie	[30]
Mikołajskie	497.9	25.9	s	eu	Yes	No	Flow-through	Masurian	[31]
Majcz Wielki	163.5	16.4	s	meso	No	No	Flow-through	Mrągowskie	[31]
Kuc	98.8	28.0	s	eu	No	No	Seepage	Mrągowskie	[31]
Ełckie	382.4	55.8	s	eu	Yes	No	Flow-through	Masurian	[32, 33]
Mutek	10.7	17.2	s	eu	No	Yes	Flow-through	Masurian	[34]
Grażyna	75.8	4.2	ns	eu	No	No	Flow-through	Drawskie	[35]
Rakowo	4.59	2.7	ns	eu	No	No	Flow-through	Drawskie	[36]
Bęskie	52.5	8.5	s	eu	Yes	No	Flow-through	Olsztyńskie	[37]

(Table 2). The sediments of 13 lakes had silica amounts higher than 50% SiO_2 in d.w., which allows us to classify these deposits as the silicate type, according to Stangenberg [39] classification.

The organic deposit type, according to [39] classification, appears when organic matter makes up over 50% of sediment's dry weight. This situation was noted in the sediments of six lakes, with the maximum amount of organic matter (81% d.w.) in the deposit of dystrophic Borowo Lake (Kashubian Lakeland). The minimum value (11.7% d.w.) occurred in the sediment of meromictic Starodworskie Lake (Table 2).

The share of calcium (calculated as $CaCO_3$) exceeded 50% $CaCO_3$ in d.w. in the deposits of two lakes: Klasztorne Górne in Strzelce Krajeńskie (63.7% $CaCO_3$ in d.w.) and Wolsztyńskie Lake in Wolsztyn (59.6% $CaCO_3$ in d.w.). Both lakes are located in the western part of Poland. According to [39] classification, the deposits of these lakes present the carbonaceous character. The lowest values of $CaCO_3$ in bottom deposits occurred in the group of Kashubian lakes, described by [12, 13], with the minimum result determined in Kiełpino Małe Lake (0.55% $CaCO_3$ in d.w.).

The magnesium content was much lower than calcium amounts in the analysed sediments. Minimum value (0.21% MgO in d.w.) was noted in the deposits of Mikołajskie Lake, whilst maximum amount – in Klasztorne Małe Lake (4.12% MgO in d.w.) (Table 2).

The amounts of iron in the analysed lake sediments ranged from 0.6% Fe_2O_3 in d.w. (Klasztorne Górne Lake) to 7.6% Fe_2O_3 in d.w. (Kiełpino Duże Lake), whist the aluminium content varied from 0.4% Al_2O_3 in d.w. (Suskie Lake in Susz) to 9.3% Al_2O_3 in d.w. (Kiełpino Małe Lake in Kashubian Lakeland) (Table 2).

Manganese amount in the analysed lakes' deposits was the lowest among all components (from 0.005% MnO in d.w. in Wolsztyńskie Lake to 0.56% MnO in d.w. in Dejguny Lake) (Table 2). This element is one of the most mobile components of sediment, because of its highest sensitivity to redox potential changes [1, 3, 40].

Table 2 Minimum, maximum, mean and standard deviation (SD) of analysed chemical properties of the profundal bottom sediment of Polish lakes (in % d.w.)

	OM	SiO_2	CaO	MnO	MgO	Fe_2O_3	Al_2O_3	N	P_2O_5	CO_2	Corg
Min	11.7	10.9	0.3	0.005	0.21	0.6	0.4	0.5	0.21	0.22	5.9
Max	81.0	79.1	35.8	0.56	4.12	7.63	9.3	3.6	2.98	28.0	40.5
Mean	33.7	39.0	8.3	0.14	1.53	3.4	3.2	1.8	0.94	7.9	16.8
SD	15.3	16.0	9.2	0.14	0.89	1.69	2.6	0.8	0.56	7.9	7.7

OM organic matter

Table 3 Minimum, maximum, mean and standard deviation (SD) of selected ratios between chemical parameters of the profundal bottom sediment in the analysed Polish lakes

	OM:N	Corg:N	Al:P	Fe:P	Ca:P	Fe:Mn	Mn:P
Min	11.6	5.8	0.5	0.8	0.3	4.5	0.02
Max	34.6	17.3	23.0	30.5	147.2	310.8	2.08
Mean	18.9	9.5	4.7	7.6	25.6	56.8	0.38
SD	5.0	2.5	3.8	5.6	35.9	70.9	0.48

OM organic matter

The nitrogen amounts in the analysed lakes' sediment ranged from 0.5% N in d.w. (Kiełpino Małe Lake in Kashubian Lakeland) to 3.6% N in d.w. (Sukiel Lake in Olsztyn). The lowest phosphorus content (0.21% P_2O_5 in d.w.) was noted in the sediment of Kashubian Przywidz Lake, whilst the maximum amount was noted in the deposit of degraded, hypertrophic Karczemne Lake in Kartuzy (2.98% P_2O_5 in d.w.) (Table 2).

The assessed ratios between selected components in the analysed bottom deposits are collated in Table 3. The maximum OM:N (34.6) was noted for the deposits of dystrophic Borowo Lake (Kashubian Lakeland), and the lowest value (11.6) was determined in the sediment of Mikołajskie Lake (Masurian Lakeland). According to [5], if this ratio exceeds 25, sediment has the polyhumic character, and values between 20 and 25 implicate the mesohumic class. According to this criterion, 14 lakes among the analysed 47 waterbodies were mesohumic in character, whereas two were polyhumic.

The Corg:N ratio gives information about the origin of organic matter in the bottom sediment [5, 6]. The values of this ratio in the analysed sediments ranged from 5.8 (eutrophic Mikołajskie Lake in Mikołajki and Skanda Lake in Olsztyn) to 17.3 (dystrophic Borowo Lake). According to [1, 5, 8] the C:N value higher that 10:1 occurs when organic matter has a higher share of terigenic origin matter (allochthonous matter). In 16 out of 47 analysed lakes, this ratio reached a value over 10 (mainly in the Kashubian lakes).

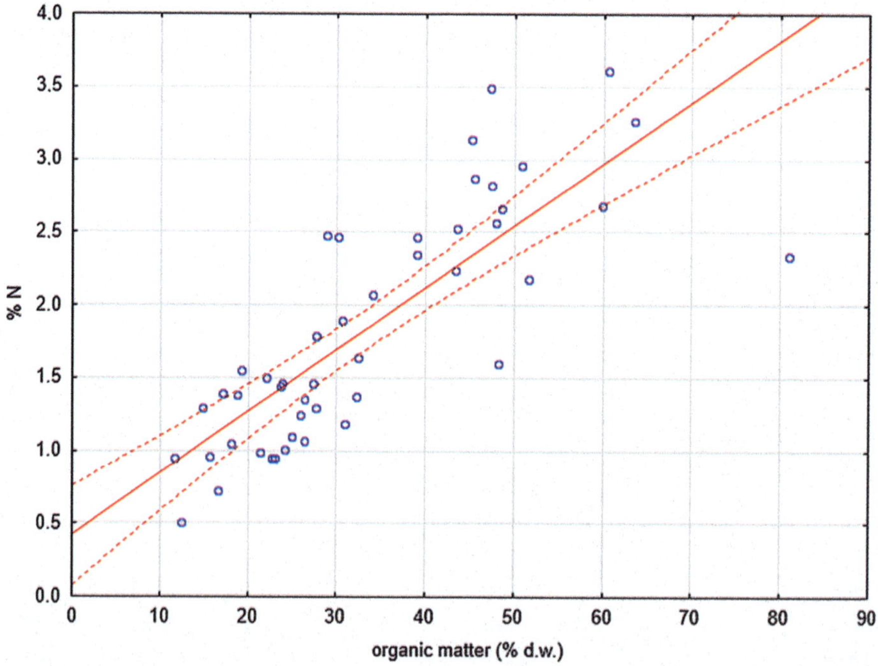

Fig. 2 Correlation between organic matter and nitrogen in the bottom sediments of the analysed Polish lakes ($N = 0.41055 + 0.04250 \times OM$)

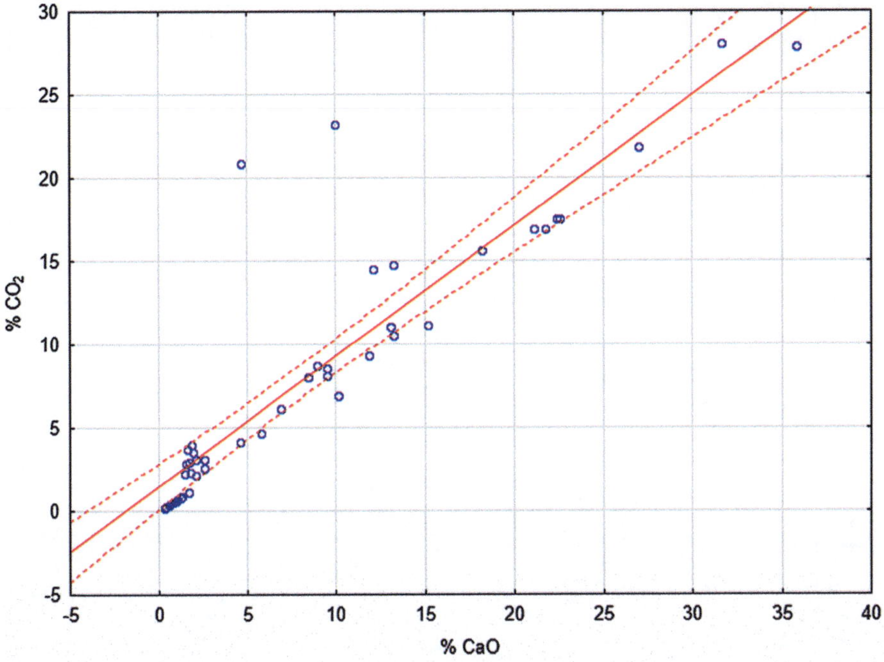

Fig. 3 Correlation between calcium and carbonates in the bottom sediments of the analysed Polish lakes ($CO_2 = 1.4094 + 0.78432 \times CaO$)

The ratios of components active in phosphorus binding in sediment (Al, Fe, Ca, Mn) to P can give information about potential phosphorus mobility in sediment and therefore its bioavailability [3, 10]. The maximum Al:P and Fe:P ratio values were noted in sediment of Kashubian Przywidz Lake (20.3 and 30.5, respectively). The lowest ratios were calculated for Borowo Lake (Al:P = 0.5) and Klasztorne Górne Lake (Fe:P = 0.8) (Table 3). The Ca:P ratio range was wider: from 0.3 (Kiełpino Duże Lake) to 147.2 (Gatno Lake), both in Kashubian Lakeland. A high Ca:P ratio was also noted in Wolsztyńskie Lake sediment (136.3). The Mn:P ratio in analysed lakes' deposits ranged from 0.02 (in 3 lakes: degraded, hypertrophic Karczemne Lake, eutrophic Mikołajskie Lake and restored Wolsztyńskie Lake) to 2.08 (in the deposit of Kashubian Grabowskie Lake) (Table 3).

The Fe:Mn ratio is treated as an indicator of redox potential changes in the water-sediment interface [1, 3, 40]. Its maximum value (310.8) was noted in the sediment of Wolsztyńskie Lake, whilst the lowest value (4.5) was determined in the sediment of Klasztorne Górne Lake.

The correlation analysis between the analysed sediment components revealed significant positive correlations between OM and N ($r = 0.81$, $n = 47$, $p < 0.05$) (Fig. 2), CaO and CO_2 ($r = 0.90$, $n = 47$, $p < 0.05$) (Fig. 3), iron and aluminium ($r = 0.62$, $n = 47$, $p < 0.05$) (Fig. 4), iron and manganese ($r = 0.47$, $n = 47$,

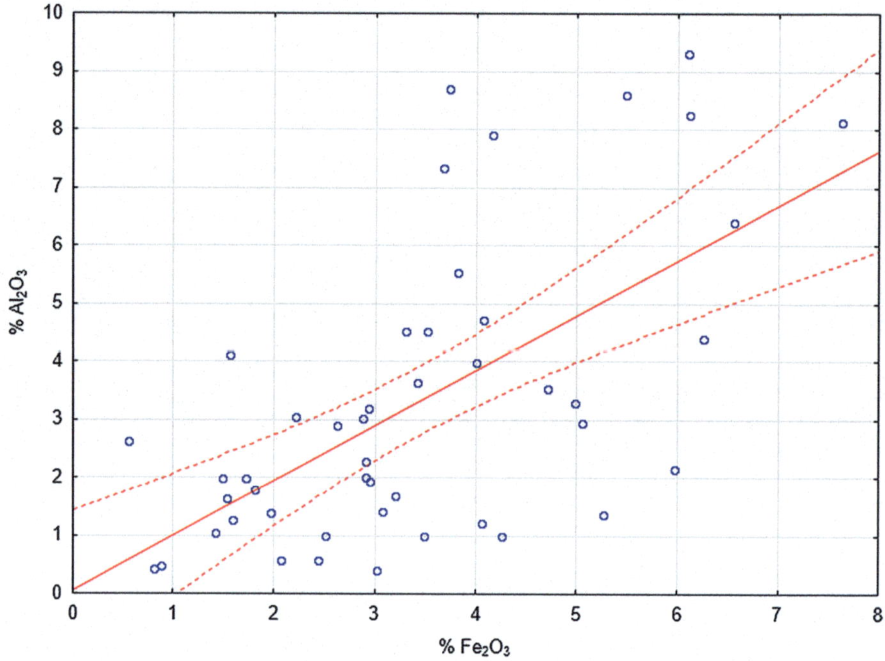

Fig. 4 Correlation between iron and aluminium in the bottom sediments of the analysed Polish lakes ($Al_2O_3 = 0.05759 + 0.94801 \times Fe_2O_3$)

$p < 0.05$) (Fig. 5), silica and aluminium ($r = 0.54$, $n = 47$, $p < 0.05$), silica and iron ($r = 0.46$, $n = 47$, $p < 0.05$).

A significant correlation between OM and nitrogen has been reported in many other studies [e.g. 3, 12, 13, 15, 19, 20, 36, 39]. It confirms that nitrogen is deposited in sediments mainly in the organic form.

The phosphorus content was positively correlated with iron ($r = 0.29$, $n = 47$, $p < 0.05$) (Fig. 6) and aluminium ($r = 0.50$, $n = 47$, $p < 0.05$) (Fig. 7), whilst a negative significant correlation was found between P and CaO ($r = -0.38$, $n = 47$, $p < 0.05$) (Fig. 8) (Table 4).

Because phosphorus is a key factor involved in eutrophication processes, an in-depth analysis was made of the relationships between phosphorus and other sediment components actively binding this element, a step recommended by [10]. Additional correlations were tested between selected components and ratios of these components to P. The results of this analysis are shown in Table 5.

The negative correlation between calcium and phosphorus (Fig. 8) and highly significant positive correlation between calcium and carbonates (Fig. 3) suggest that calcium is deposited in the sediments of the analysed lakes mainly as calcium carbonate, and the concentration of hydroxyapatite in sediment is low. A similar situation (negative correlation between calcium and phosphorus, significant positive correlation between Ca:P ratio and Ca, and simultaneous negative correlation

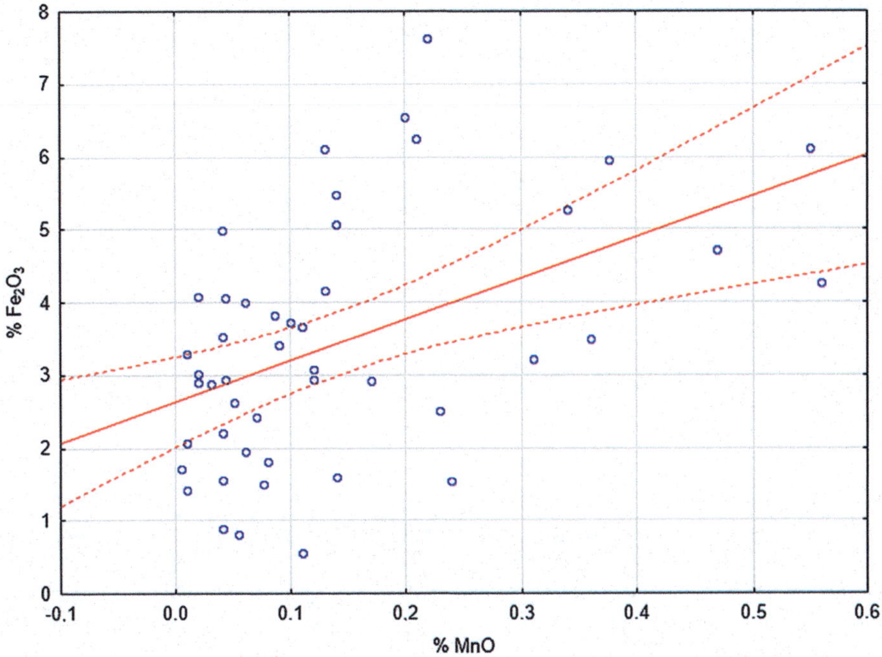

Fig. 5 Correlation between iron and manganese in the bottom sediments of the analysed Polish lakes ($Fe_2O_3 = 2.6343 + 5.6464 \times MnO$)

between Ca:P and P) (Table 5) was noted by [10]. The lack of correlation between calcium and phosphorus amounts was observed by [41, 42].

The positive correlations between iron, manganese and phosphorus obtained in our analysis of the sediments from selected lakes in Poland (Tables 4 and 5, Figs. 6 and 7) suggest that these elements play an important role in the binding of phosphorus in the investigated sediments, which agrees with other studies [3, 10, 13, 15, 17, 34, 37]. An unexpected observation, however, was the lack of simple correlation between the content organic matter and phosphorus in the analysed lakes' profundal sediments. Numerous studies have verified an important role of organic matter in phosphorus binding [e.g. 3, 10, 14, 15]. A possible explanation of this observation could be the fact that phosphorus was deposited mainly in mineral forms in the majority of the analysed lakes. A similar situation was observed by [43] in Długie Lake during restoration via an artificial aeration method.

The cluster analysis was performed in order to find similarities in the chemical composition of the bottom sediments in the analysed Polish lakes in relation to their hydrological regimes, trophic state, mixing type and, consequently, stratification, as well as sewage input presence and possible restoration with technical methods. The results of this analysis are shown in Fig. 9.

The results of this analysis clearly show similarities between lakes. The first cluster (Kiełpino Małe, Szczyczno, Okunkowo, Dzierżążno, Techlinka) consists of

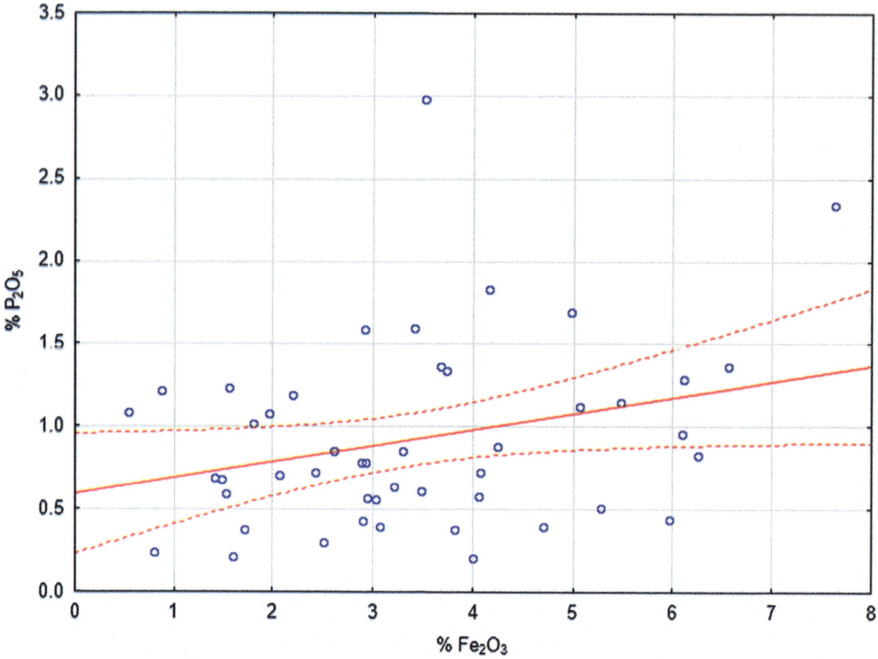

Fig. 6 Correlation between iron and phosphorus in the bottom sediments of the analysed Polish lakes ($P_2O_5 = 0.59752 + 0.09592 \times Fe_2O_3$)

non-stratified, dystrophic and eutrophic, seepage lakes with the same genesis, located in Kashubian Lakeland. They are similar to the Kashubian Kiełpino Duże and Smołdzino Lakes (stratified, eutrophic and also seepage lakes).

The second cluster includes non-stratified, calcareous, flow-through lakes located in the western part of Poland (Wolsztyńskie and Klasztorne Górne), and these lakes are similar to Kashubian eutrophic, stratified, flow-through lakes (Zagnanie, Gatno and Średnik) and also to flow-through, eutrophic and stratified lakes Mikołajskie Lake (Masurian Lakeland) and Kortowskie Lake in Olsztyn.

The third cluster consists of a group of flow-through, stratified, eutrophic lakes, also with relatively large amounts of calcium in deposits (Majcz Wielki, Bęskie, Ełckie and Wadąg) located on Masurian, Mrągowskie and Olsztyńskie Lakelands. These waterbodies are similar to flow-through, stratified lakes (Skanda, Dejguny, Mutek, Grabowskie and Przywidz lakes), located in the Olsztyńskie, Masurian and Kashubian Lakelands, respectively.

The fourth cluster includes hypertrophic, degraded by sewage inflow Kashubian lakes (Wierzysko, Klasztorne Duże, Klasztorne Małe and Karczemne). They are flow-through lakes with a high phosphorus level in sediments.

The fifth cluster gathers the flow-through, shallow, non-stratified lakes with a relatively high organic matter level, such as three Kashubian lakes (dystrophic Borowo and Karlikowo, and eutrophic Wierzchołek Lake), two lakes from Drawskie Lakeland (Grażyna, Rakowo) and Suskie Lake from Iławskie Lakeland.

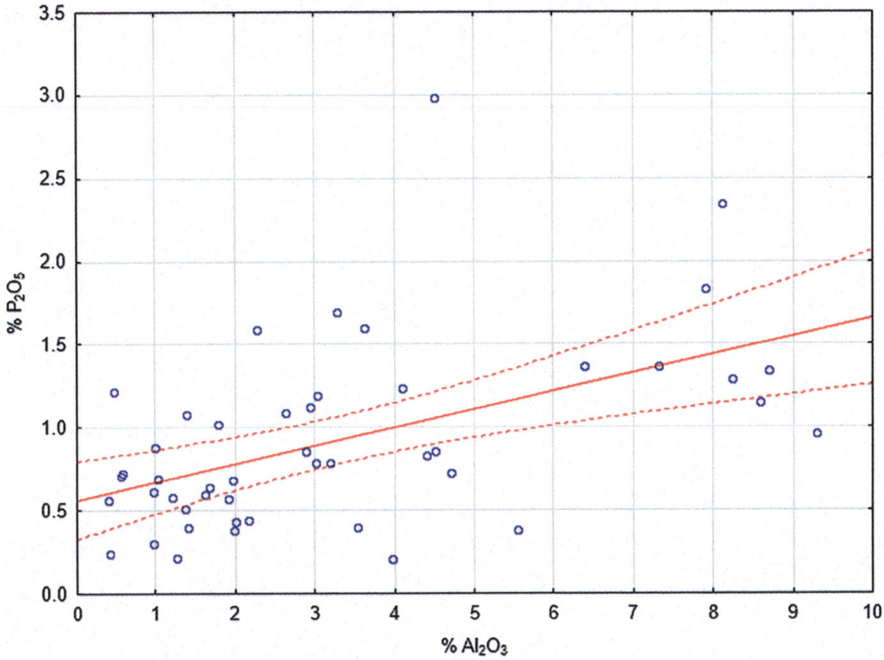

Fig. 7 Correlation between aluminium and phosphorus in the bottom sediments of the analysed Polish lakes ($P_2O_5 = 0.56331 + 0.10979 \times Al_2O_3$)

The sixth cluster includes two flow-through, stratified, eutrophic lakes (Starokiejkuckie and Gim), and another group in this cluster are seepage, stratified lakes with a high organic matter level (Kuc, Sukiel, Tyrsko, Długie, Redykajny, Głębokie).

The seventh main cluster gathers two stratified, seepage lakes (meromictic Starodworskie Lake and eutrophic Mezowskie Lake). Lakes in another group in this cluster are characterised by the seepage hydrological regime (except Mielenko Lake) and lack of stratification.

It is worth noting that the cluster analysis divided the analysed lakes into groups mainly with respect to the hydrological regime and stratification. The trophic state of lakes seems to be a less important factor which influenced the similarity of the chemical composition of bottom sediments, which is in accordance with the observations of [7]. This can confirm Grochowska's [11] observations, made for the Pasłęka river-lake system, that flow and water-sediment exchange processes are key factors in the retention of elements in bottom deposits. The local geochemistry is also important, because it can shape the physico-chemical conditions (such as pH, ionic strength or redox conditions if, for instance, the organic matter content in sediments is high) influencing deposition or sorption processes [3, 43]. However, according to the results of the cluster analysis, there are certain similarities in the chemical composition of sediments between the lakes of Kashubian and Olsztyńskie

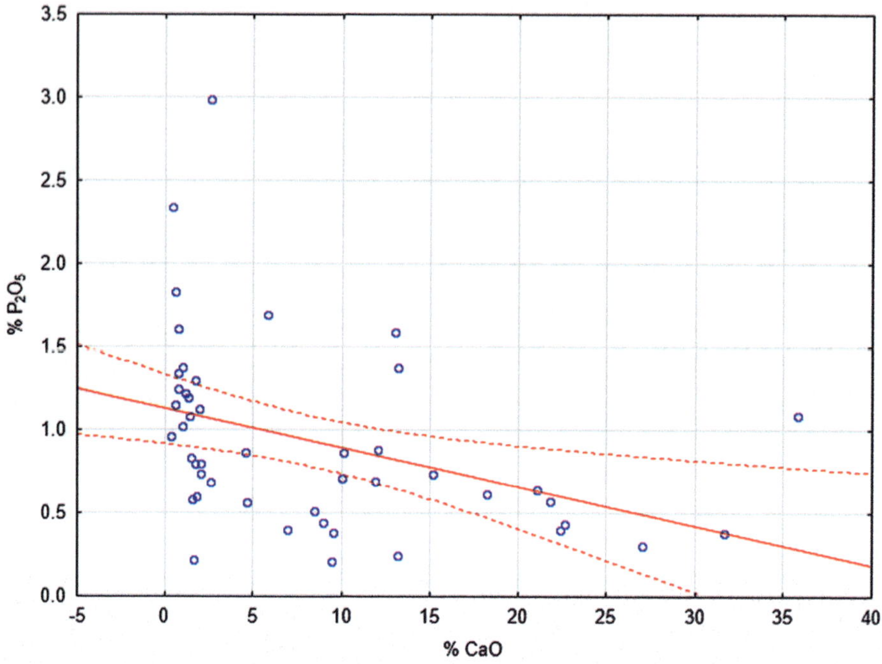

Fig. 8 Correlation between calcium and phosphorus in the bottom sediments of the analysed Polish lakes ($P_2O_5 = 1.1200 - 0.0235 \times CaO$)

Table 4 Correlation matrix between chemical components of the profundal bottom sediment in the analysed Polish lakes ($n = 47, p < 0.05$)

Variable	OM	SiO_2	CaO	MnO	Fe_2O_3	Al_2O_3	N	P_2O_5	CO_2
OM	–	−0.52	−0.35	−0.32	−0.39	ns	0.81	ns	−0.38
SiO_2		–	−0.55	ns	0.46	0.54	−0.58	ns	−0.51
CaO			–	ns	−0.29	−0.41	ns	−0.38	0.90
MnO				–	0.47	ns	ns	ns	ns
Fe_2O_3					–	0.62	−0.34	0.29	−0.31
Al_2O_3						–	−0.38	0.50	−0.51
N							–	ns	ns
P_2O_5								–	−0.43

ns non-significant

Lakelands, where silica and organic matter are quantitatively dominant deposit components and calcium compounds occurs in small amounts. This justifies the supposition that several Olsztyn-based lakes (Długie, Tyrsko, Redykajny) could have been a habitat of relict plants like *Lobelia dortmanna* in the past, but anthropopressure changed and deteriorated their water chemistry. Nevertheless, palaeolimnological investigation of sediments is needed to confirm this hypothesis.

Table 5 Correlation matrix between selected chemical components (as pure elements) of the profundal bottom sediment and the ratios of these components to phosphorus ($n = 47$, $p < 0.05$)

Variable	Fe	Mn	Ca	Al	P	Al:P	Fe:P	Ca:P	Mn:P
Fe	–	0.30	−0.30	0.59	0.29	0.32	0.36	ns	ns
Mn		–	ns	ns	0.85	ns	ns	ns	0.32
Ca			–	−0.40	ns	ns	ns	0.85	ns
Al				–	ns	0.51	ns	−0.38	ns
P					–	ns	−0.38	−0.32	ns
Al:P						–	0.69	ns	ns
Fe:P							–	0.38	0.58
Ca:P								–	ns
Mn:P									–

ns non-significant

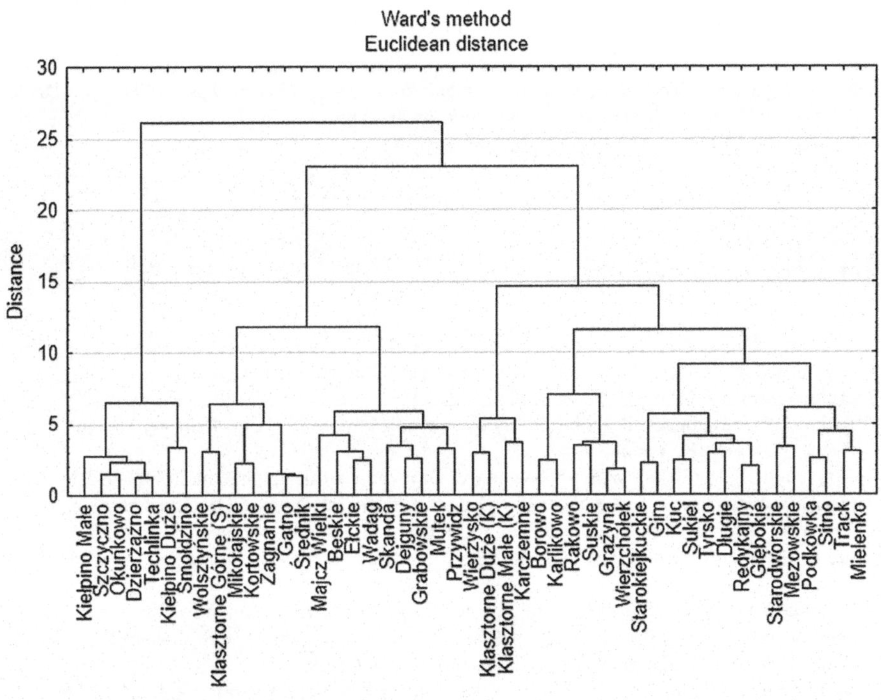

Fig. 9 Cluster analysis based on determinations of the chemical composition of bottom sediments in the analysed Polish lakes, their hydrological regime, trophic state and mixing type

4 Conclusions

In most of the investigated lakes, the dominant compounds in profundal sediments of 47 Polish lakes were silica or organic matter. Calcium carbonate dominated quantitatively in several lakes only.

Other sediment components occurred in low amounts, so that their share did not surpass a few percent of d.w. Nutrients such as N and P were deposited differently, namely, nitrogen appeared in the organic form, whist phosphorus was accumulated mostly in the form of mineral compounds. Aluminium, iron and manganese play important roles in the binding of phosphorus in sediments of the analysed Polish lakes, whist the role of calcium seems to be unimportant in this process.

The cluster analysis showed that the hydrological regime seems to be the main factor that differentiated the character of sediments in the analysed Polish lakes. The trophic state and stratification type of the lakes were less important factors in terms of the influence they had on the character of bottom sediments.

Acknowledgements The authors wish to thank to Grzegorz Wiśniewski, PhD, and Daniel Szymański, MSc, for their help in the field work, and to Marzena Karpienia, MSc, and Agnieszka Grzegorczyk, MSc, for their help in laboratory analyses.

References

1. Borówka RK (2007) Geochemiczne badania osadów jeziornych strefy umiarkowanej (Geochemical studies of lacustrine sediments of moderate zone). Studia Limnologica et Telmatologica 1(1):33–42
2. Pokojska U, Bednarek R (eds) (2012) Geochemia krajobrazu (Geochemistry of landscape). Wyd. Nauk. UMK, Toruń
3. Augustyniak R (2018) Wpływ czynników fizyczno-chemicznych i mikrobiologicznych na zasilanie wewnętrzne fosforem wód wybranych jezior miejskich (The influence of physical, chemical and microbiological factors on the phosphorus internal loading to the water of selected urban lakes). Monografie Komitetu Inżynierii Środowiska PAN, nr 140, Lublin, wyd. Perfecta info
4. Górniak A (1996) Substancje humusowe i ich rola w funkcjonowaniu ekosystemów słodkowodnych (Humic substances and their role in the functioning of freshwater ecosystems). Wyd. UW, Filia w Białymstoku, Seria Rozprawy, Białystok
5. Håkanson L, Jansson M (1983) Principles of lake sedimentology. Springer, Berlin, 316 p
6. Kaushal S, Binford MW (1999) Relationships between C:N ratios of lake sediments, organic matter sources and historical deforestation in Lake Pleasant, Massachusetts, USA. J Paleolimnol 22:439–442
7. Kentzer A (2001) Fosfor i jego biologicznie dostępne frakcje w osadach jezior różnej trofii (Phosphorus and its bioavailable fractions in the sediments of lakes with different trophic status). Wyd. UMK, Toruń
8. Koszelnik P (2009) Źródła i dystrybucja pierwiastków biogennych na przykładzie zespołu zbiorników zaporowych Solina-Myczkowce (Sources and distiribution of biogenic compounds on the example of dam reservoirs Solina-Myczkowce). Oficyna Wydawnicza Politechniki Rzeszowskiej. Rzeszów.

9. Łukawska-Matuszewska K, Voght RD, Xie R (2013) Phosphorus pools and internal loading in a eutrophic lake with gradients in sediment geochemistry created by land use in the watershed. Hydrobiologia 713:183–197
10. Bojakowska I (2016) Phosphorus in lake sediments of Poland – results of monitoring research. Limnol Rev 16(1):15–25
11. Grochowska J (2015) Obieg wybranych makropierwiastków i związków biogennych w systemie rzeczno-jeziornym na przykładzie górnej Pasłęki (The circulation of selected macroelements and biogenic compounds in the river-lake system on the example of upper Pasłęka). Wyd. Komitetu Inżynierii Środowiska PAN, Perfecta info. Lublin
12. Januszkiewicz T (1970) Charakterystyka hydrochemiczna grupy jezior na Pojezierzu Kaszubskim (Hydrochemical studies of the group of lakes on the Kashubian Lake District). Rocz. Nauk. Rol. H 92(2):61–91
13. Januszkiewicz T (1970) Skład osadów głębinowych grupy jezior na Pojezierzu Kaszubskim (Chemical composition of bottom sediments in the group of Kashubian Lake District). Rocz. Nauk. Rol. H 92(1):67–84
14. Augustyniak R, Neugebauer M, Kowalska J, Szymański D, Wiśniewski G, Filipkowska Z, Grochowska JK, Łopata M, Parszuto K, Tandyrak R (2017) Bottom deposits of stratified, seepage, urban lake (on the example of Tyrsko lake, Poland) as a factor potentially shaping lake water quality. J Ecol Eng 18(5):55–62
15. Januszkiewicz T (1978) Studia nad metodą analizy współczesnych osadów dennych jezior (The studies of method of analysis of the recent sediments of lakes). Zesz. Nauk. ART. Olsztyn 187(8):3–30
16. Grochowska J, Brzozowska R, Grzybowski M, Napiórkowska-Krzebietke A, Bigaj I, Szymański D, Zieliński R (2013) Opracowanie wyników badań fizyko-chemicznych wody i osadów dennych oraz pomiarów hydrologicznych 4 jezior: Mielenko, Karczemne, Klasztorne Duże i Małe i ich zlewni pod kątem możliwości ochrony zbiorników przed degradacją (Elaboration of the results of physical and chemical analysis of water and sediments as well as hydrological measurements of 4 lakes: Mielenko, Karczemne, Klasztorne Duże and Małe and their catchment in terms of the possibility of protecting lakes from degradation). Report for Local Government Unit in Kartuzy, manuscript
17. Januszkiewicz T (1983/1984) Chemistry of surface sediments of four lakes in Northern Poland. Acta Hydrobiol 25/26(3/4):227–242
18. Januszkiewicz T (1980) Chemistry of recent sediments of Grabowskie Lake in Kashubian Lake District in Northern Poland. Pol Arch Hydrobiol 27(3):319–336
19. Januszkiewicz T, Januszkiewicz KR (1980) Chemical composition of recent bottom sediment of the Lake Zagnanie. Acta Hydrobiol 22(2):157–178
20. Dunalska J, Wiśniewski G, Brzozowska R, Szymański D, Sieńska J (2014) Wykonanie badań fizyko-chemicznych wody i osadów dennych oraz pomiarów hydrologicznych Jeziora Wierzysko pod kątem możliwości ochrony zbiornika przed degradacją (Performing physical and chemical analysis of water and bottom sediments as well as hydrological measurements of the Wierzysko Lake in terms of the possibility of protecting the lake from degradation). Report for Local Government Unit in Kościerzyna, manuscript
21. Łopata M et al (2014) Rekultywacja jeziora Klasztorne Górne w gminie Strzelce Krajeńskie metodą chemiczną wraz z biomanipulacją oraz monitoringiem warunków środowiskowych (Restoration of Klasztorne Górne Lake in Strzelce Krajeńskie community by chemical method together with biomanipulation and environmental conditions monitoring). Report for Local Government Unit in Strzelce Krajeńskie, manuscript
22. Augustyniak R, Łopata M, Grochowska J, Tandyrak R, Parszuto K (2018) Multiannual changes in sediment phosphorus of Wolsztyńskie Lake (Wielkopolskie Lake District, Western Poland) International Multidisciplinary Scientific Geo Conference Surveying Geology and Mining Ecology Management. SGEM 18(3.1):515–521
23. Lossow K, Gawrońska H, Mientki C, Wiśniewski G, Łopata M (2005) Jeziora Olsztyna. Stan troficzny, zagrożenia (Lakes of Olsztyn. Trophic state, threats). Edycja S.C., Olsztyn

24. Tandyrak R (2018) Ewolucja miksji i stan troficzny Jeziora Starodworskiego na tle warunków klimatycznych i postępującej urbanizacji (Evolution of mixing and trophic state of Starodworskie Lake on the backgrounds of climate conditions and developing urbanization). Wyd. UWM, Olsztyn
25. Januszkiewicz T, Samulowska B (1978) Chemizm współczesnych osadów dennych jeziora Wadąg k/Olsztyna (The chemistry of recent bottom sediments of the Lake Wadąg). Zesz. Nauk. ART. Olszt., Ochrona Wód i Rybactwo Śródlądowe 8:31–58
26. Wiśniewski G (2007) Skład chemiczny osadów dennych Jeziora Kortowskiego po 50 latach rekultywacji (Chemical constitution of bottom deposits of Lake Kortowskie after 50 years of recultivation). In: Wiśniewski R, Piotrowiak J (eds) Ochrona i rekultywacja jezior, pp 191–200
27. Dunalska J, Wiśniewski G, Brzozowska R (2012) Opracowanie wyników badań fizyko-chemicznych wody i osadów dennych oraz pomiary hydrologiczne jeziora Gim i jego zlewni pod kątem możliwości ochrony zbiornika przed degradacją (Elaboration of the results of physical and chemical analysis of water and bottom sediments and hydrological measurements of Lake Gim and its basin in terms of the possibility of protecting the lake from degradation). Report for Local Government Unit in Purda, manuscript
28. Dunalska J, Grochowska J, Brzozowska R (2012) Opracowanie wyników badań fizyko-chemicznych wody i osadów dennych Jeziora Suskiego i jego zlewni pod kątem możliwości ochrony zbiornika przed degradacją (Elaboration of the results of physical and chemical analysis of water and sediments of the Suskie Lake and its basin in terms of the possibility of protecting the lake from degradation). Report for Local Government Unit in Susz, manuscript
29. Brzozowska R, Dunalska J, Zdanowski B (2005) Preliminary characteristic of the chemical composition of the top-layer bottom deposits in Lake Dejguny (Mazurskie Lake District). Limnol Rev 5:11–16
30. Dunalska J, Tandyrak R, Brzozowska R, Grzybowski M (2012) Opracowanie wyników badań fizyko-chemicznych wody i osadów dennych, składu gatunkowego makrofitów oraz pomiary hydrologiczne Jeziora Starokiejkuckiego i jego zlewni pod kątem możliwości ochrony zbiornika przed degradacją (Elaboration of the results of physical and chemical analysis of water and sediments, species composition of macrophytes and hydrological measurements of Lake Starokiejkuckie and its basin in terms of the possibility of protecting the reservoir from degradation). Report for Military Unit 2669, manuscript
31. Brzozowska R, Kang KH (2010) Powierzchniowa warstwa osadów dennych jezior Kuc, Majcz Wielki i Mikołajskiego (The surficial sediment layer of bottom deposits of Kuc, Majcz Wielki and Mikołajskie lakes). In: Dunalska J (ed) Warunki środowiskowe i stan troficzny jezior Kuc, Majcz Wielki i Mikołajskiego (Environmental conditions and trophic state of Kuc, Majcz Wielki and Mikołajskie lakes). Wyd. UWM, Olsztyn, pp 89–99
32. Marszelewski W (2005) Zmiany warunków abiotycznych w jeziorach Polski północno-wschodniej (Changes of abiotic conditions in the lakes of north-eastern Poland). Wyd. UMK, Toruń
33. Dunalska J, Wiśniewski G, Ławniczak A, Napiórkowska-Krzebietke A, Augustyniak R, Grochowska J, Korzeniewska E, Sieńska J, Szymański D (2016) Badania wody i osadów dennych oraz pomiary hydrologiczne Jeziora Ełckiego i jego zlewni wraz ze sporządzeniem projektu ochrony i rekultywacji zbiornika w celu zachowania jego potencjału rekreacyjnego i przyrodniczego (Research on water and bottom sediments as well as hydrological measurements of the Ełckie Lake and its catchment along with the preparation of a protection and restoration project to preserve its recreational and natural potential). Report for Local Government Unit in Ełk, manuscript
34. Gawrońska H (1987) Chemizm osadów dennych jeziora Mutek poddanego eksperymentowi sztucznego napowietrzania (Chemistry of bottom sediments in Lake Mutek induced by artificial aeration). Rocz. Nauk. Rol. H 101(1):39–52
35. Czerniawski R (2017) Raport z prac wykonanych w ramach zadania: "Opracowanie programu ochrony wód jeziora Grażyna w Drawnie na podstawie badań hydrochemicznych i hydrobiologicznych wód wraz z oceną możliwości rekultywacji" w okresie od kwietnia 2016

do marca 2017 (Report on the work carried out as part of the task: "Development of the water protection program of Lake Grażyna in Drawno based on hydrochemical and hydrobiological studies of water together with the assessment of restoration opportunities" from April 2016 to March 2017). Report for Local Government Unit in Drawno, Szczecin 2017, manuscript
36. Czerniawski R (2016) Raport nr 2 z prac wykonanych w ramach zadania "Opracowanie metody rekultywacji jeziora Rakowo Duże w Złocieńcu na podstawie badań fizykochemicznych i biologicznych wód" w ramach zadania "plaża nad jeziorem Rakowo Duże" w okresie od kwietnia do października 2016 (Report No. 2 on the work carried out as part of the task "Development of a method of restoration of the Lake Rakowo Duże in Złocieniec on the basis of physico-chemical and biological studies of water" as part of the task "Beach on the Lake Rakowo Duże" in the period from April to October 2016). Report for Local Government Unit in Złocieniec, Szczecin 2016, manuscript
37. Gawrońska H (1989) Skład chemiczny osadów dennych Jeziora Bęskiego (Chemical composition of bottom sediments from Lake Bęskie). Acta Academiae Agriculturae Ac Technicae Olstenensis. Protectio Aquarum et Piscatoria 17:35–43
38. Statsoft Inc. STATISTICA software package 12.5
39. Stangenberg M (1938) Skład chemiczny osadów głębinowych jezior Suwalszczyzny (Chemical composition of profundal sediments of Suwałki area lakes). Rozpr. i Spraw. Inst. Bad. Lasów Państw., A 31:1–40
40. Naeher S, Gilli A, North RP, Hamann Y, Schubert CJ (2013) Tracing bottom water oxygenation with sedimentary Mn/Fe ratios in Lake Zurich, Switzerland. Chem Geol 352:125–133
41. Brzozowska R, Gawrońska H (2005) Influence of a multi-year artificial aeration of a lake using destratification method on the sediment-water phosphorus exchange. Arch Environ Prot 31 (3):71–88
42. Fytianos K, Kotzakioti A (2005) Sequential fractionation of phosphorus in lake sediments of northern Greece. Environ Monit Assess 100:191–200
43. Orihel DM, Baulch HM, Casson NJ, North RL, Parsons CT, Seckar DCM, Venkiteswaran JJ (2017) Internal phosphorus loading in Canadian fresh waters: a critical review and data analysis. Can J Fish Aquat Sci 74:2005–2029

Characteristics of the Water Network in Postglacial Areas of Northern Poland

Elżbieta Bajkiewicz-Grabowska, Włodzimierz Golus, Maciej Markowski, and Monika Kwidzińska

Contents

Abstract The characteristics of the postglacial relief of northern Poland, which is responsible for the initial development of the water network in this area, were described. The dominant form of relief in this area is concave forms of the terrain, clusters of which form areas permanently or periodically excluded from river runoff. In the area of the basal moraine, in the beds of these depressions there are usually lakes of various sizes, including the smallest ones – postglacial ponds, some of which are periodically connected by short watercourses. On outwash plains only deep channels are occupied by lakes; all other concave forms, often with significant denivelations, are dry.

The main objects of the hydrographic network of the postglacial areas are primarily lakes, boggy areas and watercourses: those of the lower orders in the Strahler classification, usually short, often periodic, functioning as overflows, and those using valleys formed by postglacial rivers (channels, stagnant ice melt-outs, fluvioglacial valleys), which are the main drainage axis. It was pointed out that in the postglacial areas of northern Poland, three types of hydrographic network can be distinguished, differing in the degree of river runoff organization: the hydrographic network of endorheic areas, the river-lake network and the mature river network (lakeland rivers).

E. Bajkiewicz-Grabowska (✉) · W. Golus · M. Markowski · M. Kwidzińska
Department of Limnology, Institute of Geography, University of Gdańsk, Gdańsk, Poland
e-mail: bajka37@wp.pl

E. Korzeniewska, M. Harnisz (eds.), *Polish River Basins and Lakes – Part I*,
The Handbook of Environmental Chemistry 86,
https://doi.org/10.1007/978-3-030-12123-5_8

Keywords Endorheic areas · Lakeland rivers · Ponds · Postglacial landscape · River-lake network

1 Introduction

Poland is situated within an area that was covered by glaciations in the last several hundred thousand years. The subsequent periods of advance and retreat of the glacier caused continuous rejuvenation of the landscape. The discharge of surplus water from this area was therefore constantly transformed as a result of morphological, climatic and hydrological changes. The range of successive glaciations was variable. The maximum extent of the glacier (during the Weichselian glaciation) did not cover the entire area of present-day Poland; it cut it diagonally from the southwest to the northeast and embraced about 1/3 of the country's territory (Fig. 1). During the last stadial of this glaciation (Baltic glaciation), the glacier covered and remained in the area of northern Poland up to 10,250 BP (limit date of the glacial period in Poland [1]); hence the time that has passed since the withdrawal of the glacier is very short in the geological sense.

The activity of the glacier and its meltwaters together with later morphogenetic processes (accumulation and erosion) caused that the area of northern Poland, delimited from the south by the last glaciation range and from the north by the belt of the Baltic Sea Coast (Fig. 1), has a lakeland landscape. The largest longitudinal extent of the postglacial area is in the western and central part of Poland, where it reaches about 300 km in width, while in the eastern part of Poland, it narrows down to 50–70 km in width, and at the eastern border, it has only 30 km [1]. The northern Poland landscape is dominated by groups of forms of glacial, slope, fluvial and aeolian relief [2]. The most characteristic groups of forms are hypsometrically varied hills and frontal moraine belts, vast mildly waved areas of basal moraine plateaus, flat outwash areas, subglacial channels (generally deeply dissecting moraine plateaus and outwash plains), river valleys usually of longitudinal course, ice marginal streamways of latitudinal course as well as numerous concave forms resulting from the melting of buried stagnant ice bodies (kettle holes) of various sizes. In this type of landscape, denivelations can locally exceed 100 m with the slope grades up to several dozen degrees. Such a relief creates different habitat gradients related to altitude, exposure and location relative to other landscape components. The slope layout of land causes a large diversity of microclimatic, hydrological and soil conditions even in a relatively small area [3].

The high spatial diversity of the structure of the postglacial landscape is conditioned not only by the relief and geological structure but also by the richness of hydrographic objects (lakes, watercourses, swamps, springs, sapping) with a poor organization of the surface drainage network and a mosaic of natural and cultivated environments. This spatial diversity of the postglacial landscape is defined in landscape ecology [4, 5] by the term of "spotted", where various habitats, ecosystems and communities constitute "patches" of various functions in the processes of transport and exchange of matter and biological information.

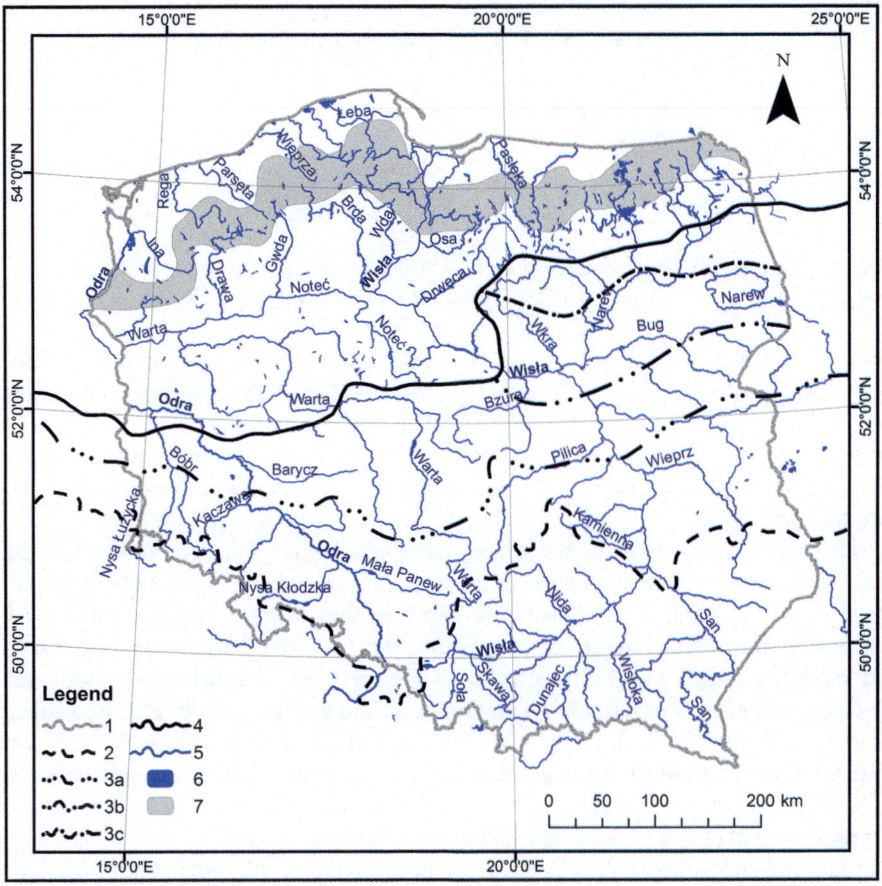

Fig. 1 Extent of Middle Poland glaciations and Vistula glaciation. (1) National border; (2) Odra glaciation; (3) Warta glaciation: (a) Pilica stadial, (b) Wkra stadial, (c) Mława stadial; (4) Vistula glaciation; (5) watercourses, (6) lakes (based on [1]), (7) lakeland hump

2 Endorheism of Lakelands

The quantitatively dominant elements of the postglacial landscape relief are endorheic depressions formed in the Holocene as a result of melting of buried stagnant ice bodies. These are terrain depressions located between hills and elevations and sometimes even small hollows on the slopes of the latter. They also occur in a valley position or in a flat area [6, 7]. The catchments of these depressions are endorheic areas, which is a local climate anomaly in the Polish Lowland [6, 8]. In the beds of endorheic depressions, there may be ponds, lakes and swamps. These are the storage depressions, most numerous on the moraine plateaus. For example, in the part of the Polish Lowland called Gdańsk Pomerania, the number of water reservoirs with an area of less than 5 ha is close to 32,000 [9], and their density in the areas of

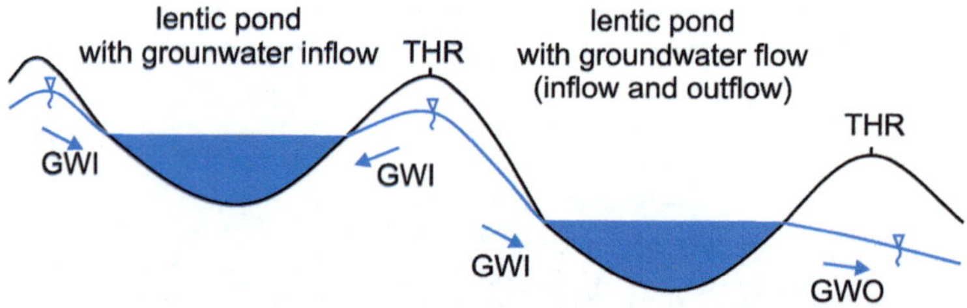

Fig. 2 Schemes of endorheic depressions with ponds in their beds (legend: *THR* threshold of endorheic depression, *GWI* inflow of groundwater, *GWO* outflow to groundwater)

moraine plateaus reaches even 80 reservoirs per 1 km^2. The bottoms of the endorheic depressions can also be dry, devoid of waterbodies – they are then absorbent depressions, most frequent on outwash plains. In the postglacial landscape, both types of endorheic depressions constitute local drainage bases performing a storage role (storing a part of precipitation water) or an infiltration role (supplying underground waters).

The catchments of endorheic depressions (both storage and infiltration ones) can be excluded from the alimentation of river systems completely (total endorheism) or periodically (periodic endorheism), or they can supply these systems only by the underground route (surface endorheism) (Fig. 2). The surface endorheism is favoured by a high variability of relief, while the total endorheism is conditioned by the geological structure and permeability of the ground related to it. Permanent surface endorheism occurs practically only in some endorheic depressions on outwash plains. Even those of them, at the bottom of which there are ponds, lakes or swamps (mostly raised bogs) due to high thresholds (Fig. 2) and topographic barriers, are not even periodically included in the river runoff.

The size of surface endorheic areas in the structure of the catchment of lakeland rivers ranges from several to several hundred km^2 (Table 1). Surface endorheism of the river catchments reaches from several to several dozen percent.

3 Hydrographic Network of Lakelands

The formation of the hydrographic network of the postglacial areas of northern Poland was influenced by the processes of permafrost degradation and melting of stagnant ice bodies. These processes were particularly intense in the Allerød and in some lakeland areas which still occurred in the Preboreal [12, 13]. The current system of the water network of northern Poland corresponds to older pre-Quaternary outflow routes, and its layout is a result of the complex nature of deglaciation, local morphological, lithological, hydrogeological conditions, climatic changes and contemporary vertical movements [14–17].

The development of the hydrographic network in lakeland areas was multiphase and varied in time and space. Its contemporary layout is the result of (1) formation of

Table 1 Proportion of endorheic areas in the structure of chosen catchments of northern Poland

The river	Total catchment area (km^2)	Extent of endorheic areas (km^2)	Share of endorheic areas (%)
Masurian and Suwałki lakelands (based on [10])			
Krutynia	700.0	301.0	43
Szeszupa	92.9	55.7	60
Czarna Hańcza	170.2	93.6	55
Wiatrołuża	176.5	74.1	42
Kamionka	94.4	32.1	34
Joka	63.3	17.7	28
Jarka	220.0	23.0	10
Gawlik	57.2	15.2	27
Dejgunka	11.7	2.8	24
(Upper) Rospuda	41.1	14.8	36
Pomeranian lakeland (based on [11])			
Słupia	1,570.5	582.2	37
Łupawa	805.6	290.7	36
Łeba	1,128.7	277.4	25
Reda	387.1	88.1	23
Wda	1,350.6	728.4	54
Wierzyca	1,559.8	706.6	45
Radunia	793.2	321.3	41
Borucinka	32.2	19.6	61

networks of proglacial rivers by subglacial runoff; (2) filling and preservation of channels by ice; (3) formation of runoff in periglacial conditions during deglaciation before melting of buried stagnant ice bodies and deglaciation of permafrost; (4) melting of buried ice bodies and formation of lakes as well as cutting of channel thresholds and formation of overflow breaches, which led to the runoff and disappearance of many lakes in the channel valleys; and (5) modern erosion and accumulation processes. Thus complex origin of the hydrographic network of the post-glaciated areas causes that the valley sections of one river system can be of different age [18–21]. This natural and multicycle development of the hydrographic network is additionally influenced by human transformations, in some areas dating back as far as the early Middle Ages.

Many modern rivers in the postglacial areas of northern Poland flow through former subglacial channels, at the bottom of which – after the melting of stagnant ice bodies or winter (hydrogenic) ice – lakes were formed (e.g. the Brda, Radunia, Krutynia). Rivers also go through melt-out areas (e.g. the Szeszupa); they also use the valleys of marginal rivers (e.g. the Noteć, Reda, Łeba) and proglacial ones (e.g. the Ełk, Rospuda, Czarna Hańcza, Lega).

The geologically young age of the water network of northern Poland means that until today it has the characteristics of an initial network, i.e. one that is under formation. This is manifested in (1) exclusion of substantial areas from the surface drainage, (2) inclusion of postglacial lakes of various sizes to river outflow, (3) periodicity of the river network, (4) unstabilized longitudinal profile of permanent watercourses and (5) frequent surface and underground incompatibility of areas supplying a given waterbody [10].

The geological and geomorphological diversity of the postglacial landscape makes it possible to distinguish three characteristic types of hydrographic network differing in the degree of organization, conditioning the discharge of water surplus outside their alimentation area in the form of river runoff. The lowest organization degree is represented by the hydrographic network of areas without outflow (endorheic). A higher organization level is displayed by a river-lake network that permanently discharges water surplus from its catchment in the form of river runoff. The highest organization degree is achieved by the lakeland river, the main river building the system of surface drainage in the postglacial area, flowing through a valley formed as a result of its erosion-accumulation activity along the route of the former outflow of postglacial rivers. It should be remembered that the mosaic of the postglacial landscape means that poorly and well-organized hydrographic networks are often found next to each other.

3.1 Hydrographic Network of Endorheic Areas

The hydrographic network in endorheic areas is in an organizational stage. It consists of swamps of small area and very numerous postglacial ponds and small lentic lakes. Watercourses, if any, are periodic, mostly episodic. The hydrographic network of the plateau endorheic areas is particularly interesting. Poor permeability of surface formations (boulder clays) that build moraine plateaus favours the storage of water in ponds located in the beds of endorheic depressions. When reservoir storage of these ponds is exceeded, they initiate runoff, thanks to which the water network develops, the number of watercourses which include subsequent ponds in the fluvial system (river runoff) via short sections increases. These watercourses function usually in early spring, creating extensive river systems that reach the 3rd and even the 4th order in the Horton-Strahler hierarchy [22] (Fig. 3). The research carried out in Gdańsk Pomerania showed that nearly 20% of ponds in endorheic areas are included in the river runoff for at least a part of the year [22], which indicates that the ponds not only store rainwater but also initiate river runoff, periodically limiting the endorheism of the postglacial areas, and that some of them supply groundwaters, constituting peculiar "hydrogeological windows" (Fig. 2) [24].

This thaw-related river network on moraine plateaus quickly disappears with decreasing catchment storage and dries up completely in late spring or early summer (Fig. 3). The development of this type of river system starts at the mouth section of the watercourse and goes upstream and the disappearance from the watercourses located topographically the highest.

The hydrographic network of endorheic areas in outwash plains landscape is characterized by a much greater stability, mainly associated with fast infiltration of

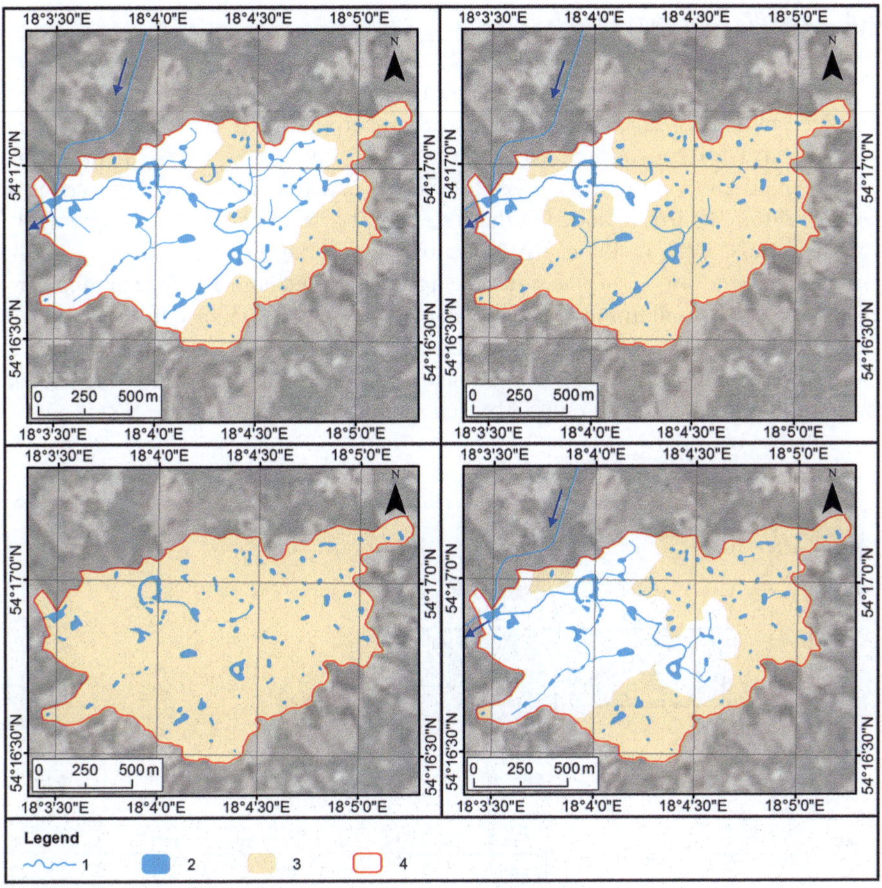

Fig. 3 Hydrographic survey of water network in endorheic area of moraine plateau during different storage states (legend: (1) watercourses, (2) ponds, (3) extent of surface endorheic areas, (4) boundary of endorheic area during low storage condition) [23]

precipitation water. At the bottom of depressions, deeply cut into outwash levels, there are often ponds and swamps, usually forming raised bogs. They are a durable and unchanging element of the water network. Due to high thresholds and topographic barriers, these endorheic areas are permanently excluded from the river runoff, but due to reservoir storage, they direct the surplus of precipitation water to aquifers. These aquifers constitute a drainage base for lakes that are part of the river-lake network or for lakeland rivers.

Studies indicate that endorheic depressions, constituting local drainage bases, play an important role in shaping the runoff of lakeland rivers [22, 24]. On the morainic plateaus of lakeland areas, the development of the hydrographic network and its durability depend on their storage state.

3.2 River-Lake Network

A higher degree of hydrographic network organization, though still initial, is demonstrated by areas with a river-lake network, incorporating into the uniform fluvial system concave forms of the area of the last glaciation, using valleys incompatible with the modern river runoff [10, 25–27]. Such a network is composed of a series of lakes of different sizes included in the river runoff system via short watercourses of various orders, often periodic. River sections connecting the lakes function on the principle of overflow of surplus water from a lake located higher to the neighbouring lake, lying below, and further to the recipient (river or lake), which constitutes a local drainage base. They may also have a character of breaches (Fig. 4). This means that the rivers flowing from the lakes included in the runoff are fed with underground water to a lesser extent and only in breach sections.

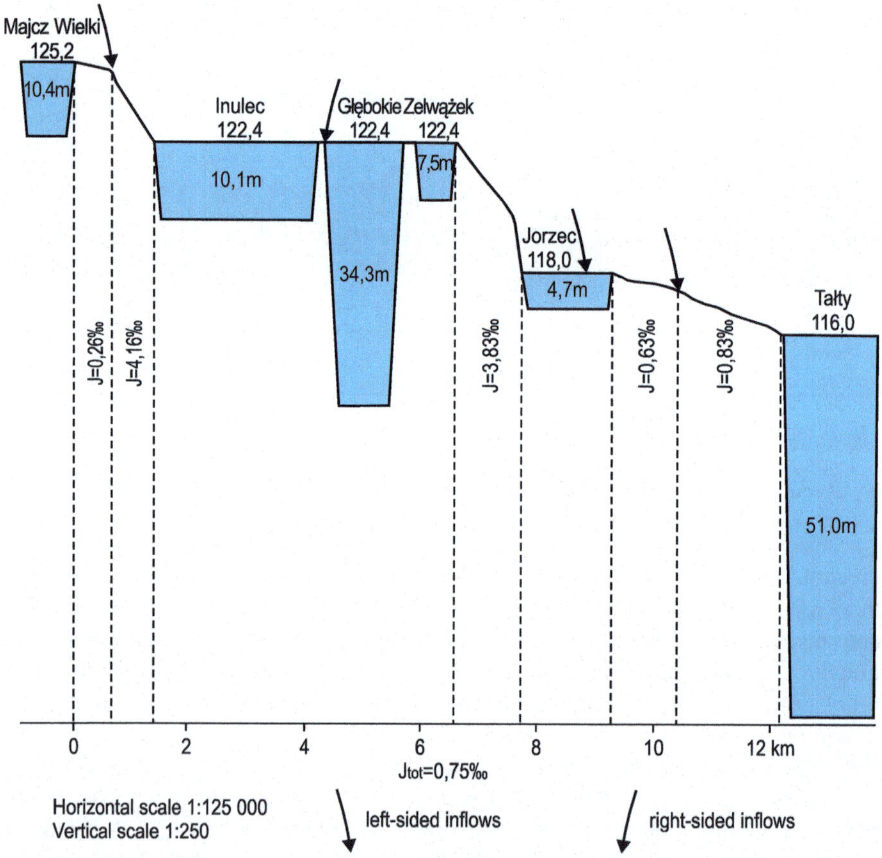

Fig. 4 Longitudinal profile of the valley of river-lake system of Jorka according to [10]

Table 2 Percentage sections of lentic water in the total length of selected river-lake networks of the North Poland

River-lake system	Length of river sections (km)	Length of lake sections (km)	Total length of the river (km)	Percentage of limnic elements in the total length of the river (%)
Krutynia	52.60	46.95	99.55	47
Szeszupa	16.24	2.52	18.76	13
(Upper) Rospuda	19.15	20.11	39.26	51
Szelmentka	6.18	11.65	17.83	65
Kamionka	11.62	3.66	15.28	24
(Upper) Lega	26.88	8.98	35.86	25
Jorka	5.80	6.30	12.10	52
Bludzia	6.56	11.21	17.77	63
Orzysza	20.67	14.96	35.63	42
(Upper) Radunia	7.39	26.16	33.55	78
Gołubska Struga	1.67	11.14	12.81	87
Czapielska Struga	0.79	4.52	5.31	85
Brda	168.24	77.27	245.51	31
Zbrzyca	28.39	20.99	49.39	43
Sępolna	30.36	13.48	43.84	31
Krówka	30.95	22.62	53.57	42
Sucha	11.46	4.57	16.03	29
Kręgiel	8.67	7.95	16.62	48
Szumionka	14.67	7.60	22.27	34

A river-lake network differs from a river network in: (1) significant share of lake sections in the river course (even to 60%) (Table 2); (2) periodicity of the river network (Table 3) mainly due to surface and subsurface supply with no contact with underground waters, which affects the seasonal variability of the runoff of the main river in the system; (3) significant share of surface endorheic areas in the structure of the area alimenting the network (Table 1); and (4) evolutionally unstable longitudinal slope of the main river of the system, which is manifested by breach sections linking lakes, indicating their adaptation to inherited valleys (Fig. 3).

The runoff stability in the river-lake system is determined by its lake elements. Basins of lakes included in this system are a drainage base not only for surface water, but also for underground water, of both shallow and deeper aquifer horizons [28–30], thanks to which lakes condition the stability of the runoff of river sections connecting these reservoirs, and affect the value of the runoff of the watercourse that forms a given river-lake system. The roles of particular lakes in forming the runoff of the watercourse that builds this system vary. Watercourses flowing from lakes depend on the level of lake basin filling. At high storage state of the reservoir, the flow of the watercourse increases, while at a low state the flow decreases or disappears completely [10, 22]. The role of the lakes lying in the middle course of the river is different. Such lakes can increase (thanks to groundwater drainage) or

Table 3 Percentage of periodic streams in catchments of selected river-lake systems of northern Poland

River-lake system	Permanent rivers (km)	Periodic rivers (km)	Total rivers (km)	Percentage of periodic streams in the catchment of the system (%)
Krutynia	220.6	367.9	588.5	63
Orzysza	20.87	14.96	35.83	42
(Upper) Rospuda	186.9	166.7	353.6	47
Bludzia	6.56	11.21	17.77	63
Szelmentka	6.18	11.05	17.23	64
Jorka	16.8	43.4	60.2	72
Jarka	95.2	296.4	391.6	76
Gawlik	40.4	37	77.4	48

reduce (as a result of alimenting groundwater) the outflow of the watercourse flowing through them. Owing to the storage capacity of the basins, they even out the outflow of the river flowing from them, which is manifested in reducing the irregularity of the flows as a result of decreasing the culmination of high water, prolonging its passage time, mitigating low water stages and shortening their duration. The influence of the lakes "closing" the river-lake system on the outflow of the watercourse building this system is already small [10]. The function of the lake in shaping the outflow depends on the storage state of the catchment, conditioned by the water content and dryness of the hydrological year.

The hydrological system of the river-lake network is influenced not only by the lakes that are part of such a network but also by the watercourse that builds this network. It often determines the ecological state of the lakes through which it flows. The river-lake network is therefore a coherent and functionally interrelated system in which there is often no continuum of flow increase along the river course, but eventually along with the increase in the catchment area, the flow of the watercourse building the river-lake network increases [10, 11, 31].

The river-lake network is formed on various levels of river runoff organization. It also occurs in all types of the lakeland landscape, although it is most often related to subglacial channels. The decisive morphological factor in its formation is the presence of low thresholds that are cut by watercourses which include lakes into the fluvial system.

The river-lake network of Polish lakelands is quite diverse. It consists of watercourses of the 2nd–3rd order, flowing through small and shallow lakes, draining areas from several to several dozen km^2, as well as large systems of 4th–6th order, with catchments of a few hundred km^2, flowing through many small and shallow lakes and large and deep lakes [10, 11, 31]. An example of a well-developed river-lake network is the Krutynia, about 100 km long, draining a catchment of 700 km^2 [11, 31]. This river flows through 17 lakes; its limnic index is 47%. Another example of such a network is the Radunia (Fig. 5), draining an area of about 213 km^2 with a limnic index of 78%. Within a section of 33.6 km, the river flows through eight lakes, including four with a maximum depth of over 20 m. Connections between the

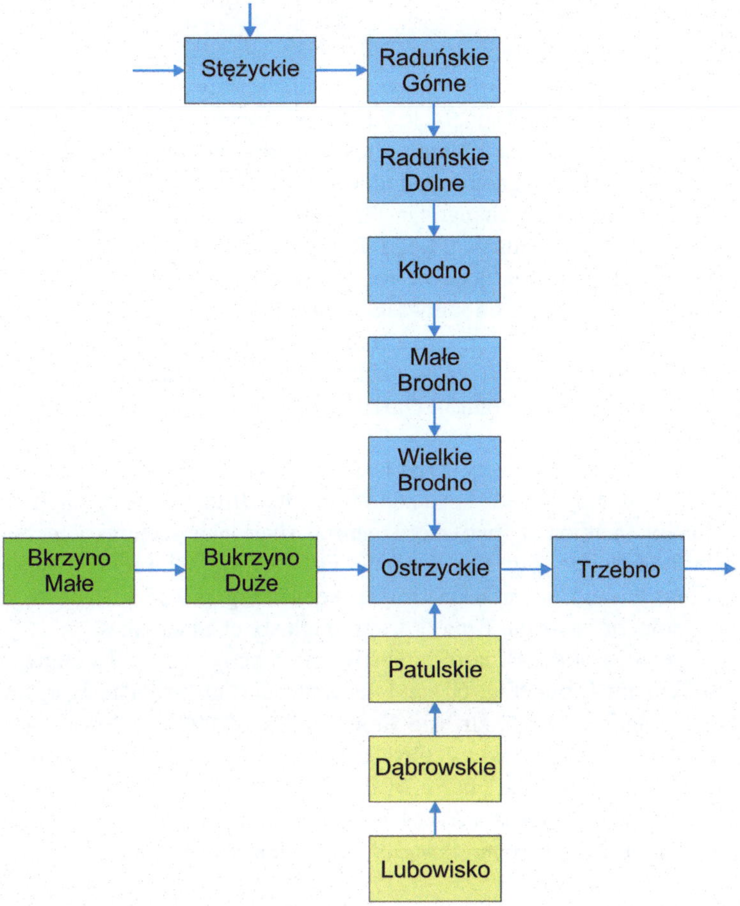

Fig. 5 Schemes of river-lake network of the upper Radunia (including lakes Stężyckie, Raduńskie Górne, Raduńskie Dolne, Kłodno, Małe Brodno, Wielkie Brodno and Ostrzyckie) and of the Czapielska Struga (including lakes Bukrzyno Małe and Bukrzyno Duże) and of the Gołubska Struga (including lakes Lubowisko, Dąbrowskie and Patulskie) [27]

lakes have a length of several metres to nearly 1 km, or they are overflows functioning on the basis of differences in water levels in neighbouring reservoirs [26].

Small river-lake networks occur widely in the postglacial landscape. Examples of such networks are the ones included in the hydrographic system of the upper Radunia (Fig. 5): the Czapielska Struga system (including lakes Bukrzyno Małe and Bukrzyno Duże) of a length of 5.3 km, alimentation area 4.81 km^2 and limnic index 85%, and the Gołubska Struga system (including lakes: Lubowisko, Dąbrowskie and Patulskie) of a length of 12.8 km, the alimentation area 28.2 km^2 and limnic index 87% [26, 27].

3.3 Lakeland Rivers

Large lakeland rivers also flow in valleys formed by postglacial rivers (fluvioglacial valleys, former subglacial channels, ice marginal streamways). Their river systems drain the slopes of the lakeland hump. The rivers draining the northern slope of the hump have beds of non-aligned longitudinal profiles whose development was to a large extent determined by changes in the level of the Baltic Sea (the Ina, Rega, Parsęta, Wieprza, Słupia, Łupawa, Reda, Radunia, Pasłęka, Łyna). The beds of the rivers draining the southern slope of the hump are developed in the bottoms of valleys formed most often on outwash routes (the Drawa, Gwda, Brda, Wda, Wierzyca, Drwęca, Pisa, Ełk, Czarna Hańcza and Biebrza) [32]. The headwater sections of these rivers have a river-lake character. An example is the upper course of the Brda (Fig. 6a) with a length of 9.019 km and a limnic index of 66%, which connects five lakes.

In the course of almost every large lakeland river, a series of typical river-lake sections can be distinguished. For example, in the Brda's course, apart from the section already mentioned, two more natural river-lake sections can be distinguished. In the first one (Fig. 6b) of a length of 15.7 km and limnic index of 42%, the Brda flows through seven lakes; in the second (Fig. 6c) of a length of 25.6 km and limnic index of 78%, the Brda flows through six channel lakes.

Many tributaries of the Brda form small river-lake networks. Examples include the Struga Siedmiu Jezior (Fig. 6d) and the Krówka (Fig. 6e). The Struga Siedmiu Jezior has a length of 9,417 km and limnic index of 80%. It flows out of Lake Ostrowite (43 m deep), flows through seven lakes and flows into Lake Charzykowskie, lying on the course of the Brda. The Krówka is 53.57 km long. Its central and lower course is a river-lake system of a length of 22.62 km (43% of the total length of the watercourse) and limnic index of 80%.

4 Conclusion

The hydrographic network of the postglacial areas is clearly different from the drainage networks of the rest of our country. It is a young network, which results from the fact that the current system of water runoff from the postglacial areas of northern Poland relies on the valleys formed by postglacial rivers. The specificity of the postglacial relief makes the degree of river network development, conditioning the discharge of water surpluses in the form of river runoff, diversified. In endorheic areas, this network is at the stage of initial development. Only with high storage state can the endorheism of these areas be periodically "destroyed" by filling the reservoir storage of some ponds located in the beds of the endorheic depressions. The most typical of postglacial areas is the river-lake network, i.e. composed of lakes linked by a network of watercourses into a coherent drainage system. The runoff stability in such a river-lake network is determined by its lake sections. These sections of

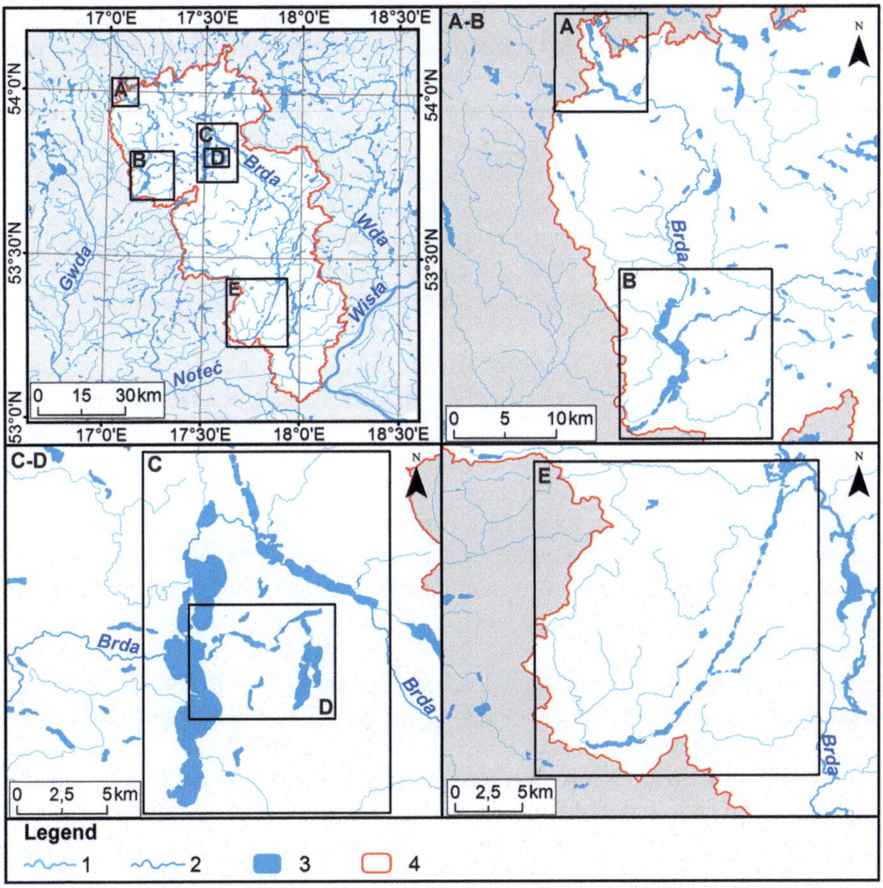

Fig. 6 Hydrographic system of the Brda based on MPHP (Map of Hydrological Division of Poland 2010); legend: (1) tributaries of the main river of a given river system, (2) main watercourses of river systems, (3) lakes, (4) the Brda catchment

river-lake systems perform various functions in the underground phase of the water circulation. They can increase river runoff by incorporating into it the water which they receive from underground drainage. They can "take" water from the river and feed it to underground waters, reducing the runoff of rivers flowing out of them. The lake elements of the river-lake network disrupt the concept of the river continuum. Due to the way they function (recipient or donor), each section of the river-lake network should be examined individually.

The catchments of rivers draining the postglacial areas are characterized by the average river network density of about 1 km km^{-2}. Differences in river network density between particular catchments are small and range from 0.35 to 1.59 km km^{-2} [10, 33]. It is worth noting that the share of periodic streams in the river network density is significant and often exceeds the length of the permanent stream network. This indicates a low level of organization of the lakeland's river network. Nonetheless, as the organization of the river network increases, its density increases. Low density of the river network and a larger proportion of endorheic areas characterize less-developed water network. The storage state of the catchment also determines the density of the river network. In the period of high catchment storage, the density of the river network many times exceeds the density observed at low catchment storage state.

Both the numerous ponds in the endorheic areas and the small river-lake systems occurring widely in the postglacial landscape contribute to the storage of the lakeland areas and therefore determine the stability of the water resources of these areas.

Areas feeding the river network of lakelands can be treated as a system of cascaded and connected reservoirs, each of which has a specific role in the formation of permanent river runoff [27, 34, 35]. Water surplus moves from higher to lower reservoirs, either in the form of surface runoff or underground by seepage through the bottom of reservoirs situated higher- to lower-lying reservoirs. The highest reservoir in the cascade is formed by catchments of endorheic areas. When the storage of this reservoir is exceeded, water surplus enters the next reservoir in the cascade initiating a permanent river runoff. At the next stage in the cascade, there is a reservoir of regional runoff, which collects surplus water from the basins initiating a permanent runoff and transfers it to the transit outflow reservoir, which in turn directs the collected water surplus to the last reservoir in the cascade – the reservoir of surplus water outflow, which is the drainage base for the whole postglacial area. The river-lake network can occur on every stage of the cascade of this model, constituting an individual link transforming the quantity and quality of water and having a significant impact on the functioning of the recipients – mature lakeland rivers.

References

1. Kosmowska-Suffczyńska D (2005) Geneza i rozwój rzeźby w czwartorzędzie. In: Richling A, Ostaszewska K (eds) Geografia fizyczna Polski. Wyd. Nauk. PWN, Warszawa, pp 32–67
2. Kostrzewski A, Zwoliński Z, Andrzejewski L, Florek W, Mazurek M, Niewiarowski W, Podgórski Z, Rachlewicz G, Smólska E, Stach A, Szmańda J, Szpikowski J (2008) Współczesna ewolucja rzeźby młodoglacjalnej Niżu Polskiego. In: Starkel L, Kostrzewski A, Kotarba A, Krzemień K (eds) Współczesne przemiany rzeźby Polski. IGiGP UJ, Kraków, pp 271–325
3. Hillbricht-Ilkowska A (2005) Ochrona jezior i krajobrazu pojeziernego – problemy, procesy, perspektywy. Kosmos, t. 54(4):285–302
4. Forman RTT, Gordon M (1986) Landscape ecology. Wiley, New York, p 622
5. Turner G, Gardner RU, O'Neill RV (2001) Landscape ecology in theory and practice. Springer, Berlin, p 398
6. Drwal J (1975) Zagadnienie bezodpływowości na obszarach młodoglacjalnych. Zesz Nauk WBiNoZ Geogr 3:7–26, Uniwersytet Gdański, Gdańsk

7. Bajkiewicz-Grabowska E, Magnuszewski A (2009) Przewodnik do ćwiczeń z hydrologii ogólnej. Wyd. nauk. PWN, Warszawa, p 195
8. Kowalska A (1971) Zagadnienie środkowoeuropejskich obszarów bezodpływowych w literaturze. Czas Geogr 42(4):353–363
9. Golus W (2014) Obieg materii w małych zbiornikach pojeziernych, PhD thesis, p 295
10. Bajkiewicz-Grabowska E (2002) Obieg materii w systemach rzeczno-jeziornych (Circulation of matter in the river-lake systems). Wydział Geografii i Studiów Regionalnych UW, Warszawa, p 274
11. Bajkiewicz-Grabowska E (1996) Zmiany wielkości przepływu w węzłach hydrograficznych systemu rzeczno-jeziornego Krutyni (Pojezierze Mazurskie). In: Hillbricht-Ilkowska A, Wiśniewski RJ (red.) Funkcjonowanie systemów rzeczno-jeziornych w krajobrazie pojeziernym: rzeka Krutynia (Pojezierze Mazurskie), vol 13, Zesz. Naukowe Kom. "Człowiek i Środowisko", pp 35–42
12. Nowaczyk B (1994) Wiek jezior i problemy zaniku brył pogrzebanego lodu na przykładzie sandru Brdy w okolicy Charzykowy. AUNC Geogr 27:97–110
13. Błaszkiewicz M (2005) Późnoglacjalna i wczesno holoceńska ewolucja obniżeń jeziornych na Pojezierzu Kociewskim (wschodnia część Pomorza). Prace Geogr 201:192
14. Sylwestrzak J (1978) Rozwój sieci dolinnej na Pomorzu pod koniec plejstocenu. Gdań. Tow. Nauk., Gdańsk, p 178
15. Brykczyński M (1986) O głównych kierunkach rozwoju sieci rzecznej Niżu polskiego w czwartorzędzie. Prz Geogr 58(3):411–440
16. Florek W (1991) Postglacjalny rozwój dolin rzek środkowej części północnego skłonu Pomorza. WSP w Słupsku, Słupsk, p 238
17. Jankowski A (1994) Geneza współczesnej sieci dolinno-rzecznej na obszarze Polski (Genesis of the Contemporary River Valley Network of the territory of Poland). In: Kołtuniak J (ed) Rzeki, kultura-cywilizacja-historia, vol 3, Wyd. Śląsk, Katowice, pp 135–155
18. Falkowski E. (1971) Historia i prognoza rozwoju układu koryta wybranych odcinków rzek nizinnych Polski. Biul Geolog UW 12. s:5–122
19. Falkowski E (1980) Problemy genezy i interpretacji ukształtowania doliny środkowej i dolnej Wisły. Prz Geolog UW 12:345–347
20. Starkel L (ed) (1991) Geografia Polski. Środowisko przyrodnicze. Wyd. Nauk. PWN, Warszawa, p 592
21. Andrzejewski L (2017) Ewolucja systemów dolinnych i zmiany obiegu wody od ustąpienia ostatniego zlodowacenia. In: Jokiel P, Marszelewski W, Pociask-Karteczka J (eds) Hydrologia Polski. Wyd. Nauk. PWN, Warszawa, pp 111–116
22. Bajkiewicz-Grabowska E, Golus W (2009) Organizacja sieci hydrograficznej w zlewni pojeziernej przy różnym stanie jej retencji. In: Bogdanowicz R, Fac-Beneda J (eds) Zasoby i ochrona wód. Obieg wody i materii w zlewniach rzecznych. Fundacja Rozwoju Uniwersytetu Gdańskiego, Gdańsk, pp 159–166
23. Golus W, Kwidzińska M, Markowski M (2017) In: Czerniawski R, Bilski P (red.) Dynamika sieci hydrograficznej obszarów pojeziernych, Funkcjonowanie i ochrona wód płynących, Szczecin, pp 257–272
24. Golus W, Bajkiewicz-Grabowska E (2016) Water circulation in the moraine ponds of northern Poland. Hydrobiologia. https://doi.org/10.1007/s10750-016-2830-7
25. Bajkiewicz-Grabowska E (2005) Rola systemów rzeczno-jeziornych w funkcjonowaniu obszarów młodoglacjalnych. Roczniki AR w Poznaniu CCCLXV, Melioracje i Inżynieria Środowiska 26:13–21
26. Bajkiewicz-Grabowska E (2007) Charakterystyka hydrograficzna. In: Borowiak D. (red.) Jeziora Kaszubskiego Parku Krajobrazowego. Badania Limnologiczne, vol 5, Katedra Limnologii UG, Gdańsk, pp 41–52
27. Bajkiewicz-Grabowska E (2012) Obieg wody i materii w małych systemach rzeczno-jeziornych obszarów młodoglacjalnych. In: Kostrzewski A, Szpikowski J (eds) Zintegrowany monitoring środowiska przyrodniczego. Funkcjonowanie geoekosystemów w różnych strefach krajobrazowych Polski. Biblioteka Monitoringu Środowiska v. XXIX, Storkowo, pp 123–138

28. Jankowska H (1985) Znaczenie jezior w kształtowaniu się odpływu podziemnego w dorzeczu górnej Raduni. Zesz Nauk Wydz BGiO UG Geogr 14:57–68
29. Nowakowski CZ, Smoleński A, Soszka H (1996) Wody podziemne i ich udział w zasilaniu i transporcie materii do jeziora. In: Hillbricht-Ilkowska A, Wiśniewski RJ (eds) Funkcjonowanie systemów rzeczno-jeziornych w krajobrazie pojeziernym: rzeka Krutynia (pojezierze Mazurskie). Zesz. Nauk. PAN "Człowiek i Środowisko", vol 13, Oficyna Wydawnicza, Instytut Ekologii PAN, pp 71–96
30. Nowiński K (2009) Rola jezior w transformacji jakości wód systemu Raduni. PhD thesis, Instytut Geografii WOiG UG, Gdańsk, p 237
31. Bajkiewicz-Grabowska E (2017) Sieć rzeczna Pojezierzy. In: Czerniawski R, Bilski P (eds) Funkcjonowanie i ochrona wód płynących. Szczecin, pp 297–306
32. Andrzejewski L, Krzemień K (2017) Doliny i koryta rzek. In: Jokiel P, Marszelewski W, Pociask-Karteczka J (eds) Hydrologia Polski. Wyd. nauk. PWN, Warszawa, pp 121–127
33. Fac-Beneda J (2009) River network of young glacial areas and its fractal dimension. River Basin Manag 351. https://doi.org/10.2495/RM090321
34. Drwal J (1982) Wykształcenie i organizacja sieci hydrograficznej jak podstawa oceny struktury odpływu na terenach młodoglacjalnych. Wyd. UG, Gdańsk, p 130
35. Fac Beneda J (2011) Młodoglacjalny system hydrograficzny. Wyd. UG, Gdańsk, p 213

The Role of Lakes in Shaping the Runoff of Lakeland Rivers

Elżbieta Bajkiewicz-Grabowska, Włodzimierz Golus, Maciej Markowski, and Monika Kwidzińska

Contents

Abstract Hydrological functions of lakes in shaping the runoff of rivers that drain them were discussed on the basis of data from the literature. The discussion indicates the role of lakes in initiating river runoff, their role in water storage and runoff equalization of a river flowing through or out of the lake, as well as the role in shaping the runoff capacity of the river: its increase due to underground water drainage by the lake basin or decrease as a result of lake waters supplying the underground water. It was pointed out that there are lakes which regardless of the character of the hydrological year (wet, average, dry) constantly increase the runoff of the river which flows through them by incorporating into it the water which they receive from underground drainage. There are also lakes that constantly feed underground waters with their waters, which reduces the runoff of rivers flowing from them. There is also a group of lakes which, depending on the level of the storage of the basin, conditioned by the water content of the hydrological year, either drain underground waters and increase river runoff or aliment them thus reducing river runoff. The above hydrological functions of lakes were discussed on the example of outflow and flow-through lakes of northern Poland.

E. Bajkiewicz-Grabowska (✉) · W. Golus · M. Markowski · M. Kwidzińska
Department of Limnology, Institute of Geography, University of Gdańsk, Gdańsk, Poland
e-mail: bajka37@wp.pl

E. Korzeniewska, M. Harnisz (eds.), *Polish River Basins and Lakes – Part I*,
The Handbook of Environmental Chemistry 86,
https://doi.org/10.1007/978-3-030-12123-5_9

Keywords Geoecosystem of lakeland river · Hydrological functions of outflow and flow-through lakes · River network of lakelands · Runoff of lakeland river

1 Introduction

The hydrographic network of young glacial areas, which consists of lakes of varying morphometry, various watercourses (generally short, often periodic), springs, sapping, and swamps, is quite unique due to the relationships among these hydrographic objects and their role in shaping river runoff [1–8]. The interrelations (morphological and hydrological) of lakes and watercourses in shaping river runoff and their arrangement in the structure of the hydrographic network justify the term river-lake network, as it is a series of lakes included in the outflow system through short watercourses [5]. The river sections connecting the lakes are short; they generally have the character of overflows, which means that the river flowing through these lakes discharges water flowing over from one reservoir to another. In such a system of surface drainage, it is not the riverbed but lake basin that is a drainage base for underground waters, both phreatic and deeper. Therefore, the volume of river runoff in such a system is not determined by river sections but by lake sections [5]. In addition to the river-lake network, the area of young glacial lakelands is also drained by lake rivers flowing either through glacial valleys (e.g., channels) or valleys formed as a result of the erosion-accumulation activity of the river on the route of the former meltwater runoff. A lake river is one that flows out of or through a lake and whose water regime clearly depends on water circulation in the lake. Therefore, it is characterized by low runoff variability [9]. The potential of the lake in shaping river runoff, both in the river-lake system and in lake rivers, is determined by the range of the annual amplitude of lake water levels, so-called active storage. Its value shapes water circulation in the lake. Understanding this circulation, and thus determining the value of the main fluxes of water exchange between the lake and its surroundings (atmosphere, topographic catchment, and underground catchment), allows to identify the function of the lake in shaping the runoff of the river draining it.

In each lake (small and large, without outflow, outflow, and flow-through), there are two directions of water flow – vertical and horizontal – which cover three phases of its exchange: atmospheric, surface, and underground. These three phases of water exchange in a lake at a given time are taken into account in the balance equation comparing the amount of water supplying the lake (so-called water intake) with the amount of water escaping from it by various ways (so-called water outgo) in the form proposed by [10]:

$(P_j - E_j)$	+	$(\Sigma I_j - O_j)$	+	ΔZ_{under}	=	ΔS_j
Atmospheric exchange flux		Surface exchange flux		Underground exchange flux		Storage difference

where $P_{j,}$ is precipitation onto the lake; E_j evaporation from the lake; ΣD_j sum of river and surface inflow from direct lake catchment; I_j river runoff from the lake; ΔZ_{under} resultant of underground supply, i.e., difference between underground inflow and underground outflow; ΔS_j difference of lake storage at the beginning and end of assessment period; and means a change in water level in the lake (in mm) corresponding to changes in its capacity (m^3).

2 Materials

Hydrological functions of lakes affecting the runoff of rivers draining them are presented on the basis of literature data, which used hydrometric data obtained from IMGW-PIB (Institute of Meteorology and Water Management – National Research Institute) of 1971–2007 and from own measurements of the authors of the cited publications. Due to the subject matter of the article and the material availability, the analysis covered Masurian and Suwałki region lakes (Fig. 1), for which monthly water balances were made in 1971–1990 [5], and Kashubian lakes (Fig. 1), which have water balances from years 1999–2007 [11].

3 Results

3.1 River Runoff Initiation

Many lakes in the lakeland areas are the initial elements of river systems. River runoff is initiated by small kettle ponds [12] and lakes of various sizes [13], as well as large lake systems, such as the bifurcating System of Great Masurian Lakes [13, 14]. All reservoirs initiating river runoff (periodic and permanent) represent the hydrological type of outflow lakes [9], i.e., those whose alimentation base ensures permanent outflow character. The size of this base is specified by the lake index.[1] Its value higher than 16.3 ensures permanent outflow character to the Masurian and Suwałki lakes [5], while for Pomeranian lakes, this index is lower, i.e., at least 7.8 [15].

While larger lakes have permanent outflow, i.e., the rivers flowing out of them (usually rivulets, more rarely streams) are permanent, smaller lakes, and in particular some kettle ponds are periodically drained by small watercourses flowing out of them (as rivulets or rills) [5–7, 12, 16]. The time of functioning of these small periodic watercourses depends on the level of the storage of the catchments of the ponds that activate their outflow. This condition is achieved when fluxes of surface

[1]Lake index (Ohle index) is a quotient of total area of lake catchment and area of the lake.

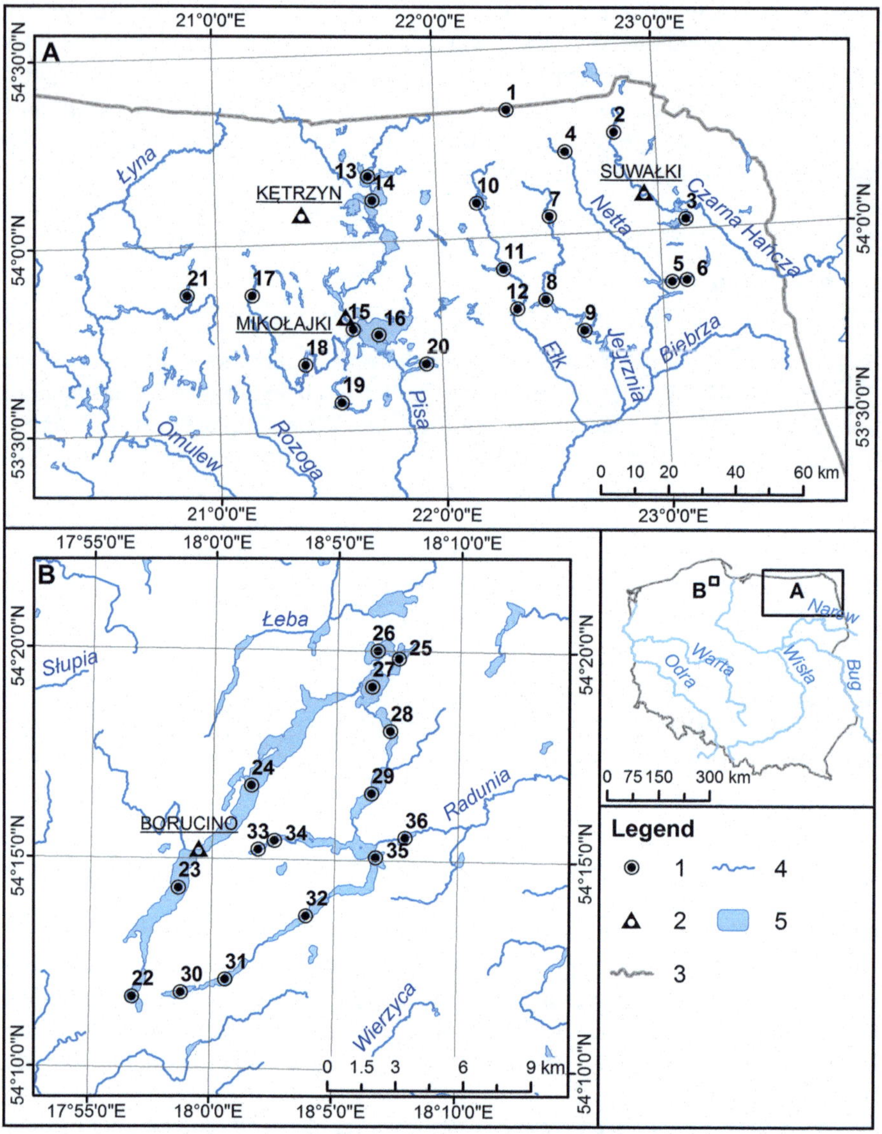

Fig. 1 Location of analyzed Masurian, Suwałki area, and Kashubian lakes. Legend: 1, Towns/villages; 2, weather stations; 3, national borders; 4, rivers; 5, lakes. Lakes: (**a**) 1, Gołdapskie; 2, Hańcza; 3, Wigry; 4, Rospuda; 5, Białe; 6, Studzieniczne; 7, Olecko Wielkie; 8, Selmęt Wielki; 9, Rajgrodzkie; 10, Litygajno; 11, Łaśmiady; 12, Ełckie; 13, Mamry Pn.; 14, Dargin; 15, Mikołajskie; 16, Śniardwy; 17, Gielądzkie; 18, Mokre; 19, Nidzkie; 20, Roś; 21, Dadaj; (**b**) 22, Stężyckie; 23, Raduńskie Górne; 24, Raduńskie Dolne; 25, Rekowo; 26, Białe; 27, Kłodno; 28, Brodno Małe; 29, Brodno Wielkie; 30, Lubowisko; 31, Dąbrowskie; 32, Patulskie; 33, Bukrzyno Małe; 34, Bukrzyno Duże; 35, Ostrzyckie; 36, Trzebno

and underground exchange begin to dominate in the water circulation structure in these kettle ponds [7, 12].

3.2 Water Storage

All lakes are natural storage reservoirs. Their storage potential is determined by the volume of water accumulated in them compared to the average water level from a long-term period, while the measure of storage capacity is active storage, i.e., that part of the lake water volume, which is within the range of the annual amplitude of water levels. The active storage capacity of the lake is conditioned by annual amplitudes of its water levels and the shape of its basin. In natural conditions, the value of water level fluctuations in lakes is affected primarily by the size of their catchments [5, 17–19]. In the case of delta and flow-through seaside lakes, an additional factor affecting the value of water level amplitudes are periodic intrusions of sea waters.

There is a large span in the amplitudes of monthly and annual water levels in Polish lakes. Small amplitudes of water levels are not always associated with large lakes and vice versa. The mean annual amplitude of lake water levels is largely dependent on the size of its catchment and the region in which the lake is located. For large flow-through lakes of northern Poland, this relationship is described by the equation [17]:

$$\Delta h = 8.9318\ \mathrm{C}^{0.4765}$$

and for Mazurian and Suwałki lakes (Fig. 1) [5]:

$$\Delta h = 0.211\ \mathrm{C} + 38.969 \text{ or } \Delta h = 9.5855\ \mathrm{C}^{0.4594}.$$

where C – lake index, Δh – annual water level amplitude.

The absolute active storage determined by periodic extremes in all hydrometrically controlled Polish lakes ranges from 0.13 to 128.2 hm^3 [20]. The largest active storage is found in lakes: Śniardwy (128.2 hm^3) drained by the Pisa, Łebsko (91.4 hm^3) drained by the Łeba, Gopło (58.8 hm^3) drained by the Noteć, and Miedwie (55.4 hm^3) drained by the Płonia. The active storage of the deepest Polish lake, Hańcza, from which the Czarna Hańcza flows, is small and amounts to 5,000 hm^3. Greater storage can be observed in the next lake on the route of this river – Lake Wigry (almost 17.0 hm^3) [5]. The largest absolute natural active storage among the studied Masurian and Suwałki lakes is found in lakes: Roś (33.8 hm^3), drained by the Pisa; Selmęt Wielki (20.1 hm^3), drained by the Jegrznia; and Orzysz (14.1 hm^3), drained by the Orzysza, tributary of the Pisa and Łaśmiady (10.7 hm^3), lying on the route of the Ełk river, while the smallest (slightly below 2 hm^3) is observed in the ribbon flow-through lakes Gołdap (drained by the Gołdapa, tributary of the Węgorapa) and Rospuda (drained by the river Rospuda, tributary of the Narew) (Fig. 1) [5].

The active storage of Polish lakes, determined by average annual amplitudes of water levels of 1961–2005, ranges from 1.25 to 1.40 km^3. This corresponds to the average fluctuations of lake water levels of 45–50 cm [20].

The research results [5] clearly indicate that local conditions, and not climatic ones, play a key role in shaping water level fluctuations in lakes.

3.3 River Runoff Equalization

Lakes included in the surface runoff system, i.e., those with an outflow and flow-through hydrological type, during periods of high river water stages, store (retain) water and, during periods of low river water stages, discharge it to the river, thus equalizing its runoff in time [21]. The effectiveness of the lake's equalization of the runoff of the river draining it is determined by the active storage capacity of its basin, and the measure of this effectiveness is the equalization index n (quotient of active lake storage capacity and the volume of outflow from the lake) [5, 21–23].

The greatest ability to equalize river runoff is found in outflow lakes, i.e., those which initiate rivers. The mean river runoff equalization index of these lakes ranges from 20% (e.g., Olecko Wielkie) to 68% (e.g., Szóstak) (Table 1). To a lesser extent, river runoff is equalized by flow-through lakes. In their case, the average river runoff equalization index, depending on the location of the lake along the course of the river, varies from 2% (e.g., Ełckie) to 22% (e.g., Orzysz). If the river flows through several lakes (Fig. 2), their equalization role decreases with its course. The lake constituting the last element in such a river-lake system, most often transfers high water stages in an unchanged form. An example could be Lake Ełckie (Fig. 2), whose average annual equalization capacity is only 2% [5]. When taking into account the maximum annual outflow from lakes, the level of equalization of river runoff by them decreases and ranges from 1% to 42% (Table 1). The equalization capacity of outflow lakes ranges from 9% (Olecko Wielkie) to 41–42% (Hańcza and Szóstak) and of flow-through lakes from 1% (Ełckie) to 17% (Orzysz).

Research [5] also indicates that the value of the river runoff equalization index decreases with the increase in the lake index C (and does not increase as often is found in literature) and decreases with the increase in exchange intensity in the lake IW.[2]

[2]Intensity of water exchange in lake (IW) is a quotient of river outflow from lake and lake capacity.

Table 1 Equalizing river runoff displayed by selected lakes of the northeastern Poland

Lake	Hydrological type of lake	Water stages amplitude, cm	Capacity of active storage, hm^3	Mean annual river runoff, million m^3	Maximum annual river runoff, million m^3	River runoff smoothing index (%)	
						n_1	n_2
Hańcza	Outflow	160	5,004	7,943	12,111	63	36
Wigry	Flow-through	85	16,958	124,460	169,607	14	10
Gołdap	Flow-through	127	1,936	73,772	86,156	3	2
Rospuda	Flow-through	56	1,891	7,598	10,562	25	18
Olecko Wielkie	Flow-through	137	3,118	15,955	34,272	20	9
Selmęt Wielki	Flow-through	154	20,082	99,297	154,940	20	12
Szóstak	Outflow	96	4,738	6,953	11,395	68	41
Litygajno	Flow-through	153	2,547	68,626	105,640	4	2
Łaśmiady	Flow-through	121	10,692	188,432	287,408	6	3
Ełckie	Flow-through	135	5,185	239,246	367,020	2	1
Dejguny	Outflow	60	4,578	12,793	19,143	36	21
Gielądzkie	Outflow	90	4,288	12,611	18,969	34	19
Mokre	Flow-through	75	6,332	127,088	172,287	5	3
Druglin	Outflow	99	4,153	8,321	18,171	50	21
Orzysz	Flow-through	131	14,089	64,406	81,041	22	13
Roś	Flow-through	187	33,800	615,773	781,830	5	3

Adapted from [5]

3.4 Shaping the Volume of River Runoff

Lakes situated on the course of a river not only equalize its runoff but also affect its volume, causing an increase in river runoff or its reduction. This process should be associated with the draining role of lake basins. While the river flowing through the lakes can only drain the level of underground water alimented either directly or indirectly by atmospheric precipitation, lakes located along its course, especially ribbon lakes, due to their depth, have hydraulic contact with deeper aquifer horizons that often lie on "far" transit underground circulation routes. Basins of these lakes can drain these aquifers, they can also aliment (supply) them. The research [24–32] indicates that the underground supply of lakes, especially ribbon lakes, is abundant and natural drainage occurs both from shallow as well as deeper aquifer horizons, which can be several (Fig. 3). In the structure of the water circulation in the lake, the flux of underground water exchange may constitute from several to several dozen percent [5, 11, 14, 33–37].

Lakes can perform various functions in the underground phase of the water cycle (Fig. 4) [5, 14]. Some lakes, regardless of the character of the hydrological year (dry, wet), always drain underground waters, increasing the runoff of the river draining them. Examples of such lakes are Olecko Wielkie, Rospuda, Gołdap, Wigry, and

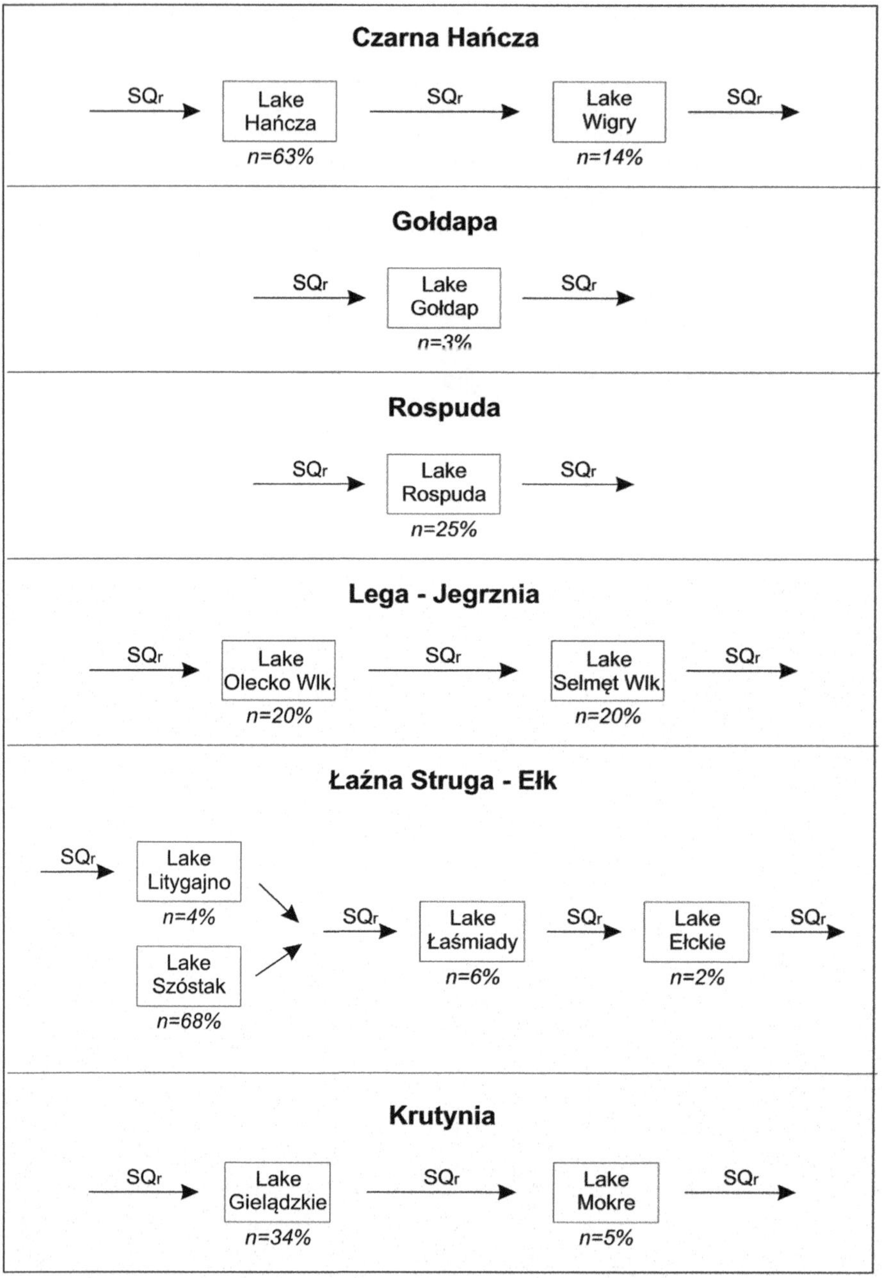

Fig. 2 Capacity of river runoff equalization by lake

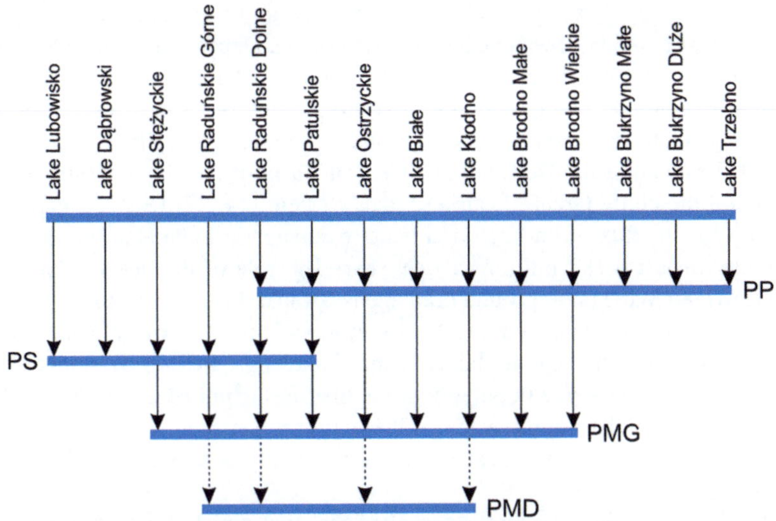

Fig. 3 Scheme of contact of lake basins with aquifer horizons (according to [32])

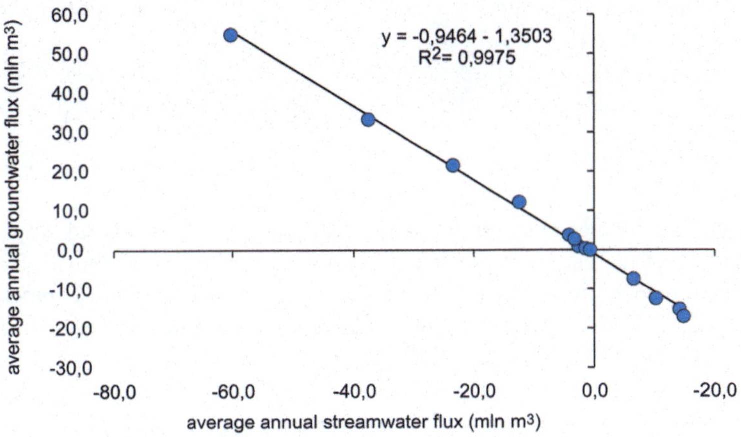

Fig. 4 Relationship between mean annual surface water exchange (mainly via river) and underground exchange in lakes of northeastern Poland [5]

Orzysz (Masurian and Suwałki area lakes) (Fig. 1) [5] and Stężyckie, Raduńskie Górne, Rekowo, Bukrzyno Małe, and Dąbrowskie (Kashubian lakes) (Fig. 1) [11]. The second group consists of lakes in which the flux of underground water exchange is dominated by an underground runoff. These lakes supply underground waters, reducing the runoff of rivers flowing out of them. These are lakes drained by the Ełk: Litygajno, Łaśmiady, Ełckie, and Selmęt Wielki drained by the river Lega (Fig. 1), i.e., lakes lying on the stretch of river courses that use the valleys of the fluvioglacial water runoff. This group is also represented by the Southern Basin of the Great Masurian Lakes System [14, 37]. The next group are lakes, whose function in the underground flux of water exchange is variable. Depending on the character of the hydrological year (wet, average, dry) they drain underground waters increasing river runoff or supply them, reducing the runoff of the rivers flowing out of them. Examples of such lakes are Dejguny, Szóstak, Mokre, Roś, Druglin, Gielądzkie, and Hańcza [5] as well as the Northern and Central Basin of the Great Masurian Lakes System and the entire Great Masurian Lakes System [14, 37] (Fig. 1). Particularly noteworthy is the flux of underground water exchange in Lake Hańcza, amounting on average to just 0.183 million m^3 per year. The role of the deepest lake in the Polish Lowland in the underground drainage is comparable to much shallower lakes. This fact should be explained by the structure of the quaternary system of the Suwałki Lakeland. The channel lakes of this Lakeland (Hańcza, Szelment Wielki, Szelment Mały), contrary to popular opinion, manifest a limited hydraulic bond with the surrounding underground waters [38]. Contact between the waters of Lake Hańcza with the deeper water-bearing levels practically does not exist, these waters flow around the lake basin [39].

The influence of the lake on the runoff of the river which drains it, and in some lakes also its direction (drainage – alimentation), depends on the amount of water which participates in the circulation in the lake in a given year. For the amount of this water primarily determines the value and direction of the flux of underground water exchange in the lake [5].

4 Conclusions

Lakeland watercourses constitute a unique geoecosystem [40, 41], to which basic properties, such as energy, matter, and information, can be attributed. Together with the alimentation area (surface and underground), they are a functionally interconnected landscape system, in which biogenic matter is transported with water, both the matter which enters the system from the catchment, and that produced in it. The main factor interfering with the transport of biogenic matter in the lakeland river system are the lakes drained by it, in which, as a result of a decrease in the kinetic energy of water, the matter accumulates in the form of bottom sediments [5]. The specificity of the young glacial areas, manifested in diversified geological structure, hydrological complexity, climatic variability, land use, and relief, causes that in lakeland watercourses (lake river, river-lake system), the matter transported with water undergoes transformation along with the course of the river and the lake sections disturb the continuity of the river runoff, the gradient nature of river zones, and the ecological processes. Lakes included in the river runoff condition the stability of rivers draining them. In shaping the river runoff, they play two main roles: equalizing and draining. The lake sections of river-lake systems and of lake rivers perform various functions in the underground phase of water circulation. They can increase river runoff (its underground component); they can reduce this runoff, directing part of the water flowing into the lake via the river into aquifers;

they can also, depending on the storage state of the catchment, increase the runoff of the river flowing out of it or reduce it. The influence of the lake on the river runoff is also determined by the rate of water exchange in the lake.

References

1. Drwal J (1975) Zagadnienie bezodpływowości na obszarach młodoglacjalnych. Zesz Nauk Wydz BiNoZ UG Geografia 3:7–26, Gdańsk
2. Drwal J (1982) Wykształcenie i organizacja sieci hydrograficznej jako podstawa oceny struktury odpływu na terenach młodo glacjalnych. Zesz Nauk Wydz BiNoZ UG Geografia 10:165–170, Gdańsk
3. Drwal J (1985) Jeziora w egzoreicznych systemach pojezierzy młodoglacjalnych. Zesz Nauk Wydz BiNoZ UG Geografia 14:7–15, Gdańsk
4. Drwal J (1990) Zastosowanie analizy sieciowej do określana wskaźnika odpływu w niekontrolowanych hydrometrycznie zlewniach pojezierzy młodo glacjalnych. In: Metody oceny i charakterystyki odpływu wód powierzchniowych i podziemnych w małych zlewniach rolniczych. IMUZ Mat Semin 29:15–21, Falenty
5. Bajkiewicz-Grabowska E (2002) Circulation of matter in the river-lake systems (Obieg materii w systemach rzeczno-jeziornych). Uniwersytet Warszawski, Wydział Geografii I Studiów Regionalnych, Warszawa, p 274
6. Bajkiewicz-Grabowska E (2007) Jeziorne ogniwa w sieci hydrograficznej obszarów młodoglacjalnych i ich rola w kształtowaniu odpływu rzeki pojeziernej. In: Michalczyk Z (ed) Obieg wody w środowisku naturalnym i przekształconym. Ser. Badania hydrograficzne w poznawaniu środowiska t.8. UMCS, Lublin, pp 39–46
7. Bajkiewicz-Grabowska E, Golus W (2009) Organizacja sieci hydrograficznej w zlewni pojeziernej przy różnym stanie jej retencji. In: Bogdanowicz R, Fac-Beneda J (eds) Zasoby i ochrona wód. Obieg wody i materii w zlewniach rzecznych. Fundacja Rozwoju Uniwersytetu Gdańskiego, Gdańsk, pp 159–166
8. Fac-Beneda J. (2011) Young-glacial hydrographic system (Młodoglacjalny system hydrograficzny). Wyd. Uniwersytetu Gdańskiego, Gdańsk, p 213
9. Bajkiewicz-Grabowska E, Mikulski Z (2006) Hydrologia ogólna. Wyd. Nauk. PWN, Warszawa, p 339
10. Bajkiewicz-Grabowska E (1982) Wymiana wody w akwenach jeziornych (na przykładzie jeziora Wigry). Prz Geof 3–4:229–240
11. Nowiński K (2009) Rola jezior w transformacji jakości wód systemu Raduni. PhD thesis, Instytut Geografii WOiG UG, Gdańsk, p 237
12. Golus W, Bajkiewicz-Grabowska E (2016) Water circulation in the moraine ponds of northern Poland. Hydrobiologia. https://doi.org/10.1007/s10750-016-2830-7
13. Mapa Podziału Hydrograficznego Polski w skali 1:10000. IMGW-PIB, Warszawa
14. Dąbrowski M (2013) Uwarunkowania naturalne i antropogeniczne obiegu wody w Systemie Wielkich Jezior Mazurskich. IMGW PIB, Warszawa, p 251
15. Lange W. (1986) Fizyczno-limnologiczne uwarunkowania tolerancji systemów jeziornych Pomorza. Zesz Nauk UG Rozprawy i Monografie 79:177, Gdańsk
16. Kwidzińska M, Markowski M, Golus W (2015) Wpływ uwarunkowań przyrodniczych na obieg wody i materii biogenicznej w systemie rzeczno-jeziornym Świętej Strugi. In: Absalon D, Matysik M, Ruman M (eds) Nowoczesne metody i rozwiązania w hydrologii i gospodarce wodnej, Komisja Hydrologiczna Polskiego Towarzystwa Geograficznego: Polskie Towarzystwo Geograficzne. Oddział Katowicki, Sosnowiec, pp 259–271
17. Pasławski Z (1975) Typologia hydrologiczna jezior Pojezierza Wielkopolskiego. Prz Geof 20(4):271–280

18. Jańczak J (1991) Fizycznogeograficzna typologia i ocena jezior na przykładzie Pojezierza Wielkopolskiego. Mat Badawcze IMGW Seria Hydrol i Oceanol 15:157
19. Borowiak D (2000) Reżimy wodne i funkcje hydrologiczne jezior Niżu Polskiego. Badania Limnol 2:164, Gdańsk
20. Borowiak D (2017) Zasoby i bilans wodny jezior. In: Jokiel P, Marszelewski W, Pociask-Karteczka J (eds) Hydrologia Polski. Wyd. Nauk. PWN, Warszawa, pp 229–235
21. Mikulski Z (1998) Gospodarka wodna. Wyd. Nauk. PWN, Warszawa, p 320
22. Chełmicki W (1999) Degradacja i ochrona wód. Część druga. Zasoby. Wyd. UJ Instytut Gografii, Kraków, p 152
23. Ciepielowski A (1999) Podstawy gospodarowania wodą. Wyd. SGGW, Warszawa, p 326
24. Glazik R (1970) Wody podziemne w dorzeczu Skarlanki i ich stosunek do rynien jeziornych. Dok Geogr IG PAN 1:70
25. Piekarek-Jankowska H (1973) Rola jezior rynnowych w kształtowaniu odpływu wód podziemnych w górnej i środkowej dolinie Rospudy. Z badań hydrogeologicznych w Polsce, vol 3, Wyd. Geol., Warszawa, pp 23–33
26. Poźniak R (1975) Wpływ warunków hydrogeologicznych na wielkość zasilania rzek nizinnych wodami podziemnymi. SGGW – rozprawy habilitacyjne, Warszawa, p 110
27. Pichiewicz S, Wróbel I (1976) Budowa geologiczna i stosunki hydrogeologiczne wybranych jezior województwa zielonogórskiego. In: Jeziora Ziemi Lubuskiej, ich wykorzystanie i ochrona przed zanieczyszczeniem. Mat Konf Łagów – Zielona Góra
28. Nowakowski CZ (1977) Charakterystyka wydajności źródeł strefy czołowo morenowej Pojezierza Suwalskiego. Biul Geol:21
29. Nowakowski CZ, Smoleński A, Soszka H (1996) Wody podziemne i ich udział w zasilaniu i transporcie materii do jeziora. In: Hillbricht-Ilkowska A, Wiśniewski RJ (eds) Funkcjonowanie systemów rzeczno-jeziornych w krajobrazie pojeziernym: rzeka Krutynia (pojezierze Mazurskie). Zesz. Nauk. PAN "Człowiek i Środowisko", vol 13, Oficyna Wydawnicza, Instytut Ekologii PAN, pp 71–96
30. Michalska M, Michalski T (1977) Perspektywy magazynowania wód powierzchniowych w strefie czołowomorenowej ostatniego zlodowacenia. Mat Konf Nauk Wydz Inż Lądowej PW, Warszawa
31. Michalska M, Michalski T (1980) Stosunki hydrogeologiczne piętrzenia wód na Pojezierzu Pomorskim. In: Stosunki wodne w zlewniach rzek Przymorza i dorzecza dolnej Wisły ze szczególnym uwzględnieniem gospodarki wodnej jezior. Mat Sesji Nauk-Tech, IMGW, Poznań
32. Jankowska H (1985) Znaczenie jezior w kształtowaniu się odpływu podziemnego w dorzeczu górnej Raduni. Zesz Nauk Wydz BGiO UG Geografia 14:57–68
33. Okulanis E (1981) Studium limnologiczne jezior raduńsko-ostrzyckich. Ossolineum-GTN, Gdańsk, p 108
34. Okulanis E (1982) Rola jezior w kształtowaniu powierzchniowych zasobów wodnych Pojezierza Kaszubskiego. Zesz Nauk UG Rozpr Monogr 37:233
35. Okulanis E (1985) Rola jezior w kształtowaniu stosunków hydrologicznych w wybranych zlewniach rzek Pojezierza Kaszubskiego. Zesz Nauk Wydz BGiO UG Geografia 14:31–44
36. Bajkiewicz-Grabowska E (1985) Factors affecting nutrients budget in lakes of the r. Jorka watershed (Masurian Lakeland, Poland). II. Hydrological budget of lakes in 1978 and 1979. Ekol Pol 33:2
37. Bajkiewicz-Grabowska E (2008) Obieg wody w systemie Wielkich Jezior Mazurskich. In: Jasser I, Robak S, Zdanowski B (eds) Ochrona i rekultywacja wód Wielkich Jezior Mazurskich narzędziem rozwoju naukowego, gospodarczego, społecznego i kulturowego regionu. Wyd. IRS, Olsztyn, pp 19–29
38. Mitręga J, Paczyński B, Płochniewski Z (1993) Wody podziemne Suwalszczyzny. Prz Geolog 41(8):569–571
39. Mitręga J, Pachla JP (1984) Hydrochemical budgeting in studiem of groundwater renevability in areas of glacial formation systems. International symposium IHP-UNESCO, Uppsala, pp 331–341

40. Kostrzewski A (1986) Zastosowanie teorii funkcjonowania geoekosystemu do badań współczesnych środowisk morfogenetycznych obszarów nizinnych Polski Północno – Zachodniej, Sprawozdania PTPN, 103 za 1984:26–28
41. Kostrzewski A (red.) (1995) Zintegrowany Monitoring Środowiska Przyrodniczego w Polsce. Propozycje programowe, Biblioteka Monitoringu Środowiska, Warszawa, p 164

Total Organic Carbon in the Water of Polish Dam Reservoirs

Andrzej Górniak

Contents

Abstract Total organic carbon (TOC) resources in Polish water reservoirs are presented as an important factor affecting water quality and ecosystem trophic state. The study is based on hydrochemical and biological data from 47 reservoirs from the years 2005–2017 and collected from the archives of the Polish National Monitoring Program, provided by the Chief Inspectorate of Environment Protection. The mean (by weight) TOC concentration in reservoirs is 6.3 mg dm^{-3}, with a range from 2.3 mg C dm^{-3} in the mountains, the Czorsztyn and Sromowce reservoirs, up to 18 mg C dm^{-3} in the hypereutrophic, lowland Siemianówka reservoir, varying according to reservoir elevation. Although reservoirs are large and deep, there is a significant negative correlation between mean reservoir depth and TOC. Seasonality and national TOC dynamics were strongly related to the rate of precipitation, with maximal concentrations in late spring and minimal in autumn or winter. The first global warming symptoms of TOC changes in reservoirs are noted, which will manifest as increased TOC and greenhouse gas emissions. Increased water retention time, which promotes water eutrophication, increases TOC resources in most Polish

A. Górniak (✉)
Department of Hydrobiology, Institute of Biology, University of Białystok, Białystok, Poland
e-mail: hydra@uwb.edu.pl

E. Korzeniewska, M. Harnisz (eds.), *Polish River Basins and Lakes – Part I*,
The Handbook of Environmental Chemistry 86,
https://doi.org/10.1007/978-3-030-12123-5_10

dam reservoirs as well as in flooded areas. Mean TOC concentrations are related to certain biological reservoir water parameters, such as phytoplankton index or diatom index. In future planning, Polish reservoirs should be placed outside lowlands, and their capacity should provide a high water exchange, in less than 2 to 3 months.

Keywords Dam · Reservoir · Total organic carbon · Water

1 Introduction

1.1 Aims of Study

Water resources are among the most frequently modified elements of the natural environment. Currently, there are more 14 million dam reservoirs in the world, with this number expected to increase in the next decades [1]. More than half of the world's global river systems are regulated by dams, which mostly lie in basins where irrigation and economic activities take place. Hydropower reservoirs and those for irrigation are dominant. The total cumulative storage of large dams is about 20% of global annual runoff [2]; in consequence of global warming and increase of deglaciated land area, a large global sea level rise of 30 mm is observed [3].

The reservoirs created by damming a preexisting river resemble a basin water storage and run-of-river system. Dam reservoirs become active modifiers of energy and mass runoff [4]. Sediment accumulation, primary productivity, and carbon mineralization along the river continuum are major effects of the perturbation of organic carbon cycling caused by damming of rivers [5]. A recent estimation shows that nearly 13% of total organic carbon global flux, carried by rivers to the oceans, has been eliminated by reservoirs, and the rate of this loss will increase with time.

European or regional studies on organic carbon cycling in dam reservoirs are not yet available. In this study, I present a first evaluation of the organic carbon resources in Polish dam reservoirs of different ages, volumes, and basin conditions. Varied environmental drivers created new organic cycle features during water storage, dependent on landscape types. Regional and hydrological conditions for organic carbon processing are presented for better reservoir management that can limit environmental disturbance and energy production costs.

1.2 Organic Matter in Water

Water, as an ideal natural solvent, contains two types of components: dissolved and not dissolved (particulate), and mineral and organic substances are found in each of these types. Next to natural organic matter (NOM), an anthropogenic organic matter (AOM) appears in waters from increased human activity, with a new, synthetic matter origin. Mineral proportions dominate over organic compounds in most of the water bodies, with the exception of dystrophic waters [6].

Organic matter in water is a mixture of organic carbon compounds, produced in the catchment soils, or terrigenous plants (allochthonous), or within water ecosystems (autochthonous). Hydrological situation, time, season and organic matter origin determine the relationship between dissolved (DOC) and particulate organic matter (POC); thus, the ratio DOC/POC, with a mean of 10:1, varies over a large range [6].

Humification of the organic matter of different origins is the most common known process, when humic and fulvic acids (HA and FA, respectively) are formed biochemically and with the active participation of bacteria, fungi, and also invertebrate organisms [7]. HA and FA, as new organic structures, together with detritus and products of incomplete decomposition, create a rich resource for autotrophs and heterotrophs [8]. Humus substance (HS) interactions with water plants and microorganisms are various, determined by humus acid concentrations, origin, pH, and mineral compound richness [8, 9]. HS are important chemical stressors in water ecosystems with effective impact on water color and photic zone density [6, 8]. Also, HS has a strong relationship with acidity and pH in the case of low water conductivity [7]. Humus substances account for 60–80% of dissolved organic carbon (DOC) [6, 7]. The presence of higher HS concentrations in waters affects the nitrogen and phosphorus cycles, because both elements are naturally associated with organic matter. In natural conditions, the increase of total organic carbon (TOC) concentration in water is connected with increasing N and P resources, primarily in particulate forms in rivers. Each high flow of the river delivers a new flux of organic matter to the reservoirs, which can become a driver of water eutrophication depending on reservoir volume and basin type. Increase of TOC concentrations often causes an increase of the easily assimilated energy pool for heterotrophs and mixotrophs, but reduces the rate of primary productivity in aquatic ecosystems [8].

1.3 Water Reservoirs in Poland

Surface water accumulation by man has had a long tradition in Poland territories, and reservoir functions change over time, according to human needs, possibilities, and technological progress. River water storage in small reservoirs was first connected with the existence of watermills, and the oldest one, from 1071, was documented in Zgorzelec on the Oder River. Also in the eleventh century the oldest Polish fishpond was built in Lower Silesia in Southwest Poland. These functions were dominant for water reservoirs up to the year 19 (XIX), when the water wheel was superseded by the water turbine for electric energy production. Hydropower reservoirs began energy production for industry and private use and at first were located in the western part of Poland. The oldest water reservoirs still functioning in recent times were built in the first years of the twentieth century as hydropower reservoirs on the Rivers Bóbr and Kwisa (Table 1), as are most of the largest Polish reservoirs. After the high

Table 1 Morphological features of the largest dam reservoirs in Poland

No.	Name	River	Year	Basin 10^3 km^2	Volume 10^6 m^3	Area km^2	Depth [m] Max	Depth [m] Mean	Stratification	*T*
1	Besko	Wisłok	1978	0.21	13.7	1.3	25.0	10.5	+	60
2	Brody Iłżeckie	Kamienna	1965	0.62	7.6	1.9	8.1	4.0	–	22
3	Bukówka	Bóbr	1987	0.06	16.8	2.0	22.4	8.4	+	194
4	Chańcza	Staszowska	1985	0.47	24.2	4.7	12.8	5.1	–	218
5	Cieszanowice	Luciąża	1998	0.08	9.1	2.6	10.4	3.5	–	106
6	Czaniec	Soła	1967	1.15	1.3	0.5	6.5	2.8	–	1
7	Czchów	Dunajec	1949	5.32	12.0	3.4	9.5	3.5	–	1.3
8	Czorsztyn	Dunajec	1997	1.20	231.9	12.3	54.5	18.9	+	116
9	Dobczyce	Raba	1986	0.77	141.7	10.7	27.9	13.2	+	146
10	Dobromierz	Strzegomka	1987	0.08	11.4	1.1	26.7	10.4	+	113
11	*Domaniów*	*Radomka*	*2001*	*0.74*	*14.4*	*5.0*	*8.6*	*2.9*	–	*31*
12	Dzierżno Małe	Drama	1938	0.18	12.6	1.7	13.1	7.4	+	24
13	*Goczałkowice*	*Wisła*	*1956*	*0.43*	*161.3*	*32.0*	*13.0*	*5.0*	–	*80*
14	*Jeziorsko*	*Warta*	*1986*	*8.39*	*202.0*	*42.3*	*11.5*	*4.8*	–	*56*
15	Klimkówka	Ropa	1994	0.18	42.6	3.1	33.3	13.7	+	148
16	Kozłowa Góra	Brynica	1939	0.14	17.6	5.8	6.5	3.0	–	307
17	Leśna	Kwisa	1907	0.29	16.8	1.4	35.8	12.0	+	38
18	Lubachów	Bystrzyca	1917	0.15	8.0	0.5	38.0	16.0	+	55
19	Łąka	Pszczynka	1986	0.17	11.2	3.5	6.9	3.2	–	80
20	Mietków	Bystrzyca	1986	0.72	71.9	9.1	15.3	7.9	+	128
21	Niedów	Witka	1962	0.32	4.9	1.9	12.5	2.6	–	13
22	Nielisz	Wieprz	2008	1.19	28.5	9.9	8.6	2.9	–	107
23	Nysa	Nysa Kłodz.	1971	3.27	124.7	20.7	13.3	6.0	–	59
24	Otmuchów	Nysa Kłodz.	1933	2.36	130.5	20.6	18.4	6.3	–	61
25	Pilchowice	Bóbr	1912	1.21	50.0	2.4	46.7	20.8	+	37

26	Pławniowice	P. Toszecki	1975	0.12	29.2	2.4	2.2	12.2	+	281
27	Poraj	Warta	1978	0.39	20.8	5.1	12.0	4.1	–	97
28	Porąbka	Soła	1936	1.10	27.2	3.3	21.2	8.2	+	22
29	Przeczyce	Przemsza	1963	0.30	20.4	4.7	12.5	4.3	–	109
30	Rożnów	Dunajec	1942	4.86	159.3	16.0	31.5	10.0	+	31
31	Rybnik	Ruda	1972	0.31	23.5	4.6	11.8	5.1	–	76
32	Rzeszów	Wisłok	1973	2.00	1.8	0.7	10.0	2.6	–	0.6
33	*Siemianówka*	*Narew*	*1991*	*1.05*	*79.5*	*32.5*	*9.2*	*2.4*	–	*198*
34	Słup	Nysa Szal.	1978	0.38	38.7	4.9	19.1	7.9	+	22
35	Solina	San	1968	1.19	472.4	22.0	60.0	21.5	+	299
36	Sosnówka	Czerwonka	2002	0.05	14.0	1.8	18.0	7.8	+	162
37	Sromowce W.	Dunajec	1994	1.30	6.4	0.9	8.5	7.1	+	3
38	Sulejów	Pilica	1973	4.90	84.3	23.8	11.3	3.5	+	38
39	Topola	Nysa Kłodz.	2003	2.14	26.5	3.4	7.8	7.8	+	Nd
40	Tresna	Soła	1967	1.03	96.1	9.6	23.8	10.0	+	90
41	*Turawa*	*Mała Panew*	*1938*	*1.42*	*106.2*	*20.8*	*13.6*	*5.1*	–	*115*
42	Wióry	Świślina	2007	0.36	35.0	4.1	23.4	8.5	+	Nd
43	Wisła-Czarne	Mała Wisła	1973	0.03	4.9	0.4	34.0	12.3	+	Nd
44	*Włocławek*	*Wisła*	*1970*	*168.9*	*453.6*	*75.0*	*12.7*	*6.0*	–	*4.5*
45	*Zegrzyński*	*Narew*	*1963*	*69.6*	*96.0*	*33.0*	*7.0*	*2.9*	–	*8*
46	Zemborzyce	Bystrzyca	1974	0.73	6.3	2.8	7.0	2.3	–	26
47	Złotniki	Kwisa	1924	0.29	12.1	1.2	27.5	10.1	+	27

T, mean water retention (days). *Italics* indicate lowland reservoirs floods in the Vistula River basin in 1920 and 1930, the next new reservoirs had an important role in limiting flood effects. The next function of reservoirs was an accumulation of potable water and for communal use for developing agglomerations in southern Poland. Water shortage in the second part of the twentieth century prompted the construction of the next large water reservoir on the main tributaries of the Vistula River, and just like on the river itself near Włocławek in the lower part of the river course. A long-time water reservoir began to become multifunctional wherein recreation use and fish farming were also significant. Among Polish water reservoirs, some of them serve for retention of mine waters, for transport, or to improve waterway quality.

In the last years in the twentieth century, nearly 140 water dam reservoirs with a total capacity greater than 10^6 m^3 and total area near 500 km^2 existed in Poland [10]. Most of these are located in the Sudety and Karpaty Mountains or in sub-mountain regions in the southern part of Poland. Only a few water reservoirs were built on lowland or upland rivers, but there are a significant number of reservoirs in the Silesia region. In the Silesia Upland an "anthropogenic Lakeland" has formed as a result of long-term surface mining of minerals, with permanent development of post-mining depression, affected by deep coal mining [11]. Around 20% of Polish reservoirs were built before the year 1945, mainly in the Sudety Mountains, but larger ones are situated in the Vistula River basin.

The dams of Polish dam reservoirs are not very high, mostly in the range of 20–30 m, but the highest are 54.5 m (Czorsztyn Reservoir on Dunajec River) and 60 m (Solina Reservoir on San River). The largest Polish water reservoirs have an area of 30–40 km^2, and only the Włocławek Reservoir on Vistula River exceeds 70 km^2. The ten largest Polish reservoirs have a capacity greater than 0.1 km^3 and only two have storage for 0.45–0.47 km^3 water (Włocławek Reservoir and Solina Reservoir) (Table 1). Reservoir morphology depends on geographic location, because mountain reservoirs are deep, with maximal depth more than 40–50 m, summer thermic stratification of the water column, and intensive sedimentation of sand and gravels. Lowland reservoirs, such as Jeziorsko, Siemianówka, and Turawa, are shallow, polymictic, with large area changes during a year, with sandy silt deposits and local bays with wetlands development. Some reservoirs, such as Goczałkowice, Nysa, and Otmuchów, have a morphology typical for lowland reservoirs, but their hydrology is connected with the mountain area.

Reservoir location on the cascade system is a specific of selected mountain valleys, formed by reservoirs Porąbka–Żywieckie–Czaniec on the Soła River (a Vistula River tributary) and another one by Topola–Otmuchów–Nysa on the Nysa Kłodzka River (Oder River tributary). Four reservoirs are situated on the Dunajec River course with a total capacity greater higher than 0.4 km^3 (nos. 7, 8, 30, and 37 in Table 1).

High multiannual variability of precipitation and available water for reservoir retention is observed in Polish territory [11, 12]. Therefore, the water volume in dam reservoirs is characterized by significant long-term variability (Fig. 1). In the past 9 years (2009–2017), extreme precipitation was noted in 2010 (including a flood in the Vistula River basin) and extremely dry years in 2012 and 2016. The variability of

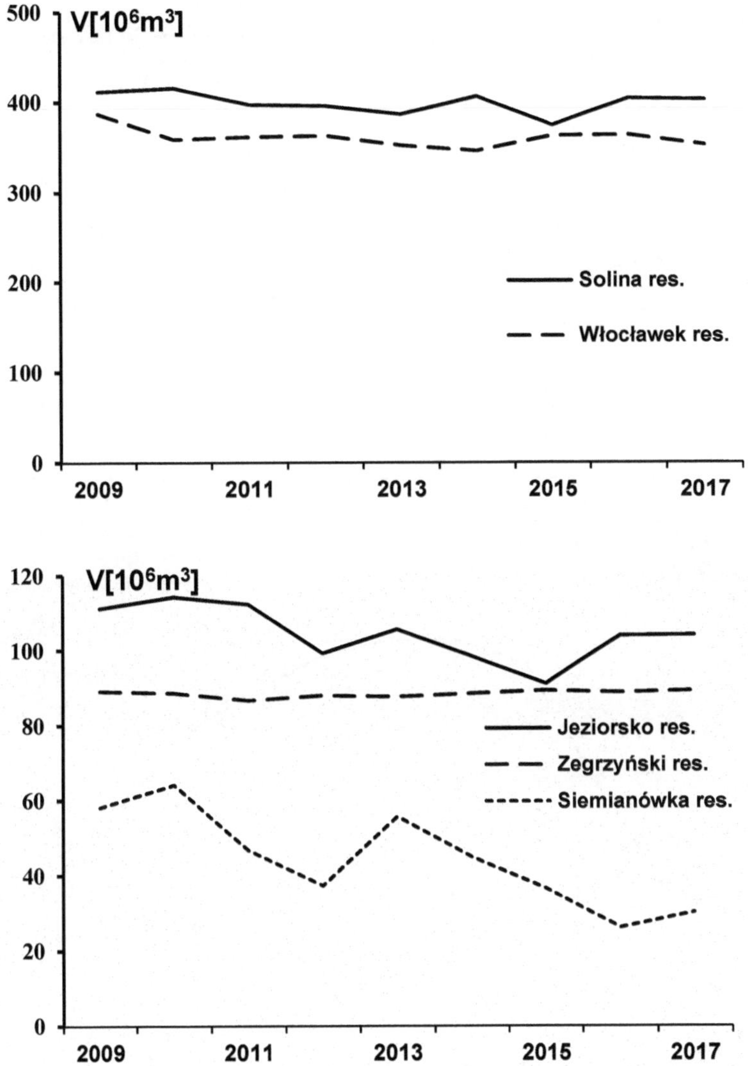

Fig. 1 Changes of mean annual water retention in the years 2009–2017 in largest (*upper panel*) and lowland (*lower panel*) reservoirs

atmospheric water supply is most visible in smaller reservoirs (<50 × 10^6 m^3) where the variation coefficient ranged from 20% to 80%. Lower variation of retained water was observed in the largest reservoirs, such as Włocławek or Solina reservoirs (Fig. 1). Low variability was also noted in reservoirs whose main function is energy production. In seasonal terms, the largest reservoir filling takes place in the spring period (April, May) and lowest level is most often in September or October (Fig. 2).

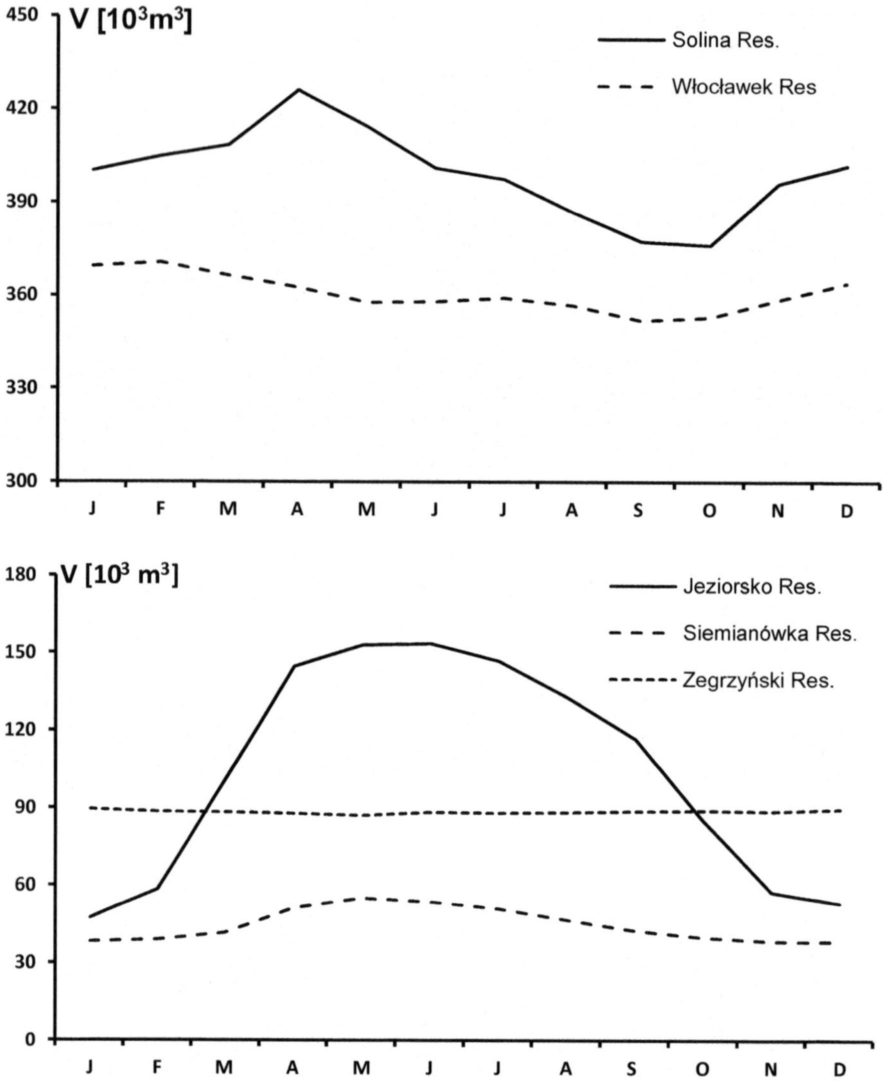

Fig. 2 Mean monthly water volume in the largest (*upper panel*) and lowland dam (*lower panel*) reservoirs in Poland in the years 2009–2017

Average water retention time is significantly differentiated in reservoir conditions, because some of them (such as reservoirs no. 6, 32, and 37 in Table 1) have a short retention time, less than 4–5 days. Among Polish reservoirs presented in Table 1 are those in which theoretical full water exchange is very long, 7–9 months. It is worth emphasizing that in the analyzed Polish reservoirs ice phenomena occur in wintertime, which is especially dangerous for reservoir dams. Extremely dangerous ice jams, affected by local floods, occur in the Włocławek Reservoir in the lower Vistula River.

2 Total Organic Carbon in Reservoirs of Poland

2.1 Data Collection and Methodological Comments

Data for this paper are a part of the data collection of the Polish National Monitoring Network and contain results from the years 2004–2015. I have collected all available data from the 47 dam reservoirs, which main features are presented in Table 1. Reservoirs with a minimum 1 year of sampling in the period 2006–2017 were selected. The number of TOC data for each reservoir varied greatly, from 4 to 120 results. In all, 1280 units of TOC data for reservoir water were analyzed, representing chiefly the largest reservoirs with capacities greater than 10^6 m^3 and located in the central and south parts of Poland. Analyzed reservoirs have about 3.2 km^3 with area of 473 km^2, which corresponds to 90% water volume cumulated in Poland. Dammed lakes functioning as a water dam reservoir and "small retention" objects were omitted in the present analysis. High water volume storage, 54%, exists in mountain or sub-mountain reservoirs; in lowlands this is 36%, and in uplands, 10%.

Sampling frequency was varied, depending on reservoir function; most of the reservoirs used for a potable water supply were sampled once per month. Reservoirs with an energy-dominated function were sampled 4 to 6 times per year, according to the National River Monitoring Protocol [13]. Samples were taken from the surface water layer (0–1 m depth) and located near the dam; on some occasions, samples were taken near the towers used for water supply. Automatic TOC measurements were made using the Polish Norm for direct analyses of TOC in water samples (not filtrated), using high-temperature fittings and measurement of CO_2 by a UV detector. Available data of water conductivity were also collected. Laboratories of the Voivodship Inspectorates of Environment Protection, where TOC analyses were provided, have a national certification of quality.

Original daily hydrological data for 25 reservoirs were provided from the national company "Polish Waters," the National Centre for Flood Protection in Warsaw. Averages of monthly water volume and outflow from reservoirs were calculated. Mean yearly water retention time for each reservoir was calculated as a ratio of the volume of water to the volume of water outflow in each calendar year. Mean TOC resources in the main hydrological regions were calculated on the basis of average TOC concentrations and available hydrological data for the period 2005–2017. When actual hydrological data were not available, the official normal water volume of the reservoir was used for calculation.

2.2 TOC Variability in Dam Reservoirs

Variability of TOC concentrations for each reservoir is presented in Table 2. The range of TOC concentrations from collected data varied from less than 1 mg C dm^{-3}

Table 2 Total organic carbon concentrations and water conductivity in water of Polish reservoirs

No.	Reservoirs	Years	*n*	TOC [mg C dm^{-3}] Min	Max	Mean	Conductivity [μS cm^{-1}]
1	Besko	2010–2015	36	2.5	8.2	4.9	294
2	Brody	2006–2015	29	3.6	54.0	11.5	334
3	Bukówka	2012, 2014	6	2.6	5.6	4.4	123
4	Chańcza	2006–2015	19	6.7	19.4	8.9	279
5	Cieszanowice	2011	4	12.1	13.9	13.3	252
6	Czaniec	2010–2015	60	2.0	4.3	2.8	196
7	Czchów	2007–2011	20	1.0	6.0	2.7	155
8	Czorsztyn	2008–2012	14	1.3	3.3	2.3	240
9	Dobczyce	2005–2014	119	2.3	5.0	3.4	266
10	Dobromierz	2011–2015	15	1.5	8.1	4.8	263
11	Domaniów	2004–2006	34	4.8	16.3	10.0	378
12	Dzierżno	2010–2015	9	4.7	7.7	5.8	526
13	Goczałkowice	2010–2016	63	2.6	6.8	4.7	201
14	Jeziorsko	2011, 2014	22	6.7	11.3	9.3	382
15	Klimkówka	2008–2012	14	1.2	5.0	3.7	202
16	Kozłowa G.	2010–2015	63	2.1	32.0	13.0	342
17	Leśna	2012, 2014	6	3.9	10.8	7.0	124
18	Lubachów	2011–2015	15	2.4	22.5	5.4	200
19	Łąka	2010–2015	9	7.6	12.0	10.3	646
20	Mietków	2012, 2014	6	3.8	7.2	5.7	435
21	Niedów	2011–2015	15	2.1	14.2	4.8	170
22	Nielisz	2012, 2015	13	4.5	7.2	5.8	463
23	Nysa	2009, 2012, 2015	9	2.1	21.0	7.5	254
24	Otmuchów	2009, 2012, 2016	9	4.1	18.2	8.1	263
25	Pilchowice	2012, 2014	6	1.8	8.6	6.2	196
26	Pławniowice	2010–2015	9	6.1	7.3	6.6	530
27	Poraj	2010–2015	9	6.2	12.0	8.6	403
28	Porąbka	2010–2015	12	2.2	4.8	2.7	201
29	Przeczyce	2010–2015	9	6.8	13.0	9.5	383
30	Rożnów	2007–2012	19	1.2	5.0	2.9	162
31	Rybnik	2010–2015	9	7.6	9.6	8.7	1210
32	Rzeszów	2005–2012	17	3.8	8.9	5.5	505
33	Siemianówka	2009–2014	47	11.5	46.6	18.1	286
34	Słup	2012, 2014	8	5.9	12.0	9.0	359
35	Solina	2010–2014	28	1.1	6.3	3.1	197
36	Sosnówka	2011–2015	15	2.8	9.3	5.0	98
37	Sromowce W.	2008–2011	12	1.2	3.7	2.3	278
38	Sulejów	2005–2015	120	3.3	26.5	9.5	319
39	Topola	2011, 2013	12	1.4	8.0	4.6	294
40	Tresna	2010–2015	18	1.9	9.2	3.5	193
41	Turawa	2009, 2011, 2014	9	7.0	21.4	11.4	299

(continued)

Table 2 (continued)

No.	Reservoirs	Years	*n*	TOC [mg C dm^{-3}]			Conductivity
				Min	Max	Mean	[μS cm^{-1}]
42	Wióry	2012–2015	10	4.5	7.1	5.8	390
43	Wisła–Czarne	2007–2014	111	0.9	7.0	3.3	74
44	Włocławek	2005–2014	89	5.1	14.0	8.6	620
45	Zegrzyński	2010–2014	89	4.9	20.6	11.7	596
46	Zemborzyce	2012–2013	8	3.9	11.6	7.8	360
47	Złotniki	2012	4	6.8	10.5	8.9	111

n, number of samples

up to 54 mg C dm^{-3}; the mean value (weight by volume) for a reservoir is 6.3 mg C dm^{-3}.

The Czorsztyn and Sromowce reservoirs had the lowest TOC levels; in the Siemianówka reservoir, TOC concentrations were ten times higher. Increase of TOC resources in the reservoirs was related to decrease of basin elevation (Fig. 3); thus, water dammed in lowland, upland, and mountain reservoirs had TOC mean (by weight) concentrations of 10.0, 8.1, and 36 mg C dm^{-3}, respectively.

These data confirm the role of soil carbon resources in the basin, because eroded soil material is the main source of the organic carbon load to surface waters [14]. Also, significant differences in reservoir TOC concentrations are derived from various characteristics of size and mass of river deposits in the upper part of each reservoir. The coarse material and that poor in organic content is buried in the upper reference part of mountain reservoirs and has a only small part in water enrichment by TOC in the reservoir during the active process of mineralization, caused by variations in dam water level. In upland, and especially in lowland, reservoirs that are more shallow, riverine material is distributed through most of the impoundment area and creates more available conditions to additional bottom TOC influx. Furthermore, lowland reservoirs with more varied flooded areas favor intensive macrophyte or other hygrophyte colonization and become "hotspots" of nutrients as well as organic compounds. Mineralized detritus of water plants and organic matter of shore-wetted soils also become sources of greenhouse gases (GHG) [15]. It is not without significance that lowland water retention usually occurs in shallow, polymictic water bodies, where multiple whole column mixing accelerates the microbiological mineralization and utilization of organic particles [4, 6], parallel to UV photobleaching of DOC components [7].

There are statistically significant relationships (negative) between TOC and mean reservoir depth (Fig. 4), but larger TOC variability (four- to sixfold) was observed in a group of shallow reservoirs than for deeper ones (two- to threefold). The same relationships, but for highly rheolimnic reservoirs, exist with a slightly lower concentration range.

Water retention time is important in the creation of TOC resources in the lowland, upland, and smaller mountain reservoirs, where statistically significant positive correlations are present (Fig. 5). Long water residence activates development of

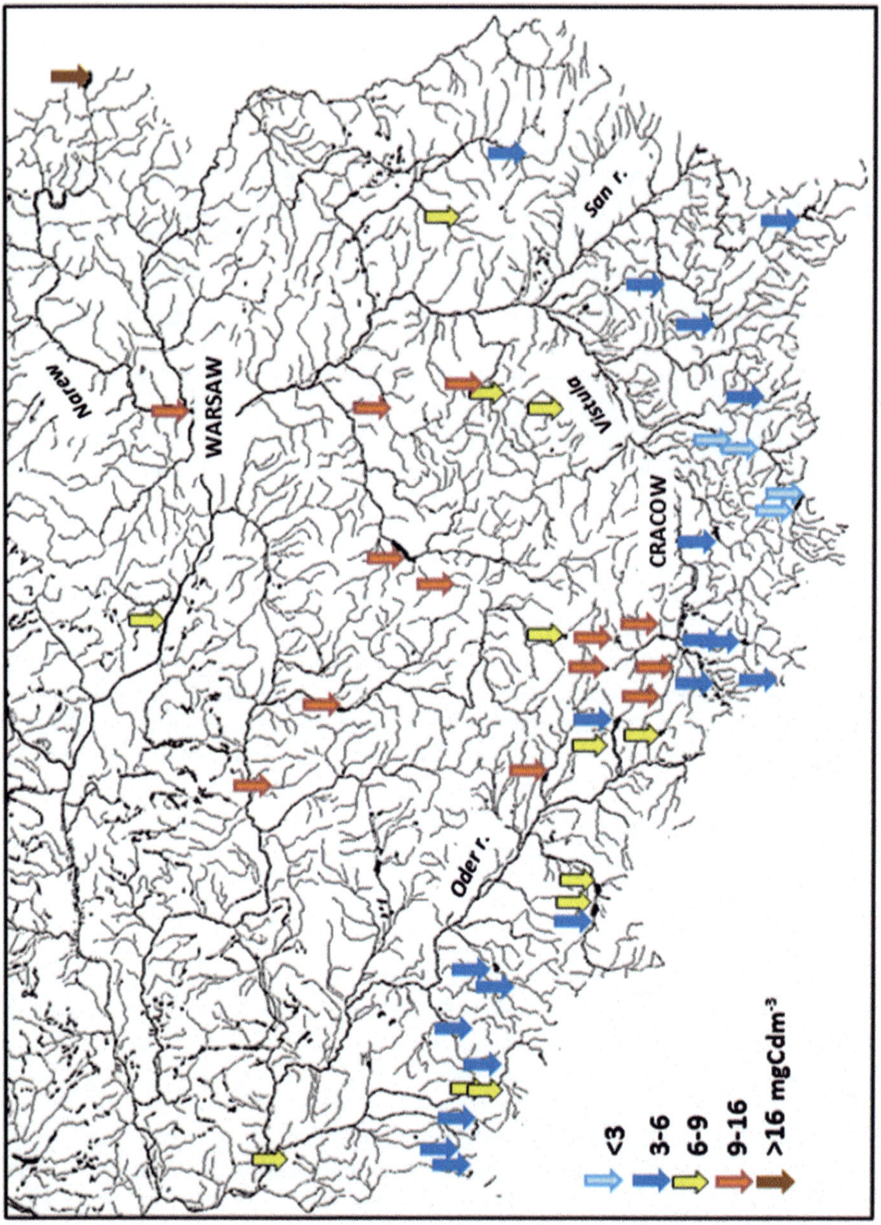

Fig. 3 Mean TOC concentrations in water of Polish reservoirs

bacteria and phytoplankton [6, 16, 17] and in consequence starts the eutrophication process gradually [18], resulting from an increase of organic matter resources not utilized by consumers in the food web [6]. An intensive allochthonous organic

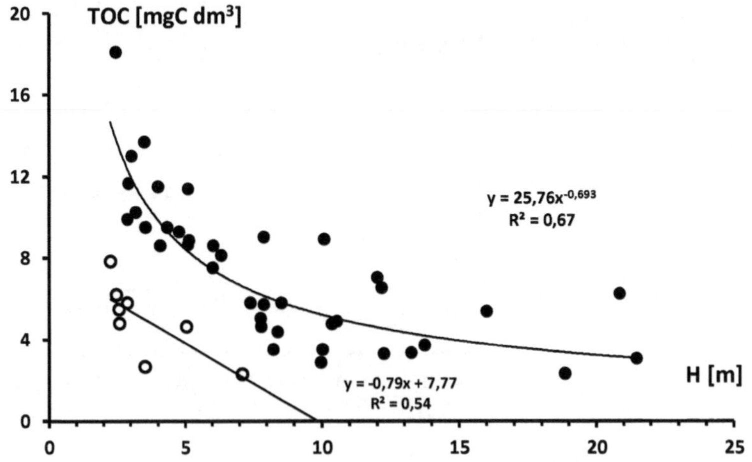

Fig. 4 Correlation between mean depth of reservoir and average TOC concentration in Polish water reservoirs; *open circles* indicate Low Silesia region

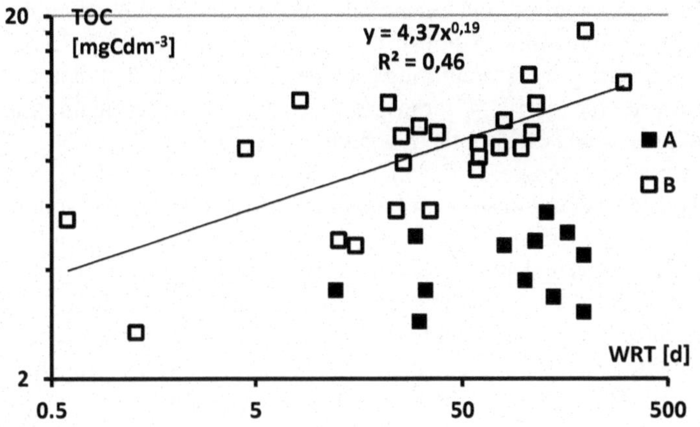

Fig. 5 Relationships between water residence time (days) and mean TOC concentrations in water of Polish reservoirs: (A) deep and large mountain reservoirs; (B) lowland, upland, and smaller mountain reservoirs

matter supply of water bodies, common with organic nitrogen and phosphorus, resulted in especially high *Cyanophyta* development. Because blue-green algae is mixotrophic, easily utilizes organic materials, and tolerates low water transparency, it becomes dominant in phytoplankton eutrophic lakes [6]. This type of eutrophic "organic" water, without additional anthropogenic water pollution, is named humeo-eutrophication [9].

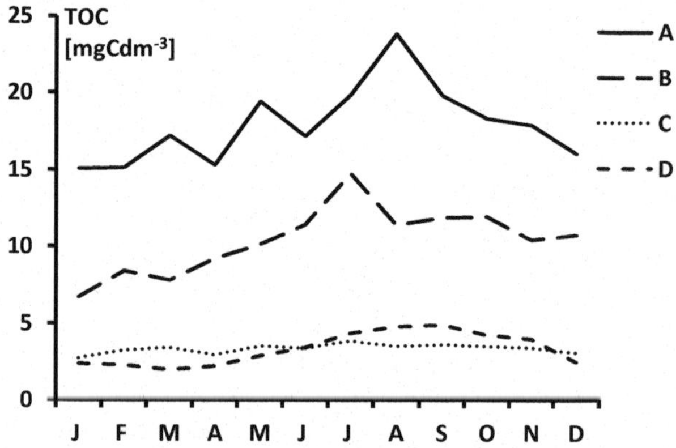

Fig. 6 Mean monthly TOC concentrations in selected reservoirs in the years 2009–2015: A, Siemianówka reservoir; B, Sulejów reservoir; C, Dobczyce reservoir; D, Wisła–Czarne reservoir

2.3 *TOC Seasonality and Multiannual Changes*

Reservoir ecosystems are very dynamic because of their strong dependence on the river hydrological regime and their main proposed function. A more objective analysis of TOC dynamics in Polish dam reservoirs is limited, because only in a few reservoirs is long-term monitoring provided. I have collected monthly data only from a few reservoirs: Wisła–Czarne and Dobczyce in the mountains and Sulejów, Włocławek, and Zegrzyński in the lowlands.

With increasing average TOC concentrations, seasonality is increased. Increase of water TOC concentrations is observed in the spring and summer months (Fig. 6). The spring period is affected by snow melting, and the summer months are associated with heavy rains and stormy season, locally inducing floods. It is noted that the spring water inflow most significantly increases TOC concentrations in lowland reservoirs, whereas in the mountains TOC increases during summer floods. A variation coefficient of TOC calculated for a few reservoirs with more frequent data indicated that mean variability variation is in the range of 22% to 55%, the highest coefficients being characteristic of basins with wetlands. Long-term data for the lowland Siemianówka reservoir showed that the organic load entering this type of reservoir with spring waters determined the resources of organic matter in reservoirs for the entire vegetation season [18].

The seasonal repetition of the TOC cycle in dam reservoirs is typical for semi-natural basins, as observed in the Wisła–Czarne reservoir with protected forest areas (Fig. 7). The greater variability of TOC resources in particular seasons was recorded in reservoirs with significant changes in basin management. Urban and agricultural areas become more reactive to rainwater supply than forests or wetlands; then, a rapid response of TOC export becomes significant [19, 20].

Despite the short period of research and analysis, characteristic elements of multiannual TOC variability in Polish reservoirs were noted. In the discussed period there were clear differences in TOC concentrations in reservoir waters under different hydroclimatic conditions. The highest TOC values were observed in all

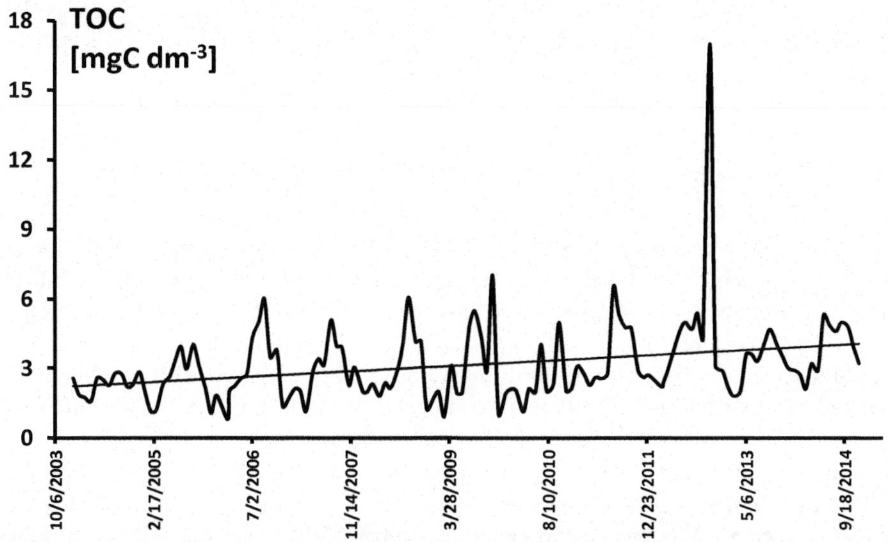

Fig. 7 Long-term variation of TOC in Wisła-Czarne reservoir in years 2003–2014

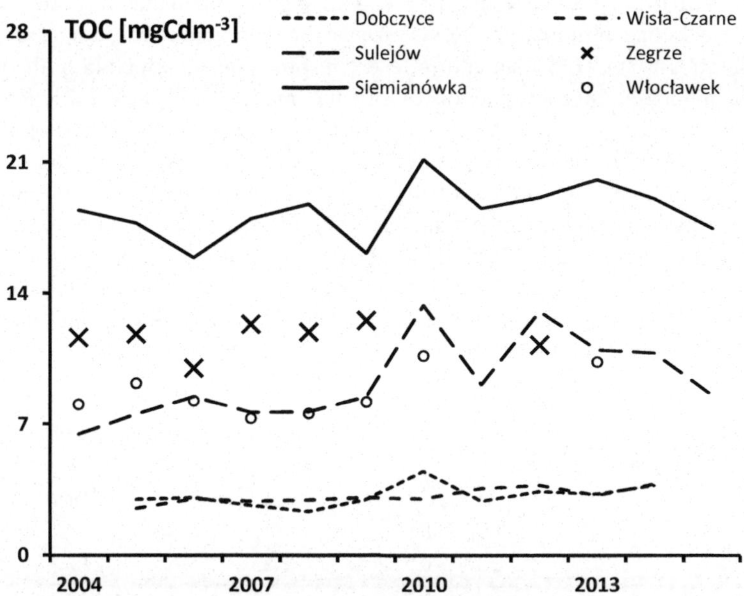

Fig. 8 Yearly mean TOC concentrations in selected Polish reservoirs

reservoirs in the year 2010 with high summer precipitation and floods (Fig. 8), when intensive soil organic carbon erosion takes place. In the dry year 2012, with low atmospheric water supply (precipitation less than 50% of multiannual mean), TOC concentrations were also low. Similar results for a drought period were presented earlier for rivers in northeast Poland [21].

2.4 Resources of Total Organic Carbon in Polish Reservoirs

Water dammed in Polish reservoirs of different regions had variable TOC mean (weight) concentrations (Table 3). Lowlands reservoirs have a TOC concentration four times higher than that observed in mountain reservoirs. Reservoirs located in the Odra River basin had significantly higher TOC concentrations than reservoirs in the Vistula River basin, 3.65 and 6.61 mg C dm^{-3}, respectively. This difference is related to the differences in the clay content in the weathered rocks, fertility, and organic carbon resources in the soils formed on them.

I have estimated that there is 1.29 t of organic carbon in the water of Polish reservoirs. In the waters of six dam reservoirs, that is, Jeziorsko, Siemianówka, Solina, Sulejów, Włocławek, and Zegrzyński, 60% of the total TOC resources of the water of Polish reservoirs is found. Additionally, nearly 40% of these resources was in the Włocławek reservoir. It should be noted that the indicated TOC resource concerns waters leaving the reservoirs and it must be assumed that they are greater in reality. TOC concentrations in the water near the dam resulted in a balance of the allochthonous basin TOC load and in-reservoir processes in which both the utilization and autochthonous production of organic matter take place [22]. Reservoir morphology, water retention time, and upstream water composition determine the real relationship between river, transitional, and lacustrine longitudinal zones usually highlighted in dam reservoirs. Globally, it is considered that in a reservoir carbon mineralization exceeds carbon fixation ($P < R$), next to the carbon burial processes [5]. Generally, longitudinal TOC concentration decrease was observed within the reservoir, but in some eutrophic reservoirs with large flooded areas TOC increase was present, as in Siemianówka reservoir [9, 18, 23], which confirms the significant role of flooded areas in the trophic state of reservoirs [5, 15, 24, 25].

Table 3 Regional differentiation of TOC and water resources in the investigated dam reservoirs in Poland

Region	n	Average ± SD [mg C dm^{-3}]	Mean (weight)	Water [%]	TOC
Mountains	4	2.75 ± 0.51	2.85	27.2	12.2
Sub-mountains	23	5.17 ± 1.61	4.68	30.4	22.4
–Vistula R. basin	10	3.91 ± 0.92	3.65	19.9	11.4
–Oder R. basin	13	6.18 ± 1.37	6.61	10.5	11.0
Uplands	11	8.52 ± 2.00	8.34	6.3	8.2
Lowlands	9	11.48 ± 2.88	10.00	36.2	57.2
Mean	47	6.97 ± 3.38	6.34		

Earlier evaluation of water TOC export along the Vistula River clearly showed the large role of Włocławek reservoir in elimination of organic carbon carried from Poland to the Baltic Sea [26], where approximately 20% of annual TOC export is reduced.

3 Perspectives

This positive aspect of functioning Polish dam reservoirs in protection of the Baltic Sea is connected also with an active "hotspot" of greenhouse gas (GHG) emission, together with other Polish dam reservoirs. During global warming, from one aspect, increase of soil organic carbon erosion is forecast, contributing to a slow increase of the TOC pool in dammed waters. The first symptoms of that increase are being observed now in selected Polish reservoirs (Fig. 8). Also forecast is an increase of air temperature that will generate additional emission of greenhouse gases from water reservoirs, while increasing the costs of electricity production. As recently globally calculated, the carbon footprint of hydropower is greater than previously assumed and in global warming this effect will increase [27]. Moreover, an increase of basin TOC flux to reservoirs affected by climatic and hydrological changes will cause an increase of water utilization costs, in consequence of changes in water treatment technologies. A more intensive organic carbon monitoring network is needed, covering water, sediments, and GHG emissions, in diverse Polish dam reservoir ecosystems, especially where TOC resources are highest.

The present evaluation of TOC resources in Polish dam reservoirs was closely related to other chemical and biological parameters of water collected by ecological monitoring from the years 2010–2015 and provided by the Chief Inspectorate of Environmental Protection. Average TOC concentrations in reservoirs were positively correlated with water conductivity ($r = 0.58, p < 0.05$). Among the biological parameters there was an inversely proportional relationship between TOC in waters and values of the phytoplankton index (IFPL) ($r = 0.79$, $p < 0.001$). A less strong correlation was observed between TOC and the benthic diatom index (IOPL) ($r = 0.51$, $p < 0.05$). Established significant correlations confirm the possibility of a wider use of the TOC parameter in assessment of the status or ecological potential of dam reservoir ecosystems.

This first evaluation of TOC assets conducted in Polish dam reservoirs offers conclusions for location of the next planned water reservoirs. The reservoirs should be located outside the lowland areas, where there are vast areas of peat soils abundantly supplying water to the TOC. The same problem of high TOC concentrations in lowland dam reservoirs exists in large reservoirs in the Volga River basin [28] or in dammed rivers in Lithuania, Latvia, or Estonia.

Most of the lowland Polish dam reservoirs have low water quality and natural high TOC resources, accelerating the eutrophication process with late summer heavy blooms of toxic *Cyanophyta*, examples of which are the Sulejów, Siemianówka, or Zemborzyce reservoirs [18, 29]. Future Polish reservoirs should be placed on rivers where relationships between the catchment areas over a reservoir area have rates higher than 1000 (data not presented), ensuring high water exchange, and preventing intensive phytoplankton development, especially of *Cyanobacteria.*

References

1. Lehner B et al (2011) High-resolution mapping of the world's reservoirs and dams for sustainable river-flow management. Front Ecol Environ 9:494–502
2. Vörösmarty CJ, Sharma KP, Fekete BM, Copeland AH, Holden J, Marble J, Lough A (1997) The storage and aging of continental runoff in large reservoir systems of the world. Ambio 26 (4):210–219
3. Chao BF, Wu YH, Li YS (2008) Impact of artificial reservoir water impoundment on global sea level. Science 320(5873):212–214. https://doi.org/10.1126/science.1154580
4. Hayes NM, Deemer BR, Corman JR, Razavi NR, Strock KE (2017) Key differences between lakes and reservoirs modify climate signals: a case for a new conceptual model. Limnol Oceanogr Lett 2:47–62
5. Maavara T, Lauerwald R, Regnier P, Van Cappellen P (2017) Global perturbation of organic carbon cycling by river damming. Nat Commun 8:15347. https://doi.org/10.1038/ncomms15347
6. Wetzel RG (2001) Limnology lake and reservoir ecosystems. Academic Press, San Diego
7. Thurman LJ, von Wachenfeldt E (2009) Interactions of dissolved organic matter and humic substances. In: Likens GE (ed) Biogeochemistry of inland waters. Academic/Elsevier, Oxford, pp 464–470
8. Williamson CE, Morris DP, Pace ML, Olson OG (1999) Dissolved organic carbon and nutrients as regulators of lake ecosystems: resurrection of a more integrated paradigm. Limnol Oceanogr 44:795–803
9. Górniak A (1996) Humic substances and their role in freshwater ecosystems functioning. Diss. Univ. Vars. 448, Białystok, p 151 (in Polish)
10. Environment in 2017 (2017) Polish Statistical Office, Warszawa
11. Jokiel P et al (eds) (2017) Hydrology of Poland. Polish Scientific Press Office, Warsaw, p 342 (in Polish)
12. Gutry-Korycka M, Sadurski A, Kundzewicz ZW, Pociask-Karteczka J, Skrzypczyk L (2014) Water resources and their use. Nauka 1:77–98 (in Polish)
13. Błachuta J, Picińska-Fałtynowicz J (2010) Methodological guidance for conducting of monitoring and evaluation of ecological potential of reservoirs in Poland. IMGW, Wroclaw, p 58 (in Polish)
14. Lal R (2003) Soil erosion and the global carbon budget. Environ Int 29:437–450
15. Deemer BR, Harrison JA, Li S, Beaulieu JJ, Delsontro T, Barros EN, Bezerra Neto JF, Powers SM, Dos Santos MA, Arie Vonk J (2016) Greenhouse gas emissions from reservoir water surfaces: a new global synthesis. Bioscience 66:949–964
16. Tranvik LJ et al (2009) Lakes and reservoirs as regulators of carbon cycling and climate. Limnol Oceanogr 54:2298–2314. https://doi.org/10.4319/lo.2009.54.6_part_2.2298
17. Cole JJ, Prairie YT, Caraco NF, McDowell WH, Tranvik LJ, Striegl RG, Duarte CM, Kortelainen P, Downing JA, Middelburg J, Melack J (2007) Plumbing the global carbon cycle: integrating inland waters into the terrestrial carbon budget. Ecosystems 10:111–184
18. Górniak A, Zieliński P, Jekatierynczuk-Rudczyk E, Grabowska M, Suchowolec T (2002) The role of dissolved organic carbon in a shallow lowland reservoir ecosystem – a long-term study. Acta Hydrochim Hydrobiol 30:179–189
19. Stanley EH, Powers SM, Lotting NR, Buffam I, Crawford JT (2012) Contemporary changes in dissolved organic carbon (DOC) in human-dominated rivers: is there a role for DOC management? Freshw Biol 57(Suppl 1):26–42

20. Ford WI, Fox JF (2014) Model of particulate organic carbon transport in an agriculturally impacted stream. Hydrol Process 28:662–675
21. Zieliński P, Górniak A, Piekarski MK (2009) The effect of hydrological drought on chemical quality of dissolved organic carbon concentrations in lowland rivers. Pol J Ecol 57:217–227
22. Thornton KW, Kimmel BL, Payne FE (eds) (1990) Reservoir limnology: ecological perspectives. Wiley, New York
23. Górniak A (2017) Total organic carbon in the waters of Vistula and Niemen basin rivers. Gosp Wodn 1:19–24 (in Polish)
24. Barros N, Cole JJ, Tranvik LJ, Prairie YT, Bastviken D, del Giorgio PA, Roland F, Huszar VLM (2011) Carbon emission from hydroelectric reservoirs linked to reservoir age and latitude. Nat Geosci 4:593–596
25. Catalan N, Marce R, Kothawala DN, Tranvik LJ (2016) Organic carbon decomposition rates controlled by water retention time across inland waters. Nat Geosci 9:501–504
26. Górniak A (2017) Spatial and temporal patterns of total organic carbon along the Vistula River course (Central Europe). Appl Geochem 87:93–101
27. Scherer L, Pfister S (2016) Hydropower's biogenic carbon footprint. PLoS One 11(9): e0161947. https://doi.org/10.1371/journal.pone.0161947
28. Mineeva NM, Bikbulatova EM (2008) The contribution of primary production to the overall organic matter pool in Volga reservoirs. Water Res 35:562–569
29. Grabowska M, Mazur-Marzec H (2016) The influence of hydrological conditions on phytoplankton community structure and cyanopeptide concentration in dammed lowland river. Environ Monit Assess 188:488. https://doi.org/10.1007/s10661-016-5506-x

The Great Masurian Lakes: Hydrological Regime and Summer Phytoplankton

Agnieszka Napiórkowska-Krzebietke, Bogusław Zdanowski, Elżbieta Bajkiewicz-Grabowska, Konrad Stawecki, and Bartosz Czarnecki

Contents

Abstract The lakes of the Great Masurian Lakes System create the largest lake complex in Poland hydrologically divided into three basins – Northern, Central, and Southern. The water flow direction depends on the position of the watershed between Vistula and Pregoła Rivers which can vary depending on the water level and the magnitude of the water outflow from the Pisa and Węgorapa Rivers. The studies aimed to analyze the changes in the hydrological regime and the ecological state of these lakes in 2008 and 2010–2012. In June 2010, the bifurcation zone occurred in the central part of the system due to the high quantity of precipitation. A positive heterograde oxygen curve in summer was in less eutrophic lakes, while a negative heterograde oxygen curve in more eutrophic lakes and a clinograde oxygen curve in highly eutrophic lakes. Phosphorus content was typical of medium or highly eutrophic lakes; therefore classifications based on physicochemical elements indicated below good ecological potential and status of all lakes. The least intensive phytoplankton growth was in the Northern Basin whereas the most intensive in the Southern Basin. Biological assessments based on phytoplankton indicated maximum potential in two lakes and good potential and status in five lakes, whereas the

A. Napiórkowska-Krzebietke (✉) · B. Zdanowski · K. Stawecki · B. Czarnecki
Department of Ichthyology, Hydrobiology and Aquatic Ecology, Inland Fisheries Institute in Olsztyn, Olsztyn, Poland
e-mail: a.napiorkowska-krzebietke@infish.com.pl

E. Bajkiewicz-Grabowska
Faculty of Oceanography and Geography, Institute of Geography, University of Gdańsk, Gdańsk, Poland

E. Korzeniewska, M. Harnisz (eds.), *Polish River Basins and Lakes – Part I*,
The Handbook of Environmental Chemistry 86,
https://doi.org/10.1007/978-3-030-12123-5_11

remaining lakes had moderate, poor, and bad potential and status. In 2010, when oxygen conditions improved significantly, the classification based on PMPL indicated significant deterioration up to bad ecological potential and status, in the lakes within the bifurcation zone and in several lakes in the Southern and Northern Basins.

Keywords Bifurcation zone · Ecological classification · Phytoplankton · Stratification · Trophy · Watershed

1 Introduction

The Great Masurian Lakes System comprises the largest lake complex in Poland and on the North European Plain. These lakes lie in the Masurian Valley [1], and the central valley is occupied by the largest among them, which are Śniardwy Lake (109.7 km^2 surface area) and the Mamry lakes complex (102.3 km^2 surface area). The Great Masurian Lakes catchment basin has a surface area of approximately 3,200 km^2 [2], of which approximately 20% is in the Węgorapa River Basin and approximately 80% is in the Pisa River Basin (Fig. 1). The Great Masurian Lakes occupy approximately 9.7% of the surface area of this catchment, and they store nearly 2.5 km^3 water, which is 12.7% of the lake water reserves in Poland [3].

The Great Masurian Lakes System includes basins with varying surface water areas (from 19 to 11,340 ha) and maximum depths (from approximately 3 to 51 m) (Table 1). The shallowest are lakes Szymon, Kotek, and Kirsajty, while the deepest are Tałty-Ryńskie, Bełdany, and Mamry Północne (Fig. 2). The smaller, shallower lakes and Śniardwy Lake, the largest, (Fig. 3) are mixed to the bottom, while the deeper lakes are usually fully stratified dimictic lakes. The direction in which water masses move determines the biological and trophic conditions in individual lake systems [4, 5].

Based on results of physicochemical and biological tests conducted in 2008 and in the 2010–2012 period, changes in the hydrological regime, physicochemical variables, and phytoplankton in the lakes of the Great Masurian Lakes System were analyzed. Physicochemical and biological data collected by the Department of Ichthyology, Hydrobiology, and Aquatic Ecology, Inland Fisheries Institute in Olsztyn, and hydrological information from data collection stations of the Institute of Meteorology and Water Management–National Research Institute were used for these analyses. The lakes in this system were examined at the height of the summer stagnation period in August.

Sampling sites were designated each time in the deepest part of the lakes. The in situ measurements taken were as follows: water temperature and oxygen content in vertical profiles from the surface to the bottom at intervals of 1 m with a YSI ProODO optical dissolved oxygen meter, water electrolytic conductivity and pH with an OAKTON pH/Con model 300 meter, and water transparency with a Secchi disk. Water samples for physicochemical analyses were collected with a Toń-2 water sampling dipper from the surface water layer, but water samples for biological

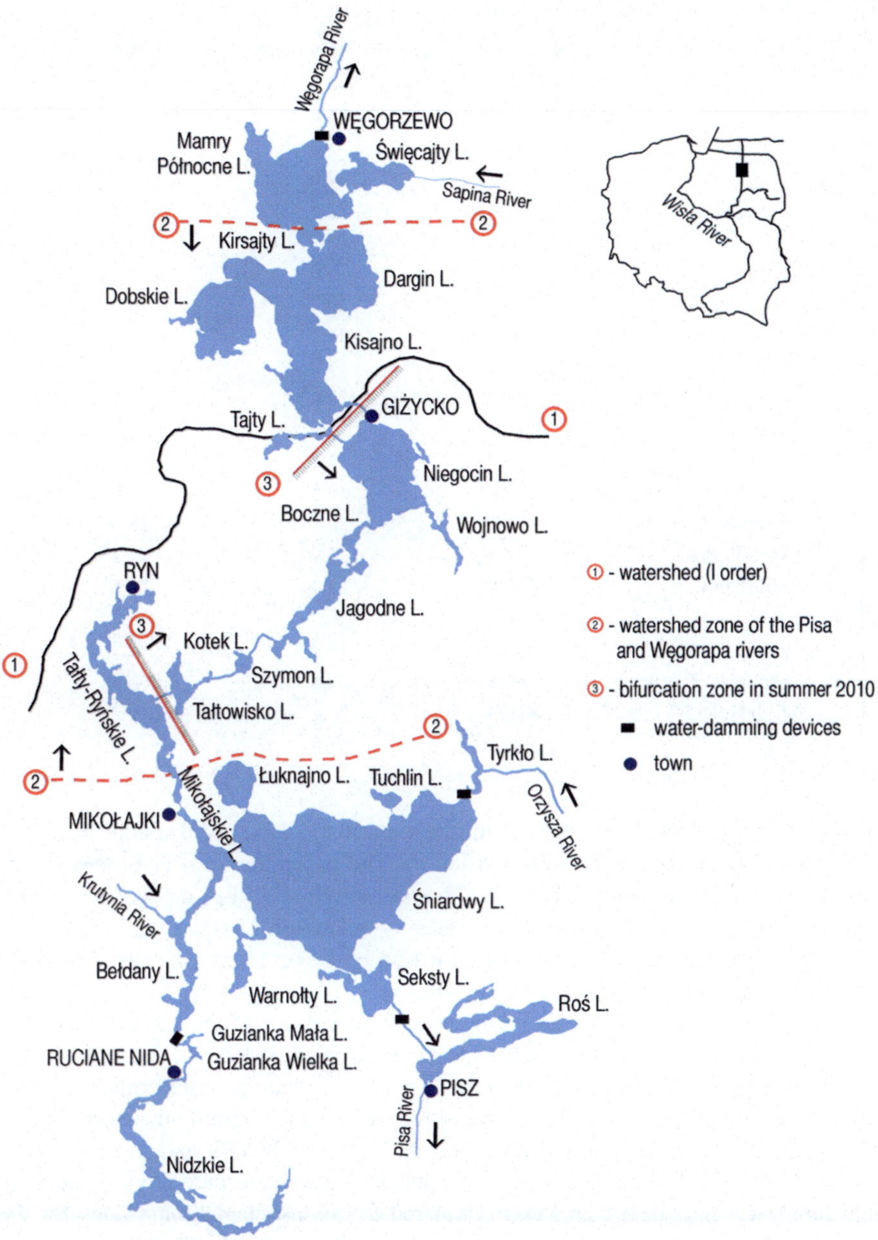

Fig. 1 The Great Masurian Lakes System (northeastern Poland)

Table 1 Morphometric characteristics of the Great Masurian Lakes

Lake		Surface area (ha)	Max depth (m)	Mean depth (m)
↑	Mamry Północne	2,504.0	43.8	11.7
	Święcajty	968.4	28.0	8.7
	Kirsajty	207.0	5.8	3.2
	Dargin	3,030.0	37.6	10.6
	Dobskie	1,776.0	22.5	7.8
	Kisajno	1,896.0	25.0	8.4
↓	Tajty	265.1	34.0	7.5
	Niegocin	2,600.0	39.7	9.9
	Wojnowo	176.3	14.2	6.3
	Boczne	183.3	17.0	8.6
	Jagodne	942.7	37.4	8.7
	Szymon	154.0	2.9	1.1
	Kotek	19.1	3.2	1.6
	Tałtowisko	326.9	39.5	14.0
	Tałty-Ryńskie	1,831.2	50.8	13.6
	Mikołajskie	497.9	25.9	11.2
	Śniardwy	11,340.4	23.4	5.8
↑	Bełdany	940.6	46.0	10.0
	Guzianka Mała	36.8	13.3	2.7
	Guzianka Wielka	59.6	25.5	6.5
	Nidzkie	1,818.0	23.7	6.2
	Roś	1,888.0	31.8	11.4

analyses were collected from the epilimnion (stratified lakes) and from the surface to the bottom water layer (non-stratified lakes). The basic chemical parameters were determined with standard methods [6, 7], chlorophyll *a* (Chl *a*) content was determined with the Lorenzen method [8], and phytoplankton biomass was determined with the Utermöhl method [9] and the method based on measuring the volume of cells using verified patterns in reports [10].

In accordance with the Water Framework Directive, six of the lakes of the Great Masurian Lakes System that were examined (Wojnowo, Kotek, Guzianka Mała, Guzianka Wielka, Nidzkie, Roś) were categorized as natural water bodies (NWB), while the remaining lakes were classified as heavily modified water bodies (HMWB). The assessment of the ecological status of NWBs and the ecological potential of HMWBs was based on the summer values of selected physicochemical and biological parameters and was compared to quality class limit values for the averages for the entire growth season in accordance with the Regulation of the Minister of Environment of July 21, 2016, on classification methods for water bodies and environmental quality standards for priority substances. This assessment applies only to extreme summer situations.

Fig. 2 Mamry Północne Lake

2 Hydrology of the Great Masurian Lakes

The hydrological system of the Great Masurian Lakes in its current state originated in the nineteenth century, when a connection was dug between Lakes Śniardwy, Niegocin, and the Mamry lakes complex. This is divided into three parts, or basins. The Northern Basin system includes the Mamry lakes complex of Lakes Mamry Północne, Święcajty, Kirsajty, Dargin, Dobskie, and Kisajno. The Central Basin comprises Lakes Niegocin, Boczne, Wojnowo, Jagodne, and Tajty. The Southern Basin includes Lakes Szymon, Kotek, Tałtowisko, Tałty-Ryńskie, Mikołajskie, Bełdany, and Śniardwy.

Fig. 3 Śniardwy Lake

The system's main tributaries are the Sapina and Krutynia Rivers. The water supply for the lakes of the Northern Basin is from groundwater to a significant degree. The Northern Basin is drained by the Węgorapa River, while the Middle and Southern Basins are drained by the Pisa River [4, 11]. The Great Masurian Lakes System has a two-way outflow of water that is regulated by weirs at the outflow. The direction of water exchange between these waters depends on the position of the Vistula River and Pregoła River watershed. The ordinate water level in the system is maintained at 116.0 m above sea level. The location of the first-order watershed determines the distribution of waters between the Węgorapa and Pisa Basins (Fig. 1). This is described by Bajkiewicz-Grabowska [4, 12] and Dąbrowski [11, 13, 14] and can vary depending on the state of the water level in the system, and it depends on the magnitude of the water outflow from the Pisa and Węgorapa Rivers [15]. As a rule, the first-order watershed runs at the mouth of Giżycko Canal from Niegocin Lake. It can change within the borders of Dargin Lake to Mikołajskie Lake.

The surface area of the endorheic basin (bifurcation) when at an average water level can lie between Lakes Kisajno and Jagodne and, in extreme cases, between Lakes Kisajno and Mikołajskie [16] (Fig. 1). Water exchange among lakes in the bifurcation zone is very limited. Waters from above its northern border flow from the Great Masurian Lakes into the Węgorapa River, while those to the south of it flow into the Pisa River. The northern end of the linear bifurcation usually corresponds to

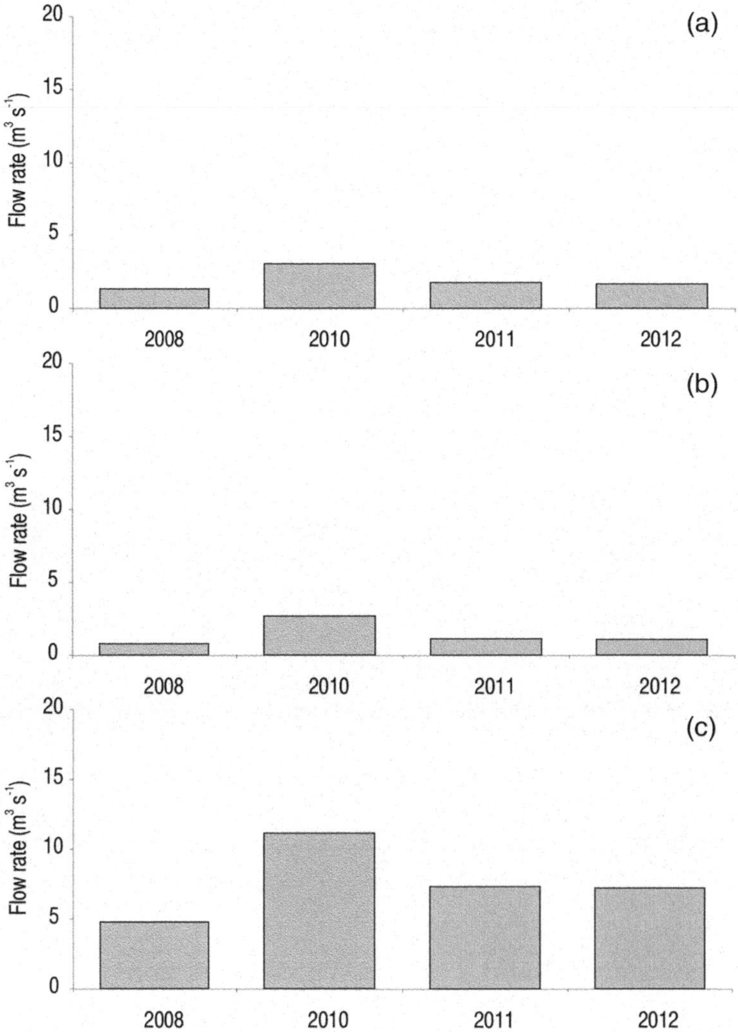

Fig. 4 The inflow of river waters to the Northern (**a**), Central (**b**), and Southern (**c**) Basins of the Great Masurian Lakes in June in 2008–2012

the highest water table level in the system and is identified with the position of the first-order watershed [2].

The position of the first-order watershed can change in different years depending on the quantity of precipitation. One such extreme situation was noted in 2010 from May to June. The most water was accumulated in the southern lakes (Fig. 4). The first-order watershed was then located in Mikołajskie Lake. Consequently, the most water flowed out via the Pisa River, while decidedly less flowed through the

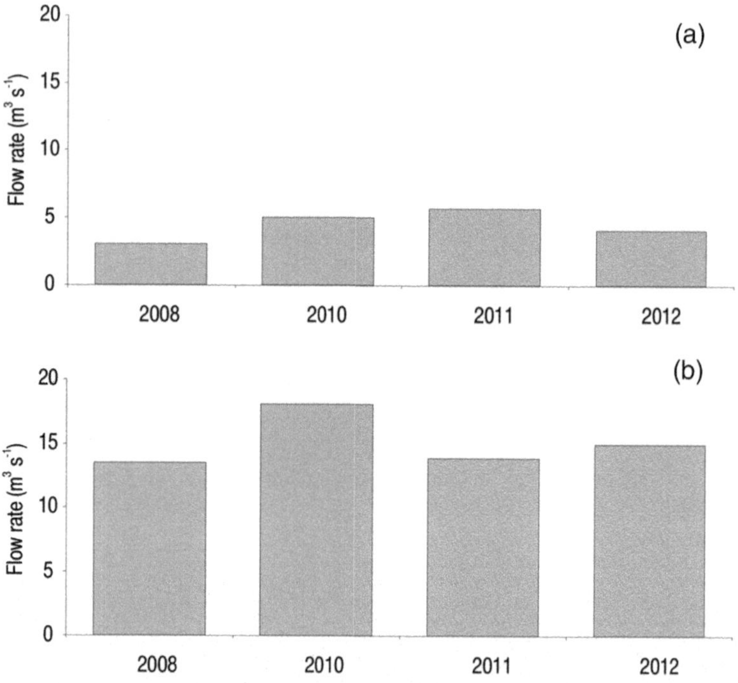

Fig. 5 Intensity of the river outflow from the Great Masurian Lakes on the Węgorapa (**a**) and Pisa (**b**) Rivers in June in 2008–2012

Węgorapa River (Fig. 5). The bifurcation zone that favors water retention was the central part of the system in June 2010 (Fig. 1).

3 Physicochemical and Trophic Water Conditions

The water surface layer temperatures of the Great Masurian Lakes fluctuated during the summer during the study period from approximately 20 to 26°C. The waters of Northern Basin were usually cooler, while those of the Southern Basin were warmer. The temperature of the near-bottom water layers in stratified lakes usually ranged from about 6 to 9°C (Figs. 6, 7, and 8). The warmest growth season was noted in 2010. The oxygen content in the summer in the epilimnion ranged from approximately 7 to 9 mg O_2 dm^{-3}, and the water oxygen saturation was from 70% to 110%.

The changes noted in water temperatures and oxygen contents in Śniardwy Lake were characteristic of strongly mixed lakes. A positive heterograde oxygen curve which describes the oxygen content distribution in the vertical profile in summer in less eutrophic lakes occurred primarily in Mamry Północne Lake (Fig. 6). A negative heterograde oxygen curve described more eutrophic lakes. In other lakes the changes in water temperatures and oxygen contents were characterized by clinograde typical

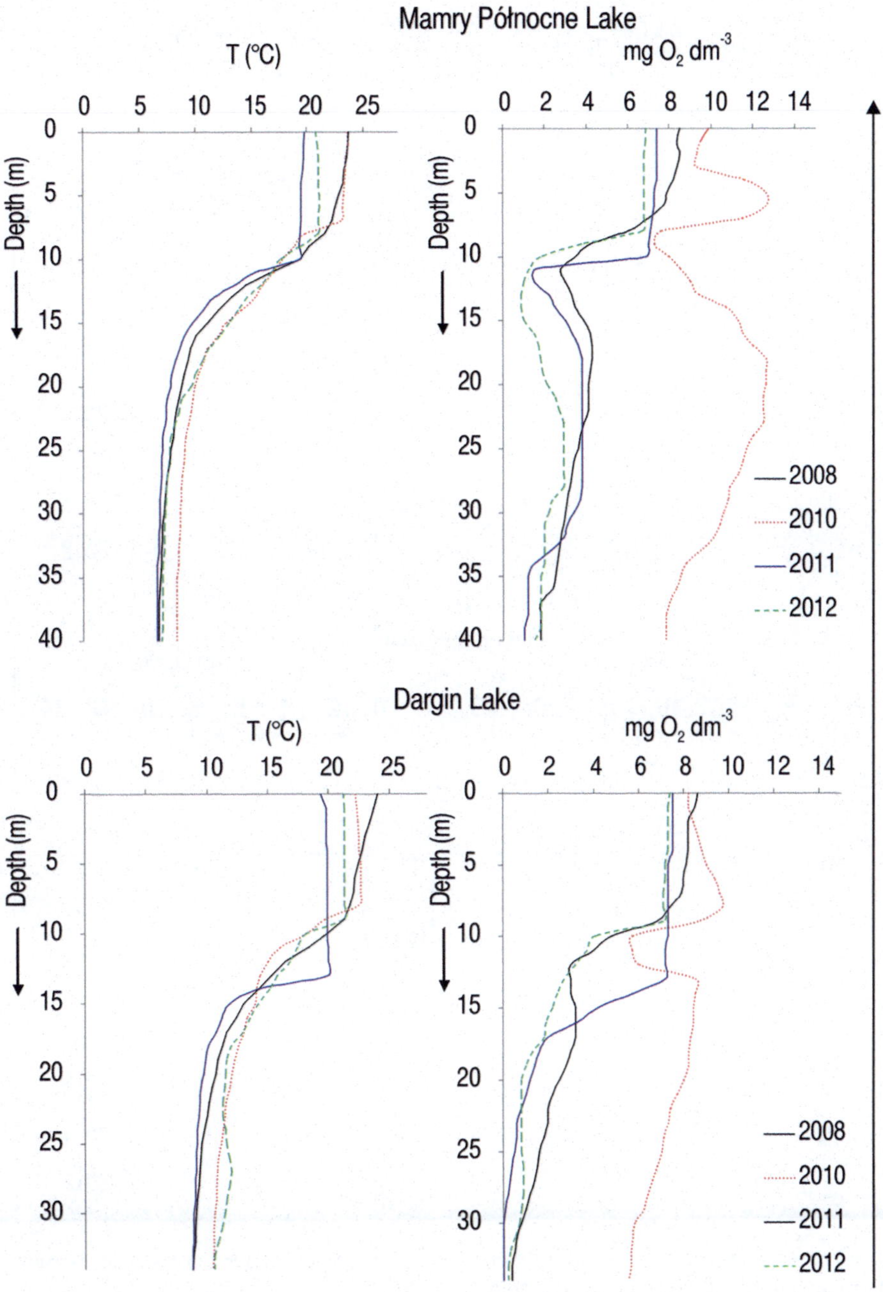

Fig. 6 Temperature and oxygen stratification in the lakes of Northern Basin of the Great Masurian Lakes during summer stagnation (August) in 2008–2012

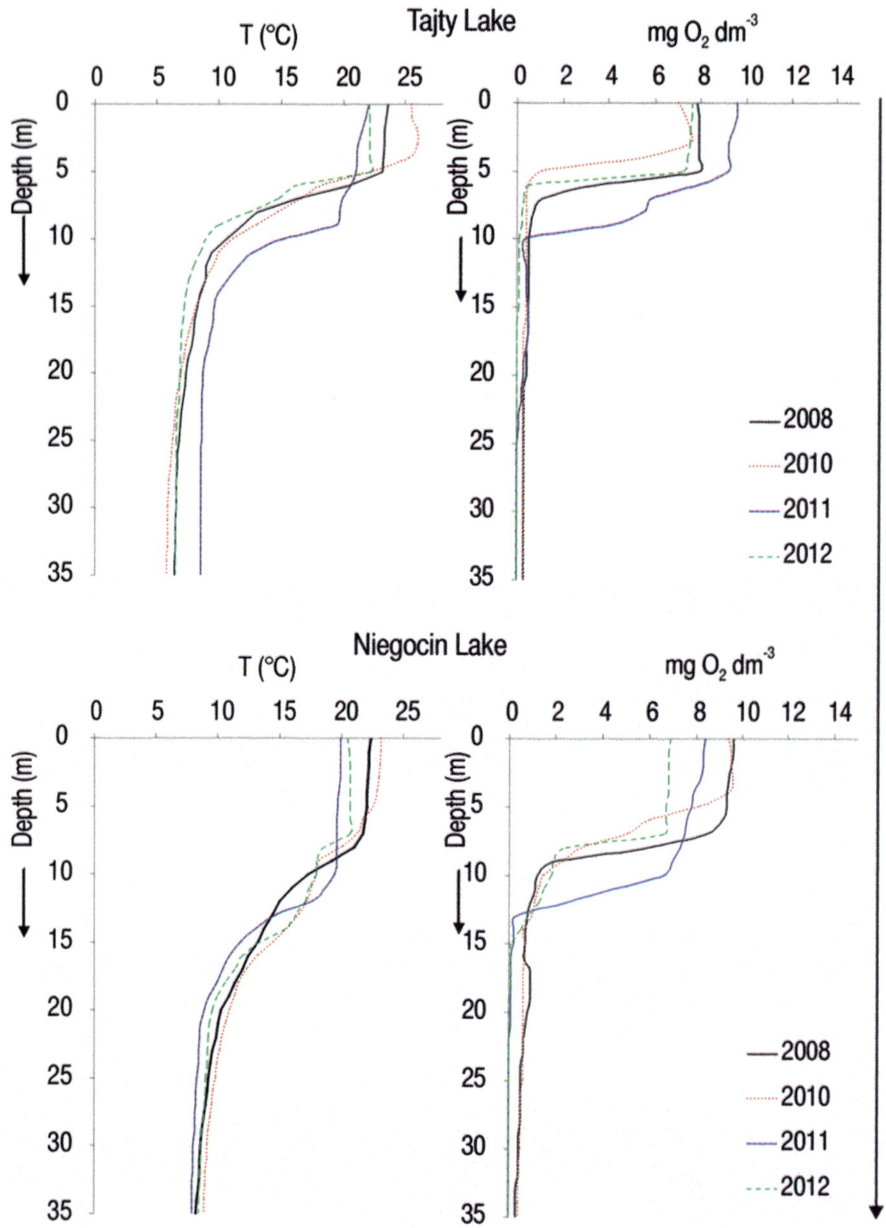

Fig. 7 Temperature and oxygen stratification in the lakes of Central Basin of the Great Masurian Lakes during summer stagnation (August) in 2008–2012

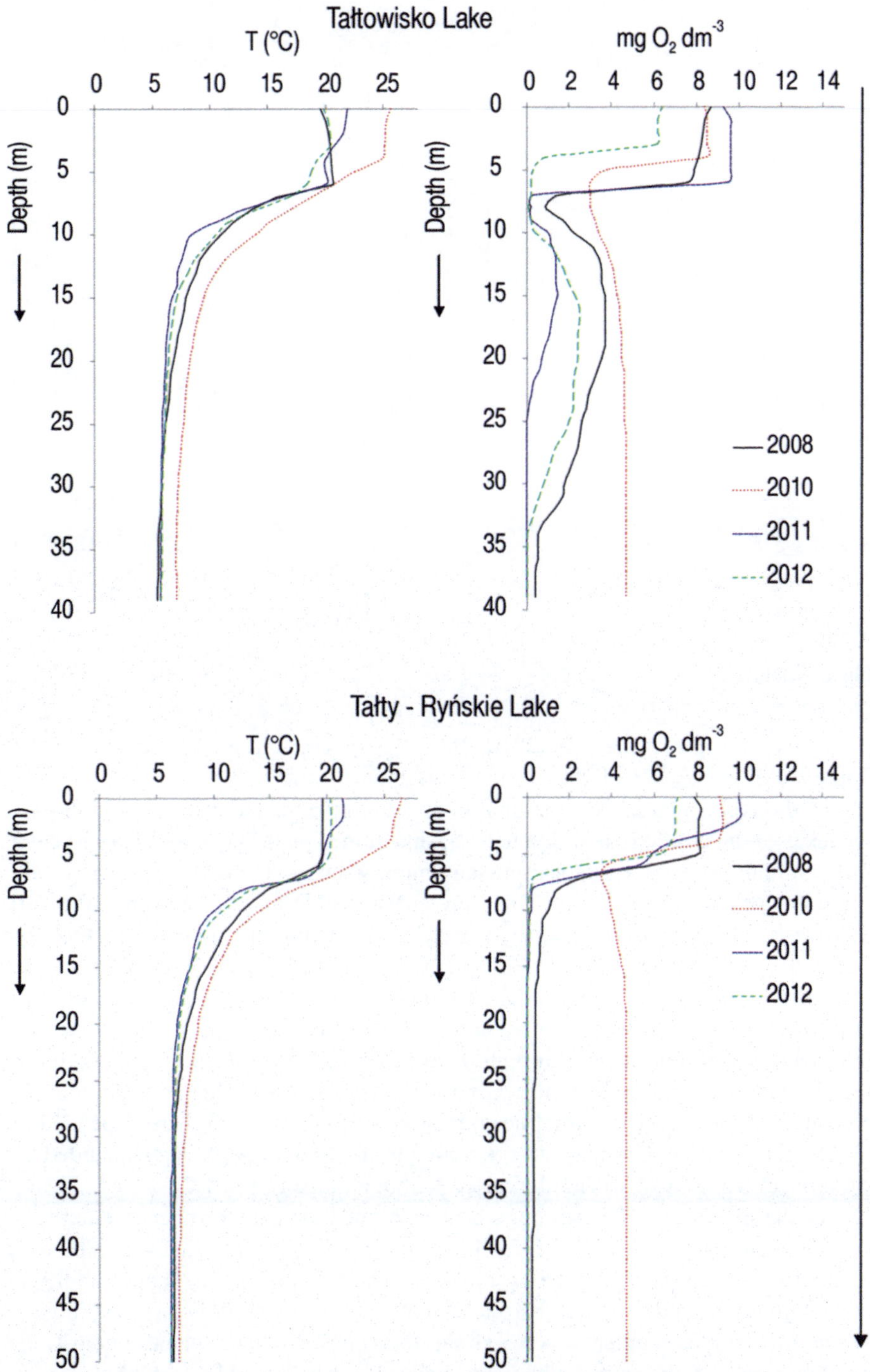

Fig. 8 Temperature and oxygen stratification in the lakes of Southern Basin of the Great Masurian Lakes during summer stagnation (August) in 2008–2012

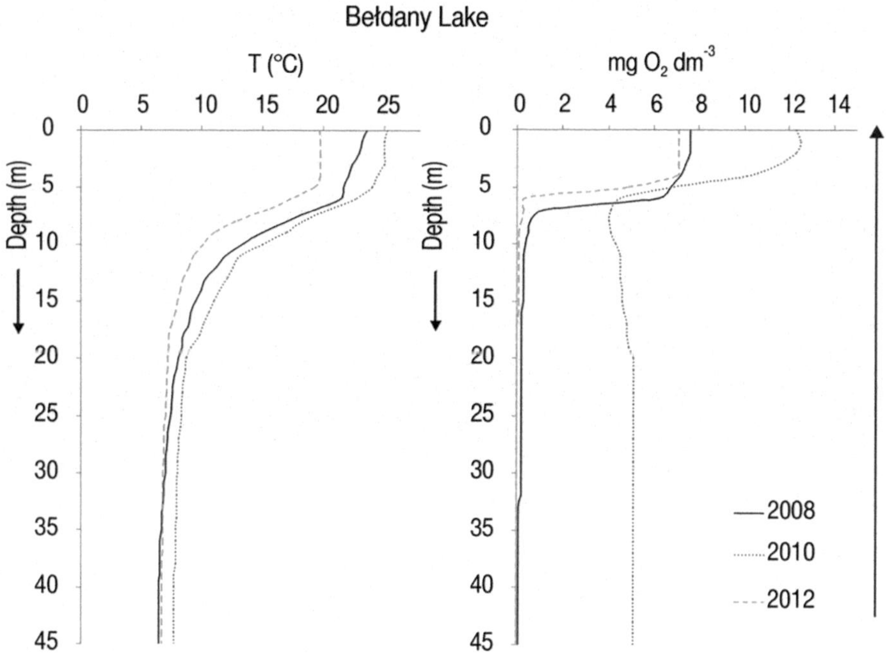

Fig. 8 (continued)

of highly eutrophic lakes. In 2010, the oxygen content in the vertical profile improved in most lakes of Northern and Southern Basins.

Good oxygen conditions in hypolimnion were noted in summer 2010 in most of the lakes of the Northern and Southern Basins of the Great Masurian Lakes System. The minimum oxygen content in metalimnion, which is typical for b-mesotrophy, occurred primarily in Lakes Święcajty, Mamry Północne, Dargin (Fig. 6), Tałtowisko, Tałty-Ryńskie, and Bełdany (Fig. 8). Below the thermocline, the oxygen content did not usually decrease <4.0 O_2 mg dm^{-3} and 40% water oxygen saturation. This unusual vertical distribution of oxygen content in summer in these lakes resulted from intense rainfall in the spring and summer, when seston was intensely washed out of flow-through lakes. Part of the seston accumulated in the metalimnion, and, subject to mineralization, it contributed to lowering the oxygen content in this layer. The rate of accumulation of oxygen deficit here depended on the sharpness of the thermal gradient, which in the flow-through lakes included the water layer from 5 to 9 m and fluctuated within a range of 9–14°C.

The thermoclines of flow-through lakes exhibited the Gliwicz phenomenon [17], i.e., the formation of a refugium for filter-feeding zooplankton that provided easy access to food and protection from planktivorous fish predation. The durability of this refugium depended on the sharpness of the thermal gradient in the metalimnion. When the oxygen content decreases in this layer to trace values, cold-water planktivorous fish (European whitefish, vendace, smelt) probably do not have access to this refugium, because they remain in the cooler, more oxygenated zones of the hypolimnion and, as a rule, forage less well. They are in worse condition and do not start breeding in fall [18]. In stagnating lakes, the metalimnion thermal gradient was less sharp at 5–9°C.

The salinity of the lakes is typical of basins located in the Polish Lowlands; hence they are bicarbonate-lime, where the composition of dissolved ions is dominated by Ca^{2+} and HCO_3^- [19–23]. The water pH in most of the lakes is alkaline because of intense phytoplankton photosynthesis. The lakes of the Southern Basin are less mineralized, while those in the Central Basin are much more strongly so (Table 2). The electrolytic conductivity of the surface water layers could change from 226 μS cm^{-1} in Nidzkie Lake situated in the forest catchment to 449 μS cm^{-1} in Wojnowo Lake situated in the agricultural catchment, while that of calcium could change from 31.5 to 69.6 mg Ca^{2+} dm^{-3}, respectively, and that of bicarbonate from 90.6 to 229 mg HCO_3^- dm^{-3}, respectively. The contents of magnesium and sodium did not, as a rule, exceed 12.3 mg Mg^{2+} dm^{-3}, potassium did not exceed 5.7 K^+ mg dm^{-3}, and carbonates did not exceed 16.5 mg CO_3^{2-}. The chloride concentration in the surface water layers in the lakes located in the lower part of the Southern Basin varied within the range of 4.1–5.2 mg Cl^- dm^{-3}, while in the central part of the system, they increased several times up to 22.4 mg Cl^- dm^{-3} in Niegocin Lake. The concentration of sulfates changed similarly from 8.3 to 33.3 mg SO_4^{2-} dm^{-3}, respectively.

The Great Masurian Lakes System was characterized by significant resources of phosphorus and nitrogen that are typical of medium or highly eutrophic lakes [24, 25]. The phosphorus content in the surface water layers was lower in the lakes of the Northern Basin and ranged from 0.045 mg TP dm^{-3} in Kirsajty Lake and up to 0.091 mg TP dm^{-3} in Święcajty Lake, and it was higher in the lakes of the Southern Basin, where it ranged from 0.071 to 0.128 mg TP dm^{-3} (Table 3). The concentration of total nitrogen in the surface water layers of the system's lakes in summer system ranged from 0.70 to 1.3 mg TN dm^{-3}. Lower contents of phosphorus and nitrogen were noted in summer 2010 in lakes located at the northern and southern limits of the system, where intense phosphate co-precipitation with calcite could occur. Water transparency was lower in the lakes of the Central and Southern Basins (Tables 3 and 4). The lowest Secchi depths were noted in Kotek Lake (0.4 m) and Lakes Boczne, Bełdany, Roś (0.9 m), and Śniardwy (1.4 m). The waters of Lakes Mamry Północne, Dargin, Dobskie, Kisajno, and Kirsajty were more transparent. Transparency in summer 2010 decreased in the Northern Basin to a maximum of 1.7 m in Dargin Lake and to 2.0 m in Kirsajty Lake, while in the Southern Basin the maximum was 0.9 m. Only in Śniardwy Lake was it higher.

4 Phytoplankton

The summer phytoplankton assemblages in the lakes of the Central and Southern Basins of the Great Masurian Lakes System in 1973 and 1976 were dominated primarily by large dinoflagellates *Ceratium hirundinella* (O.F.Müller) Dujardin

Table 2 Chemical composition (mean values) of surface water in the Great Masurian Lakes System in the summer period in 2008–2012

Lake	Ca^{2+}	Mg^{2+}	Na^{+}	K^{+}	CO_3^{2-}	HCO_3^{-}	Cl^{-}	SO_4^{2-}	pH	Conductivity
	mg dm^{-3}									µS cm^{-1}
Mamry Północne	43.3	11.0	8.4	3.9	5.2	138.7	13.2	22.9	8.62	311
Święcajty	42.1	10.5	8.5	4.0	4.4	150.4	13.1	21.6	8.64	323
Kirsajty	38.3	10.5	8.2	3.1	2.6	133.8	11.4	22.6	8.28	289
Dargin	43.1	9.4	8.5	4.2	4.6	144.7	13.0	20.0	8.56	330
Dobskie	43.4	9.3	8.4	4.1	5.4	147.9	15.3	22.1	8.63	332
Kisajno	43.5	9.6	8.3	3.8	5.3	140.7	15.9	23.1	8.63	329
Tajty	48.9	10.6	8.6	4.5	6.1	166.1	15.9	24.3	8.53	371
Niegocin	38.1	9.4	10.9	4.3	8.7	145.2	18.0	26.7	8.63	376
Wojnowo	63.1	12.3	9.2	4.4	10.3	204.2	12.4	20.3	8.53	435
Boczne	51.3	11.7	12.3	5.3	9.3	148.5	17.0	23.3	8.47	381
Jagodne	50.0	11.5	12.0	4.9	7.9	153.3	18.3	24.4	8.59	374
Szymon	44.6	11.2	11.2	4.8	7.1	156.2	15.0	21.7	8.39	375
Kotek	47.6	11.2	11.0	4.7	4.0	178.0	15.5	25.9	8.35	384
Tałtowisko	48.5	11.3	10.8	4.8	9.4	153.5	16.5	23.2	8.58	382
Ryńskie	47.7	11.5	10.4	4.5	11.5	129.0	13.3	19.4	8.62	363
Tałty	44.1	11.2	10.6	4.4	9.0	136.3	17.8	26.7	8.65	367
Mikołajskie	47.3	11.8	11.2	4.7	5.7	144.7	18.3	25.9	8.58	348
Śniardwy	49.6	11.3	10.0	3.7	6.8	147.1	11.8	16.8	8.65	341
Bełdany	44.8	9.3	7.9	2.9	9.0	137.0	9.0	15.6	8.63	323
Guzianka Mała	37.3	6.4	5.6	1.2	7.7	110.0	4.6	8.7	8.69	253
Guzianka Wielka	36.3	6.4	5.8	1.2	6.8	108.9	4.7	10.0	8.66	248
Nidzkie	34.4	6.2	5.6	1.3	4.4	110.7	5.8	11.4	8.61	229
Roś	51.2	10.8	9.1	3.4	5.4	151.5	11.3	16.7	8.59	348

Table 3 Assessment of the lake ecological status/potential in the Great Masurian Lakes System in 2008, 2011, 2012

Lake	Hypolimnion oxygenation %	Conductivity in 20°C µS cm^{-1}	Total nitrogen mg dm^{-3}	Total phosphorus mg dm^{-3}	Secchi disk visibility m	**Physicochemical assessment**	PMPL	**Biological assessment**
Mamry Północne	24.5	310	0.94	0.075	3.3	BGP	0.83	I
Święcajty	1.4	321	1.16	0.085	1.1	BGP	3.46	IV
Kirsajty	*5.4*[a]	284	1.03	0.068	3.2	BGP	0.73	I
Dargin	4.5	327	1.09	0.074	2.2	BGP	1.79	II
Dobskie	2.9	329	1.13	0.085	1.9	BGP	2.14	III
Kisajno	9.7	325	1.02	0.089	1.9	BGP	2.24	III
Tajty	1.5	365	1.10	0.075	1.4	BGP	1.96	II
Niegocin	2.9	371	1.26	0.078	1.6	BGP	1.88	II
Wojnowo	2.2	430	1.75	0.095	0.9	BGS	3.48	IV
Boczne	4.9	376	1.14	0.078	1.5	BGP	2.19	III
Jagodne	0.2	370	1.25	0.081	0.9	BGP	3.14	IV
Szymon	*7.9*[a]	371	1.38	0.086	1.3	BGP	2.65	III
Kotek	*7.4*[a]	381	1.41	0.082	0.9	BGS	2.44	III
Tałtowisko	11.1	378	1.32	0.093	1.2	BGP	3.15	IV
Ryńskie	1.1	359	1.34	0.085	1.2	BGP	3.94	IV
Tałty	1.1	365	1.26	0.085	1.4	BGP	3.11	IV
Mikołajskie	1.7	347	1.21	0.088	1.5	BGP	3.07	IV
Śniardwy	*6.3*[a]	341	1.24	0.094	1.6	BGP	2.21	III
Bełdany	0.9	321	1.26	0.093	1.3	BGP	3.50	IV
Guzianka Mała	1.9	251	1.48	0.096	1.3	BGS	3.91	IV
Guzianka Wielka	2.2	243	1.27	0.091	1.2	BGS	3.97	IV
Nidzkie	1.9	231	1.36	0.118	0.9	BGS	4.16	V
Roś	1.8	345	1.26	0.099	1.4	BGS	3.60	IV

PMPL Phytoplankton Metric for Polish Lakes

[a]*Dissolved oxygen content, mg dm*$^{-3}$

Legend:

Biological assessment

class I	class II	class III	class IV	class V	potential	status

Physicochemical assessment

class I	for TP and SDV
classes I-II	class II for TP and SDV
BGP	below good potential
BGS	below good status

Table 4 Assessment of the lake ecological status/potential in the Great Masurian Lakes System in 2010

Lake	Hypolimnion oxygenation %	Conductivity in 20°C µS cm^{-1}	Total nitrogen mg dm^{-3}	Total phosphorus mg dm^{-3}	Secchi disk visibility m	**Physicochemical assessment**	PMPL	**Biological assessment**
Mamry Północne	96.5	311	0.85	0.076	1.5	BGP	3.10	IV
Święcajty	47.2	328	0.95	0.085	0.8	BGP	4.16	V
Kirsajty	*9.0*[a]	305	0.83	0.060	2.0	II	1.72	II
Dargin	60.8	339	0.76	0.061	1.7	BGP	1.23	II
Dobskie	39.6	340	0.72	0.065	1.5	BGP	1.61	II
Kisajno	22.5	339	0.76	0.065	1.4	BGP	2.53	III
Tajty	2.7	387	1.08	0.060	1.4	BGP	2.43	III
Niegocin	4.4	390	1.15	0.075	1.0	BGP	2.80	III
Wojnowo	3.8	449	1.34	0.082	0.9	BGS	3.41	IV
Boczne	4.2	396	1.09	0.119	0.9	BGP	2.99	III
Jagodne	39.6	385	1.28	0.063	0.9	BGP	4.27	V
Szymon	*8.0*[a]	387	1.35	0.095	0.6	BGP	4.14	V
Kotek	*4.7*[a]	390	0.87	0.173	0.4	BGS	4.04	V
Tałtowisko	38.5	394	1.48	0.075	0.8	BGP	4.89	V
Ryńskie	39.2	374	1.07	0.078	0.5	BGP	4.27	V
Tałty	33.0	372	1.05	0.092	0.5	BGP	4.66	V
Mikołajskie	31.1	350	1.13	0.116	0.7	BGP	4.06	V
Śniardwy	*4.0*[a]	342	0.92	0.077	1.4	BGP	1.50	II
Bełdany	42.9	327	1.16	0.072	0.9	BGP	3.35	IV
Guzianka Mała	46.3	265	0.87	0.088	0.5	BGS	3.98	IV
Guzianka Wielka	48.4	252	0.84	0.068	0.5	BGS	3.74	IV
Nidzkie	40.2	204	0.98	0.087	0.5	BGS	4.28	V
Roś	34.4	355	1.20	0.075	0.9	BGS	4.42	V

PMPL Phytoplankton Metric for Polish Lakes

[a]*Dissolved oxygen content, mg dm^{-3}*

Legend:

Biological assessment

class I	class II	class III	class IV	class V	potential	status

Physicochemical assessment

class I	for TP and SDV
classes I-II	class II for TP and SDV
BGP	below good potential
BGS	below good status

[26, 27]. However, cyanobacterium *Planktothrix agardhii* (Gomont) Anagnostidis & Komárek co-dominated the phytoplankton in Lakes Guzianka Wielka, Jagodne, Kotek, and Szymon. Other cyanobacteria taxa of the genera *Aphanizomenon*, *Planktolyngbya*, and *Microcystis* were also present in most of the lakes.

In 2008 and 2011–2012, in turn, the summer phytoplankton assemblages in all the lakes of the Great Masurian Lakes System were dominated mainly by cyanobacteria [28]. Significant differences among the lakes concerned the biomass-based domination within the Cyanobacteria class and mainly of species representing three orders. Chroococcales order included primarily *Aphanocapsa incerta* (Lemmermann) G. Cronberg & Komárek, and *Aphanothece* sp. Nostocales order included primarily *Aphanizomenon gracile* Lemmermann; *Cuspidothrix issatschenkoi* (Usachev) P. Rajaniemi, Komárek, R. Willame, P. Hrouzek, K. Kastovská, L. Hoffmann & K. Sivonen; *Dolichospermum lemmermanii* (Richter) P. Wacklin, L. Hoffmann & J. Komárek; *D. flosaquae* (Brébisson ex Bornet & Flahault) P. Wacklin, L. Hoffmann & J. Komárek; and *Gloeotrichia echinulata* P. G. Richter. Oscillatoriales order included primarily *Limnothrix redekei* (Goor) Meffert, *Planktolyngbya limnetica* (Lemmermann) Komárková-Legnerová & Cronberg, *Pseudanabaena limnetica* (Lemmermann) Komárek, and *Planktothrix agardhii*. The share of Oscillatoriales increased with increasing total biomass and cyanobacteria biomass in the lakes from the north to the south, while the share of Chroococcales decreased. This was an evidence of the generally increasing trophic gradients in the lakes to the south.

The large share of the biomass of filamentous *Planktolyngbya limnetica* and *Pseudanabaena limnetica* of the order Oscillatoriales and trichomes *Aphanizomenon gracile* or *Gloeotrichia echinulata* of the order Nostocales also signaled progressing eutrophication. These species are well recognized as toxic or potentially toxic [29–32]. For example, in Niegocin Lake microcystin concentrations, presented as the sum of MC-LR, MC-RR, and MC-YR, were confirmed at a maximum of 0.159 μg dm^{-3} [33], while in Jagodne Lake they were approximately 7.16 μg dm^{-3} and in Lakes Szymon and Tałtowisko they were 4.79 and 4.94 μg dm^{-3}, respectively [34].

In the 1970s and 1980s, the phytoplankton biomass was the lowest (up to 2 mg dm^{-3}); this also included the lakes of the Northern Basin [26, 27, 35, 36]. The share of cyanobacteria in the total biomass ranged from 7% to 15%, which was similar to the share of dinoflagellates. The phytoplankton was dominated then by nanoplanktonic forms.

Higher biomass of phytoplankton, which was already dominated by cyanobacteria, was noted in the lakes in the Central Basin, primarily in Wojnowo Lake at approximately 37 mg dm^{-3}. The phytoplankton biomass in Niegocin Lake in the 1970s was similar with a mean of approximately 4 mg dm^{-3}. However, the domination structure in summer differed with significantly lower shares of dinoflagellates, and the share of cyanobacteria increased to 74%. In the 1990s, a significantly higher total biomass of phytoplankton dominated by filamentous cyanobacteria and high concentrations of chlorophyll *a* and very low Secchi depth were noted in Niegocin Lake [19, 21, 37–39]. The phytoplankton biomass in

Jagodne Lake was relatively low in the 1970s but with high share of cyanobacteria (up to 70%).

The phytoplankton biomass in the lakes in the Southern Basins was the highest. In the 1970s, in Lakes Szymon, Tałty-Ryńskie, and Kotek, the biomass reached maximum values of approximately 37 mg dm^{-3}, 34 mg dm^{-3}, and 24 mg dm^{-3} and large shares of cyanobacteria, dinoflagellates, or diatoms [26, 27]. Significantly lower biomass and cyanobacteria shares were noted then in Tałtowisko Lake. The phytoplankton biomass in Nidzkie Lake was sixfold smaller, and in Tałty-Ryńskie Lake it was fourfold smaller. The distinct domination of dinoflagellates was also noted. In Mikołajskie Lake, with its large share of dinoflagellates in the phytoplankton, and in Lakes Guzianka Mała and Guzianka Wielka, with their co-dominant dinoflagellates and cyanobacteria, the differences in biomass size were twofold. Significant changes also occurred in the structure of phytoplankton from the domination or co-domination of dinoflagellates toward the notable dominance of cyanobacteria.

Similar to previous studies, the lowest average biomasses of 1.7–4.1 mg dm^{-3} (2008, 2011–2012) were noted in the phytoplankton in the Northern Basin, primarily in Lakes Mamry Północne, Kirsajty, Dargin, Dobskie, and Kisajno (Fig. 9). Relatively low chlorophyll *a* concentrations of a mean of approximately 7.7 μg dm^{-3} were also noted in these lakes (Fig. 10). Higher biomass and chlorophyll content were noted in Święcajty Lake (17.0 mg dm^{-3} and 17.8 μg dm^{-3}, respectively), which was more intensely supplied by the Sapina River. In 2010, an increase in phytoplankton abundance was noted, especially in Mamry Północne Lake (eightfold higher biomass and fourfold higher chlorophyll concentrations).

In 2008 and 2011–2012, in the lakes of the Central Basin (Niegocin, Boczne, Wojnowo, Jagodne, and Tajty), the biomass and concentration of Chl *a* noted were

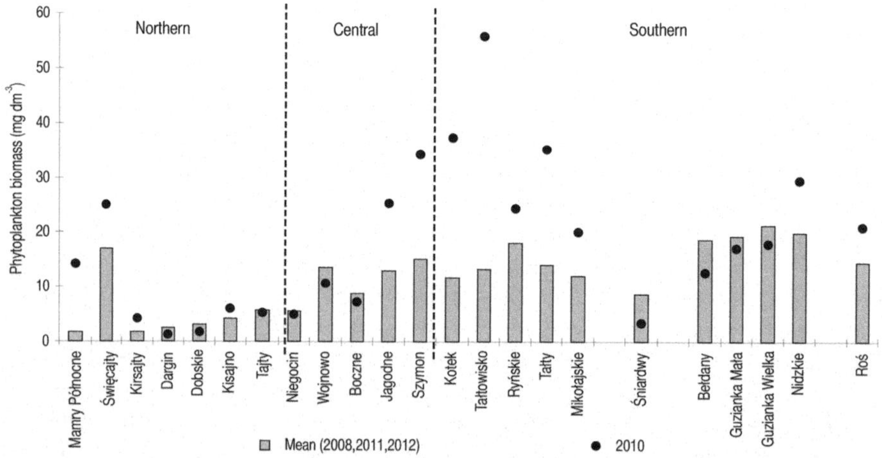

Fig. 9 Summer phytoplankton biomass in the Great Masurian Lakes System with the division into basins of hydrographic system (according to Napiórkowska-Krzebietke [28], modified)

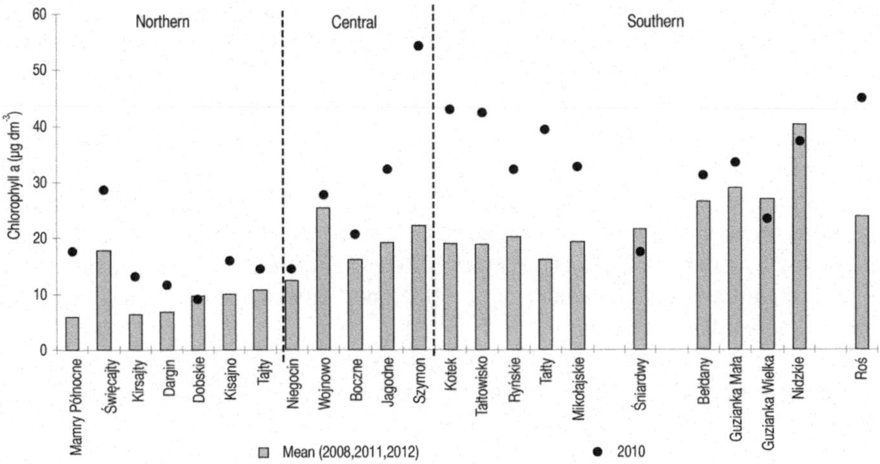

Fig. 10 Summer chlorophyll *a* concentration in the Great Masurian Lakes System with the division into basins of hydrographic system

generally higher than in the lakes of the Northern Basin. The mean values noted were 5.5–13.5 mg dm^{-3} and 10.6–25.3 μg dm^{-3} (Figs. 9 and 10), respectively. These values were similar to those noted in 2010, with the exception of the twofold higher biomass and concentration of chlorophyll *a* in Jagodne Lake, which was then located in the bifurcation zone of the first-order watershed. Currently, however, the biomass noted in Wojnowo Lake was approximately threefold lower at a mean of approximately 12 mg dm^{-3}.

The highest phytoplankton biomass in 2008 and 2010–2012 was also noted in the lakes of the Southern Basin at means that ranged from 8.6 mg dm^{-3} to 21.2 mg dm^{-3} (2008, 2011–2012) (Fig. 9).

The biomasses noted in Lakes Szymon, Kotek, and Tałtowisko that were in the bifurcation zone in summer 2010 and in Tałty-Ryńskie and Mikołajskie that were below this zone were approximately twofold to fourfold higher, but in the remaining lakes these differences were smaller. In Lakes Śniardwy, Bełdany, Guzianka Mała, and Guzianka Wielka, biomasses in 2010 were even lower than the means from 2008 and 2011–2012. Chl *a* concentrations were also the highest in these lakes with similar trends in this parameter and in the total phytoplankton biomass in 2010 (Fig. 10). The maximum Chl *a* of 54.3 μg dm^{-3} was noted in Szymon Lake.

5 Conclusions

Classifications based on physicochemical elements averaged for the study years of 2008, 2011, and 2012 (Table 3) showed that the ecological potential and status was below good of waters in all the lakes mainly because of increased total phosphorus content. The decided majority of lakes had also bad oxygen conditions in the hypolimnion and low water transparency (measured with Secchi disk visibility) with the exceptions of Lakes Mamry Północne, Kirsajty, Śniardwy, and Szymon. Simplified biological assessments based on the phytoplankton index (PMPL – Phytoplankton Metric for Polish Lakes) indicated class I, which is the maximum potential, in only two of the lakes – Mamry Północne and Kirsajty (Table 3). The deep, stratified Lakes Dargin, Tajty, and Niegocin and the shallow non-stratified Lakes Szymon and Kotek were designated as class II, which indicates good potential and status. The remaining lakes were designated as classes III and IV, which indicate moderate and poor potential and status, while only Nidzkie Lake was designated as class V, which is poor status.

In summer 2010, no significant changes were noted in total nitrogen or total phosphorus or in conductivity, which led to similar water classifications as in the other years of the study (Table 4). However, oxygen conditions improved significantly compared to 2009 and 2011–2012, with the exception of the lakes in the Central Basin, but water transparency in almost all of the lakes deteriorated. Precipitation was significantly higher in the spring and summer in 2010 than in the other years of the study. The bifurcation zone, which had a direct impact on the quality of the lakes studied, separated then the lakes of the Central Basin. Those lakes consist of an approximate 14% share of the total area and an 18% share of the total volume of the Great Masurian Lakes System.

The classification of biological elements based on PMPL also indicated significant deterioration in water quality class, i.e., ecological potential and status, in 2010 compared to the other years of the study. The lakes within the bifurcation zone and several lakes in the Southern Basin, from Lakes Jagodne to Mikołajskie, and Nidzkie and Roś, had then bad potential or status. Water quality also deteriorated in the Northern Basin lakes of Święcajty (bad potential), Mamry Północne (poor potential), and Kirsajty (good potential). The exception was Śniardwy Lake, the area of which is 34% of the total area and 24% of the total volume of the Great Masurian Lakes System, where water quality was better and which was designated as class II with good ecological potential.

Acknowledgments The material was collected and analyzed primarily as a part of statutory research (S-009) conducted at the Stanisław Sakowicz Inland Fisheries Institute in Olsztyn.

References

1. Kondracki J (1972) Polska Północno-Wschodnia (Northeastern Poland). PWN, Warszawa (in Polish)
2. Atlas Podziału Hydrograficznego Polski (Hydrographic Atlas of Poland) (2005) praca zbiorowa pod kier. H. Czarneckiej, IMGW, Warszawa (in Polish)
3. Choiński A (2006) Katalog jezior Polski (Catalog of Polish lakes). Wyd. Nauk. UAM, Poznań, p 381 (in Polish)

4. Bajkiewicz-Grabowska E (2008) Obieg wody w systemie Wielkich Jezior Mazurskich (Water circulation in the Great Masurian Lakes System). In: Jasser I, Rybak S, Zdanowski B (eds) Ochrona i rekultywacja wód wielkich Jezior Mazurskich narzędziem rozwoju naukowego, gospodarczego, społecznego i kulturowego regionu (Conservation and restoration of the waters of the Great Masurian Lakes as tools for the scientific, economic, social, and cultural development of the region). Wyd IRŚ, Olsztyn, pp 19–29 (in Polish)
5. Zdanowski B, Wierzchowska M, Wołos A (2009) Change patterns in the trophic state of Lake Mamry Północne and Lake Niegocin (Masurian Lake District, Poland). Limnol Rev 9(1):39–60
6. Standard Methods for Examination of Water & Wastewater (1999) Am Publ Health ASN., New York, p 1325
7. Hermanowicz W, Dożańska W, Dojlido J, Koziorowski B, Zerbe J (1999) Fizyczno-chemiczne badanie wody i ścieków (Physicochemical studies of water and wastewater). Arkady, Warszawa, p 555 (in Polish)
8. Lorenzen CJ (1967) Determination of chlorophyll and pheo-pigments: spectrophotometric equations. Limnol Oceanogr 12:343–346
9. CEN EN 15204 (2006) Water quality – guidance standard for the routine analysis of phytoplankton abundance and composition using inverted microscopy (Utermöhl technique)
10. Napiórkowska-Krzebietke A, Kobos J (2016) Assessment of the cell biovolume of phytoplankton widespread in coastal and inland water bodies. Water Res 104:532–546
11. Dąbrowski M (2013) Uwarunkowania naturalne i antropogeniczne obiegu wody w Systemie Wielkich Jezior Mazurskich (Natural and anthropogenic conditions of water circulation in the Great Mazurian Lakes System). Wyd Instytut Meteorologii i Gospodarki Wodnej – Państwowy Instytut Badawczy, Warszawa, pp 1–251 (in Polish)
12. Bajkiewicz-Grabowska E (1992) Czy zaszły zmiany w położeniu działu wodnego w systemie Wielkich Jezior Mazurskich? (Have changes occurred in the location of the watershed in the Great Masurian Lakes System?). Prace i Studia Geograficzne UW, p 12 (in Polish)
13. Dąbrowski M (2002) Położenie głównego działu wodnego w systemie Wielkich Jezior Mazurskich (Location of the main watershed in the Great Mazurian Lakes System). Gosp Wodna 6:238–243 (in Polish)
14. Dąbrowski M (2002) Changes in the water level of lakes in northeastern Poland. Limnol Rev 2:85–92
15. Mikulski Z (1966) Kształtowanie się działu wodnego na Wielkich Jeziorach Mazurskich (Watershed formation in the Great Mazurian Lakes). Przegląd Geograficzny 38:381–392 (in Polish)
16. Bajkiewicz-Grabowska E, Mikulski Z (2006) Hydrologia ogólna (General hydrology). Wyd Nauk PWN, Warszawa, wyd. IV, p 339 (in Polish)
17. Gliwicz M (1979) Metalimnetic gradients and trophic state of lake epilimnia. Mem Ist Ital Idrobiol 37:121–145
18. Świerzowski A, Doroszczyk L (2004) Seasonal differences in situ measurements of the target strength of vendace (Coregonus albula L.) in lake Pluszne. Hydroacoustics 7:217–226
19. Zdanowski B, Korycka A, Zachwieja J (1984) Thermal and oxygen conditions and the chemical composition of the Great Masurian Lakes. Ekol Pol 32:651–677
20. Zdanowski B, Hutorowicz A, Świątecki A (1993) Zmienność wskaźników trofii i czystości wody wielkich Jezior Mazurskich (Variability in trophic and quality indicators of the waters of the Great Masurian Lakes). Zintegrowany Monitoring Środowiska Przyrodniczego, PIOŚ Warszawa (in Polish)
21. Zdanowski B, Hutorowicz A (1994) Salinity and trophy of Great Masurian Lakes (Masurian Lakeland, Poland). Ekol Pol 42:317–331
22. Marszelewski W (2005) Zmiany warunków abiotycznych w jeziorach Polski Północno-Wschodniej (Changes of abiotic conditions in lakes of North-Eastern Poland). Wyd UMK, Toruń, p 288 (in Polish)
23. Korycka A (1991) Characterization of the chemical composition of the water in the lakes of northern Poland. Rocz Nauk Rol H-102(3):1–112 (in Polish)

24. Zdanowski B (1982) Nitrogen and phosphorus in lakes and their eutrophication. Pol Arch Hydrobiol 29:541–597
25. Kufel L (2001) Uncoupling of chlorophyll and nutrients in lakes – possible reasons, expected consequences. Hydrobiologia 443:59–67
26. Spodniewska I (1978) Phytoplankton as the indicator of lake eutrophication I. Summer situation in 34 Masurian Lakes in 1973. Ekol Pol 26(1):53–70
27. Spodniewska I (1979) Phytoplankton as the indicator of lake eutrophication II. Summer situation in 25 Masurian Lakes in 1976. Ekol Pol 27(3):481–496
28. Napiórkowska-Krzebietke A (2015) Cyanobacterial bloom intensity in the ecologically relevant state of lakes – an approach to Water Framework Directive implementation. Oceanol Hydrobiol Stud 44(1):97–108. https://doi.org/10.1515/ohst-2015-0010
29. Kobos J, Błaszczyk A, Hohlfeld N, Toruńska-Sitarz A, Krakowiak A, Hebel A, Stryk K, Grabowska M, Toporowska M, Kokociński M, Messyasz B, Rybak A, Napiórkowska-Krzebietke A, Nawrocka L, Pełechata A, Budzyńska A, Zagajewski P, Mazur-Marzec H (2013) Cyanobacteria and cyanotoxins in Polish freshwater bodies. Oceanol Hydrobiol Stud 42(4):358–378. https://doi.org/10.2478/s13545-013-0093-8
30. Bukowska A, Bielczynska A, Karnkowska A, Chróst RJ, Jasser I (2014) Molecular (PCR-DGGE) versus morphological approach: analysis of taxonomic composition of potentially toxic cyanobacteria in freshwater lakes. Aquat Biosyst 10:2
31. Bukowska A, Kaliński T, Koper M, Kostrzewska-Szlakowska I, Kwiatowski J, Mazur-Marzec H, Jasser I (2017) Predicting blooms of toxic cyanobacteria in eutrophic lakes with diverse cyanobacterial communities. Sci Rep 7:8342. https://doi.org/10.1038/s41598-017-08701-8
32. Napiórkowska-Krzebietke A, Hutorowicz A (2015) The physicochemical background for the development of potentially harmful cyanobacterium *Gloeotrichia echinulata* J. S. Smith ex Richt. J Elem 20(2):363–376. https://doi.org/10.5601/jelem.2014.19.4.756
33. Jakubowska N, Zagajewski P, Gołdyn R (2013) Water blooms and cyanobacterial toxins in lakes. Pol J Environ Stud 22(4):1077–1082
34. Mankiewicz J, Komárková J, Izydorczyk K, Jurczak T, Tarczynska M, Zalewski M (2005) Hepatotoxic cyanobacterial blooms in the lakes of northern Poland. Environ Toxicol 20:499–506. https://doi.org/10.1002/tox.20138
35. Napiórkowska-Krzebietke A, Hutorowicz A (2005) Long-term changes of phytoplankton in Lake Mamry Północne. Oceanol Hydrobiol Stud 34(3):217–228
36. Napiórkowska-Krzebietke A, Hutorowicz A (2007) Long-term changes in the biomass and composition of phytoplankton in a shallow, flow-through lake Kirsajty (Masurian Lakeland, Poland). Pol J Nat Sci 22:512–524
37. Napiórkowska-Krzebietke A, Hutorowicz A (2006) Long-term changes of phytoplankton in Lake Niegocin, in the Masurian Lake Region, Poland. Oceanol Hydrobiol Stud 35(3):209–226
38. Hutorowicz A, Napiórkowska-Krzebietke A (2007) Changes in phytoplankton assemblages after the reduction of sewage discharge into Lake Niegocin (Mazurian Lake District, Poland). Oceanol Hydrobiol Stud 36:137–145
39. Napiórkowska-Krzebietke A, Wierzchowska M, Błocka B, Hutorowicz J, Hutorowicz A, Zdanowski B (2007) Changes in the trophic state of Lake Niegocin after the modernization of a local wastewater treatment plant. Limnol Rev 7(3):153–159

Trophic State, Eutrophication, and the Threats for Water Quality of the Great Mazurian Lake System

Waldemar Siuda, Karolina Grabowska, Tomasz Kaliński, Bartosz Kiersztyn, and Ryszard J. Chróst

Contents

Abstract One of the greatest threats to water quality is accelerated eutrophication, resulting from human activity, like the high intensity of tourism, surface runoffs from fertilized fields, and municipal pollution. Water eutrophication manifests as excessive growth of phytoplankton caused by overabundant nitrogen, phosphorus, and other nutrient supply which causes deterioration of water quality related to the amount of bacterial biomass in eutrophicated water reservoirs. The Great Mazurian Lake System (GMLS) is a chain of lakes located in mesoregion of the Great Mazurian Lakes in the Northeastern Poland. All lakes of the GMLS are connected by natural or artificial channels built in the eighteenth and nineteenth centuries and nowadays create widely spilled, long (the easiest route from northern to southern edge is about 110 km) gutter unique on the scale of the continent. The lakes of GMLS are of glacial origin. During the last five decades, all lakes of the GMLS passed different levels of eutrophication, thus significantly changing their trophic states. This report describes past and present trophic conditions of lakes of GMLS and analyzes environmental factors responsible for eutrophication of their waters.

W. Siuda · K. Grabowska · T. Kaliński · B. Kiersztyn · R. J. Chróst (✉)
Department of Microbial Ecology and Environmental Biotechnology, Faculty of Biology, University of Warsaw, Warszawa, Poland
e-mail: rjchrost@me.com

E. Korzeniewska, M. Harnisz (eds.), *Polish River Basins and Lakes – Part I*,
The Handbook of Environmental Chemistry 86,
https://doi.org/10.1007/978-3-030-12123-5_12

Eutrophication processes are not only responsible for high nutrients levels in lakes, extensive growth of phytoplankton biomass and productivity, cyanobacterial predominance, etc., but eutrophication is also responsible and connected to several threats for water quality. Presence of pathogenic bacteria, as well as the potential presence of many antibiotic-resistant bacteria in lakes of the GMLS, is discussed.

Keywords Antibiotic resistance · Eutrophication · Lakes · Pathogenic bacteria

1 Introduction

1.1 The Great Mazurian Lake System

The Great Mazurian Lake System (GMLS) is a chain of lakes located in mesoregion of the Great Mazurian Lakes in the Northeastern Poland, stretched between mesoregions: Sępopolska Lowland, Mrągowo Lake District, Mazury Forest, Ełckie Lakeland, and Borecka Forest (Fig. 1). All lakes of the GMLS are connected by natural or artificial channels built in the eighteenth and nineteenth centuries and nowadays create widely spilled, long (the easiest route from northern to southern edge is about 110 km) gutter unique on the scale of the continent. The lakes of the GMLS are of glacial origin. Of these, several basic types are distinguished. In addition to mostly round lakes of head moraines and bottom moraine, the system consists of numerous similar to them, but incomparably shallower, floodplain lakes and deep gutter lakes. From the morphological point of view, the GMLS can be roughly divided into three subsystems – complex of Lake Mamry located at the northern outskirts of the system connected through melt-type Lake Niegocin with deep gutter lakes (Jagodne, Tałtowisko, Ryńskie, Tałty, Mikołajskie, and Bełdany) and finally with complex of Lake Śniardwy. From the hydrological point of view, the whole GMLS, which comprises about 20% of surface water resources of Poland, consists of various trophic types of lakes belonging to two different river basins. The northern meso-eutrophic part of the system including lakes Przystań, Mamry, Dargin, Łabap, and Kisajno drains water to the Pregoła River, whereas the southern eutrophic one comprising lakes Niegocin, Boczne, Jagodne Szymoneckie, Szymon, Kotek, Tałtowisko, Ryńskie, Tałty, Mikołajskie, Bełdany, and Śniardwy channels water to Vistula River (Fig. 1).

The total catchment area of the GMLS covers about 3,645 km^2. Although the water content in the southern part of GMLS is only a little more than that in the northern one (1,258 and 1,024 million m^3, respectively), their catchment areas differ substantially (3,030 and 615 km^2, respectively). This results in a relative stability of trophic conditions in northern lakes and a much greater susceptibility of southern lakes to eutrophication processes [1]. The basic limnological parameters of the GMLS lakes are included in Table 1.

Fig. 1 The Great Mazurian Lakes System, Northeastern Poland

Table 1 Basic limnological parameters of lakes of the Great Mazurian Lake System

Lake	Area (km^2)	Volume (10^6 m^3)	Depth (m)		Shoreline length (km)	Catchment area (km^2)	Schindler's coefficient
			Maximum	Average			
Mamry[a]	25.04	298.300	43.8	11.7	34.00	31.3	0.192
Dargin	30.3	322.100	37.6	10.6	32.80	87.1	0.354
Kisajno	18.96	159.264	25.0	8.4	50.10	nd	nd
Niegocin	26.00	258.522	39.7	10.0	35.40	51.7	0.302
Jagodne	9.43	82.705	37.4	8.7	35.41	89.5	1.206
Tałtowisko	3.27	45.831	39.5	14.0	11.50	72.1	1.647
Ryńskie	6.71	67.406	50.8	10.0	27.7	53.4[b]	0.289[b]
Tałty	11.6	180.856	44.7	15.6	31.0		
Mikołajskie	4.98	55.740	25.9	11.2	15.10	14.1	0.342
Śniardwy	113.4	660.212	23.4	5.8	97.15	15.0	0.195
Bełdany	9.41	94.848	46.0	10.0	34.4	46.1	0.590

[a]Together with Lake Przystań
[b]Summarized for lakes Tałty and Ryńskie

The destination of the border of the watershed, which separates both parts of the system, was a serious problem. From the mid-nineteenth century, its location was changed twice. For over 60 years, it ran through the Kula Channel at the southern end of Lake Niegocin. The studies of Skibniewski and Mikulski [2] as well as hydrometric measurements and field studies of Mikulski [3] determined its location at the southern end of Lake Kisajno. This location confirms the Hydrographic Division of Poland in force since 1980 [4]. The analysis of current research supplemented with current observational materials now allows the separation of the bifurcation area within the GMLS (placed between Lake Kisajno and Lake Niegocin), which, depending on the method of water management, variously supplies the systems of both river basins.

The Council Directive 75/440/EEC of 16 June 1975 (http://eur-lex.europa.eu/legal-content/EN/TXT/PDF/?uri=CELEX:31975L0440&from=en) defines surface water quality in terms of 46 biological and chemical parameters. It should be pointed out that data concerning their concentrations in majority of Mazurian lakes are scarce or not available. However, one can expect that in the GMLS lakes, which are not exposed to direct chemical pollution generated by heavy industry or to waste imported from waters flowing from other parts of Poland, concentrations of most of them (such as heavy metals, surfactants, phenols, polycyclic aromatic hydrocarbons, and many others) are probably below dangerous level and thus practically have no diagnostic significance. In contrast to chemical threats, the GMLS is primarily responsible for biological hazards such as raw and only partially purified domestic sewages, which contain biogenic substances, pathogenic bacteria, and viruses, overfertilization by artificial fertilizer residues, overproduction of organic matter, and related disturbances of matter and energy flow through lake ecosystems. Because of that to characterize water quality of the GMLS in this study, only basic physicochemical and biological parameters were chosen. The most important among them were water transparency (Secchi disc visibility – SD), pH, specific conductivity, oxygen concentrations (O_2) and its spatial distribution in depth profiles, concentrations of chlorophyll$_a$ (Chl$_a$), total phosphorus (TP), total nitrogen (TN), and dissolved organic carbon (DOC). The trophic conditions in lakes of GMLS were quantitatively expressed as average Carlson's trophic state index (TSI_{Avg}) [5] calculated as a mean value of TSI based on chlorophyll$_a$ (TSI_{Chl}), Secchi disc visibility (TSI_{SD}), and total phosphorus concentration (TSI_{TP}), which is directly related to intensity of many biological processes such as gross primary production (GPP), bacterial number (BN), bacterial production (BP), enzymatic activity [6], and planktonic respiratory activity (dark oxygen consumption – RSP) [7]. The uniqueness of GMLS as an experimental object for research of eutrophication and evolution of lakes of Central and Eastern Europe lies in the diversity of the lakes that make up the system. Lakes of each part of the system share a comparable agricultural, wetland, and urban areas. Therefore, they are, to some extent, similar in terms of external input of mineral and organic matter and finally in their trophic conditions (Fig. 2). On the other hand, both parts of the GMLS substantially differ in respect to their trophic status, the size of catchment area, and the level of anthropopressure [8]. In contrast to the catchment area of the southern lakes, where

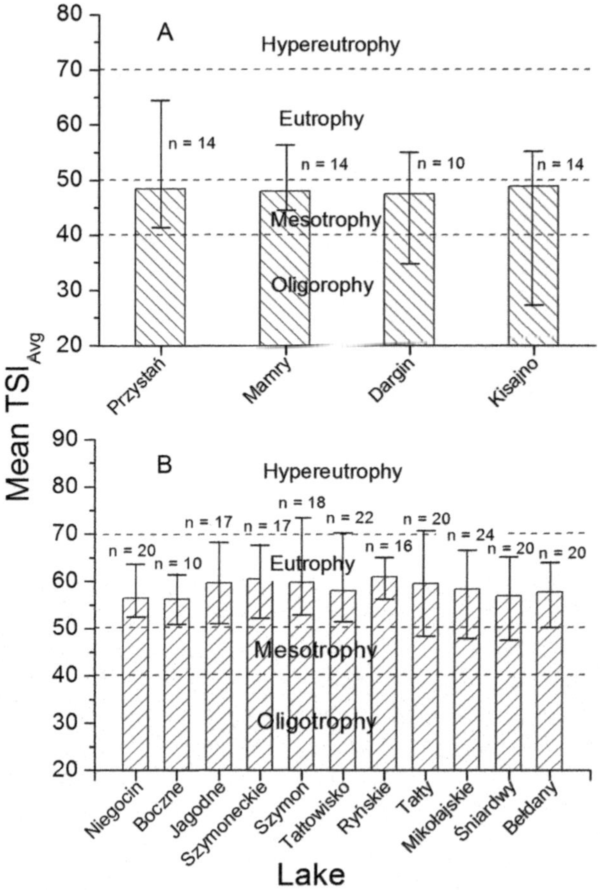

Fig. 2 Mean trophic state index of lakes of the northern (**a**) and southern (**b**) part of GMLS in summer seasons (July–August) of years 2005–2017. $TSI_{Avg} = (TSI_{Chl} + TSI_{TP} + TSI_{SD})/3$, *n* – the number of determinations, thin bars – the range of variability

the largest cities in the region are located, in a much less populated drainage basin of the northern lakes, there are no large urban centers, villages, and tourist centers. Moreover, in opposite to southern part of the GMLS, the shores of the northern lakes are low and in 80% overgrown with reeds. Their littoral zones that, with areas of reeds and bulrushes, create an effective barrier protecting them from the external input of biogenic substances are much wider and richer with submerged vegetation. Because of that northern lakes still represent various stages of meso- and meso-eutrophy, whereas the present trophic status of southern ones varies from low to advanced eutrophy. The only exception together to this rule is Lake Śniardwy, the largest Polish lake located on the southern edge of the system. In respect to trophic conditions, its northern part is similar to meso-eutrophic northern lakes, whereas the trophy of southern area resembles that of the southern lakes of the GMLS.

1.2 *Eutrophication of the Great Mazurian Lake System: An Overview*

According to the OECD definition from 1982, eutrophication is "increasing water content in food substances, which stimulates a series of symptomatic changes, among which an increase in the production of algae and macrophytes and lowering (dropping) of water quality are considered harmful and unfavorable for a human economy." The nutrients that particularly strongly stimulate the growth of fertility of aquatic ecosystems are mineral and organic compounds of phosphorus and nitrogen [9]. Apart from primary production or phosphorus and nitrogen concentrations, good determinants of trophic status of lakes are also water turbidity, concentration of dissolved organic carbon, secondary production, bacterial organic carbon demand (defined as the sum of bacterial production and respiration), bacterial number, and activity of some extracellular enzymes, such as aminopeptidase, esterase, and summer alkaline phosphatase [6].

According to Odum [10], it is necessary to distinguish natural eutrophication as a manifestation of the natural evolution of water reservoirs, and anthropogenic eutrophication evoked and stimulated by human activity. Human influence on lakes of GMLS (anthropogenic eutrophication) has started in the fifteenth century. At the time at least, some of GMLS lakes were oligo- or mesotrophic [11, 12]. Further colonization of Mazuria region resulting from human settlement, deforestation, and agriculture development changed the GMLS catchment area, intensified external input of biogenic substances to the lakes, and finally led to continuous but slow rising in their productivity [13]. Since the beginning of the 1960s of the last centuries, the impact of civilization on Mazuria region has significantly increased and has become more diverse and complex. Therefore nowadays, natural component of evolution of trophic status of the GMLS lakes has only of purely theoretic significance because anthropogenic eutrophication processes became incomparably faster and more important than the natural ones. This is particularly true for southern lakes, whose drainage area has changed much more drastically than drainage area of the northern lakes.

Present trophic status of the GMLS was shaped by four basic factors: (1) geographical location, which determines bifurcation nature of the system and its division into two separate parts carrying waters into two watersheds and evolving in a diverse way; (2) relatively low anthropogenic impact on catchment areas of northern lakes and strong anthropopressure exerted simultaneously on the catchment areas of southern lakes, which resulted in accumulation of substantial amounts of phosphorus in their waters and sediments; (3) political economic and social changes in Poland in the years 1980–1990, which initiated a drastic decline in the external input of nutrients caused by the collapse of nonrational agriculture and tourism in the region; and (4) climate changes occurring in the last few decades, resulting in an increase in mean daily temperatures in winter, shorter ice cover [14], and also shortening of autumn and spring homothermal periods.

In theory, lake eutrophication should be directly related to rates of phosphorus and nitrogen loading from drainage area [15]. In practice however, this relationship can be significantly modified by a variety of other factors such as weather conditions, morphometric characteristics of the lake (i.e., depth, extent of the shoreline, and time of water retention), human activity within the lake (i.e., fishery and degradation of ecotone zones), and mechanisms of carbon and nitrogen cycling specific for a given lake [16–18].

In order to understand the mechanisms that influenced the current trophic status of individual lakes of the GMLS and to predict the direction of their evolution in the future, it is useful to analyze changes in phosphorus concentration in Lake Niegocin which is the key element of the system. Due to the location of border of the watershed and direction of the water flow through both parts of the GMLS, fluctuations in the trophic conditions and water quality of this lake primarily affected the present trophic status of the southern part of the GMLS. However, they did not affect, or influenced only to a small extent, trophic conditions of northern lakes.

Lake Niegocin is one of the largest lakes of the GMLS. The surface of the lake is 2,600 ha, average depth 10 m, and the maximum depth 39.7 m. The bottom has numerous depressions and shallows. Water from Lake Niegocin flows through Kula Channel to the south supplying directly Lake Jagodne and indirectly all other lakes of the southern part of the GMLS. About 25% of its direct catchment area covers arable land; similar part is overgrown with forests. Urban areas occupy about 20%. On the northern shore of the lake is situated the city of Giżycko – the largest urban agglomeration in the region. In year 2016, the city was inhabited by 29,642 permanent residents but during the 4 months of the tourist season (July–September) visited by 19,849 tourists. Until 1995, in which in Giżycko the first sewage treatment plant was established, Lake Niegocin was a receiver of untreated municipal sewage and pollutants from galvanizing dairy and fish canning factory. As a result of that, in the 1980s of the last century, this lake became one of the most polluted and degraded water reservoirs in Europe.

Analysis of total phosphorus in surface and bottom zones of this lake (Fig. 3) showed that after dramatic changes in the years 1975–2005 [1], during the last decade, concentrations of this biogen stabilized on acceptable level. Such trends were not observed in the waters of the northern lakes, where the total concentration of phosphorus as well as other determinants of the trophic state remained almost unchanged for 30 years.

2 Trophic State of the Great Mazurian Lake System

2.1 Northern Lakes

Similarly, as in the case of the southern part of the GMLS, present trophic conditions of northern lakes were affected, to some extent by external input of biogenic substances from northwestern suburbs of Giżycko City, which affected directly

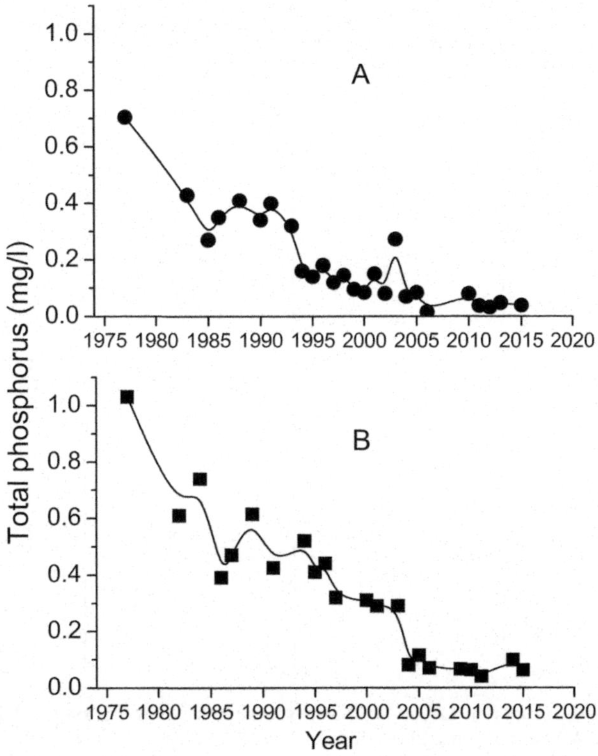

Fig. 3 Changes in total phosphorus concentration in surface (**a**) and profundal (**b**) waters of Lake Niegocin in summer periods (July–August) 1977–2015. Based on published data [19–22] and data collected by the Department of Microbial Ecology and Environmental Biotechnology, University of Warsaw, in unpublished GMLS database

relatively deep Lake Kisajno and indirectly all other northern lakes. However, anthropogenic impact on the trophic status of Lake Kisajno was, in comparison to Lake Niegocin, much smaller and less visible because the majority of wastes produced by Giżycko City were discharged to the southern part of the GMLS. Although there is no available data concerning phosphorus load collected in bottom sediments of Lake Kisajno, one can presume that it is relatively low. Additionally, because water column of this lake is relatively well oxygenated (Fig. 4a), insoluble phosphates accumulated in its bottom sediments seem to be at least partially protected against re-solubilization and penetration into the profundal zone, as in lakes in which deep anoxia is developed. Consequently, during mixing periods only small amounts of phosphorus compounds liberated from bottom sediments of Lake Kisajno are exported to Lake Dargin and further to other lakes of the northern part of the GMLS.

Variations in concentrations of P but also N forms available for phytoplankton in the photic zone of lakes of the northern part of the GMLS are primarily dependent on the rate of their regeneration from organic compounds in water column and to a lesser extent on external input determined by weather conditions or poor efficiency of their translocation from profundal waters during spring water overturn [7]. The characteristics of the northern part of the GMLS in terms of selected physicochemical and biological factors are presented in Table 2.

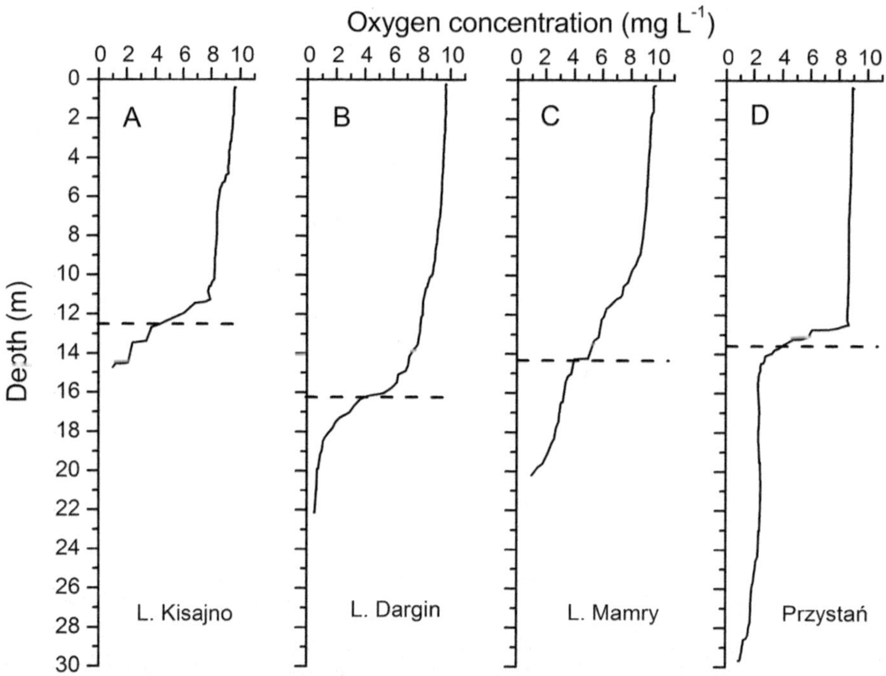

Fig. 4 Depth profiles of dissolved oxygen concentrations in lakes of the northern part of the GMLS in August 2009. Lakes are arranged according to the direction of water flow from the south (**a**) to the north (**d**). Dashed line – the depth below which oxygen concentration is critical for fish (<4 mg/L)

Table 2 Mean values ($n = 10$–14) of basic physicochemical and biological parameters of surface waters of lakes of the northern part of the GMLS during summer months (July–August) of years 2005–2015

	Lake			
Parameter	Przystań	Mamry	Dargin	Kisajno
Trophic state index	49.7	50.7	49.3	50.6
Secchi disc (m)	3.2	2.6	2.8	2.7
pH	8.2	8.2	8.3	8.2
Specific conductivity (μS/cm)	326	308	345	329
Total phosphorus (mg P/L)	0.021	0.025	0.025	0.027
Total nitrogen (mg N/L)	0.655	0.708	0.762	0.807
DOC (mg C/L)	9.7	9.9	10.3	10.1
Chlorophyll$_a$ (μg/L)	17.2	12.8	21.9	13.8
Bacterial number ($\times 10^6$ cells/ml)	4.0	3.5	4.8	4.2
Primary production[a] (mg C/L/day)	0.621	0.770	0.631	0.790
Respiration[a] (mg C/L/day)	0.331	0.462	0.252	0.460

[a]Mean values ($n = 6$) for summer seasons 2009–2011

The ratio between nitrogen and phosphorus concentrations is one of most important factors, which determines species composition and phytoplankton production of freshwater environments. The productivity of majority of oligo- and mesotrophic lakes, in which TN/TP ratio is relatively high, is mainly limited by phosphorus concentrations, whereas in eutrophic or hyper-eutrophic environments with much higher TN/TP ratio, planktonic primary production is commonly limited by nitrogen concentrations [23–25]. Moreover, low TN/TP ratio is commonly used as indicator of cyanobacterial dominance [26]. It has been the topic of many studies, but their results are equivocal. The molar TN/TP ratios, which were reported as promoting cyanobacterial growth, varied from 11 [27] to 64–66 noticed by Smith [28] in 17 lakes throughout the world and by Nöges et al. [29] in two Estonian lakes. According to Kauppinen [30] and Siuda et al. [7], during summer period (July–August) years 2010–2012 in northern lakes, mean TN/TP ratio (molar) was 46 and varied from 11 in Lake Przystań to 113 in Lake Mamry. Therefore, although in lakes of the northern part of the GMLS phytoplankton was probably limited alternately by dynamically changing availability of P and N resources, P limitation was more frequent. Another consequence of variations in N/P ratio in all lakes of the northern part of the GMLS was only weak dominance of cyanobacterial over eukaryotic phytoplankton biomass (Fig. 5).

One of the most evident symptoms of eutrophication of lakes is the increase in concentration of dissolved organic matter (DOM) in their waters caused by growing difference between net primary production summarized with external DOM input and DOM respiration rates. Such a difference defines the trophic type of the lake as net autotrophic or net heterotrophic [31] and allows drawing conclusions concerning its ecological stability and eutrophication/deterioration rate. Although there is no

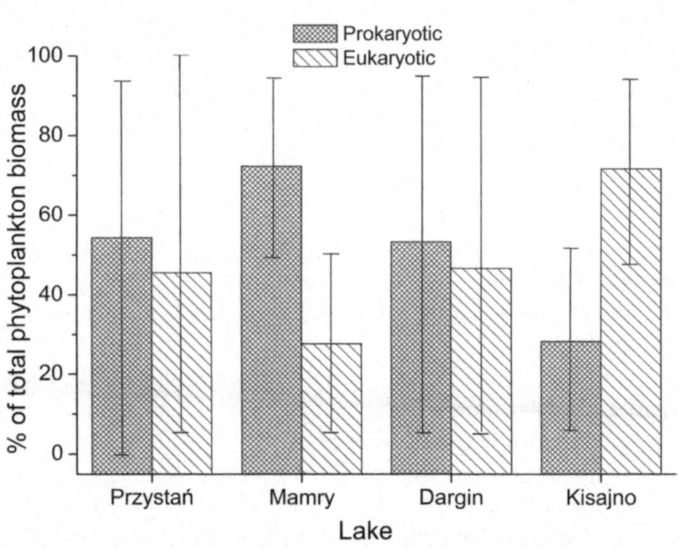

Fig. 5 Average ($n = 5$) percentage of prokaryotic (cyanobacteria) and eukaryotic phytoplankton (algae) in the total phytoplankton biomass in lakes of the northern part of the Great Mazurian Lake System during summer periods of 2009–2011. Thin bars represent standard deviations

information on the scale of external DOM input to lakes of the northern part of GMLS, one can assume that it is probably small, compared to the DOM produced by phytoplankton.

During investigations carried out in years 1994–2002, Chróst and Siuda [6] found that phytoplankton primary production in northern lakes (Przystań and Mamry) reached 0.65 mg C/L/day. Later research done in years 2009–2011 by Siuda et al. [7] showed that gross primary production calculated for whole photic zone of northern lakes was about two times lower than in lakes of the southern one and varied from 0.24 mg C/L/h in Lake Dargin to 1.172 mg C/L/h in Lake Mamry. They also speculated that this apparent increase in primary production rates was rather an effect of differences in methodology used than proof of the increase in the trophy conditions of the studied lakes. An additional evidence of stability and still relatively low trophic status of northern lakes is the fact that in contrast to the southern part of the GMLS, where cyanobacterial production dominated, in lakes of the northern part of the GMLS, also contribution of eukaryotic phytoplankton to overall primary production rates was quantitatively significant [7].

2.2 *Southern Lakes*

The present trophic conditions of lakes, which form the southern part of the GMLS were strongly affected by four major factors: (1) intensive fertilization of Lake Niegocin in years 1960–1995; (2) the collapse of agriculture and tourism in Mazuria Region caused by economic crisis in Poland, which lasted until the end of the twentieth century; (3) translocation of biogenic substances (mainly phosphorus nitrogen and dissolved organic matter) down the system in years 1995–2005 and now; and (4) mainly by internal loading of phosphorus from bottom sediments of deep gutter lakes and rapidly increasing touristic movement. Until the early 1990s of the last century, Lake Niegocin was a main receiver of raw or insufficiently cleaned domestic sewage from Giżycko City and pollution of agricultural origin, mainly the rests of lavishly used artificial fertilizers and manure from animal farms. After year 1995, large amounts of phosphorus and organic matter collected in this lake overflowed first to relatively deep Lake Jagodne which, as a kind of natural "settler" for Lake Niegocin, received about 30 mg P/m^3/year [1] and then to the other lakes placed down the system (Fig. 6), where "wave of the pollution" underwent gradual dilution.

A distinct improvement in water quality in Lake Niegocin has been observed since 2002, after the extension of the sewerage network and significant modernization of the municipal sewage treatment plant in Giżycko City. At present, it operates in accordance with the highest standards of Polish and EU environmental law. In years 2005–2016, during summer months the mean trophic state index of Lake Niegocin oscillated around 56.2 and changed from 52.6 to 63.7. Analysis of other physicochemical and biological parameters showed that at the same time rapid de-eutrophication of this lake observed in years 1997–2005 slowed down or even stopped.

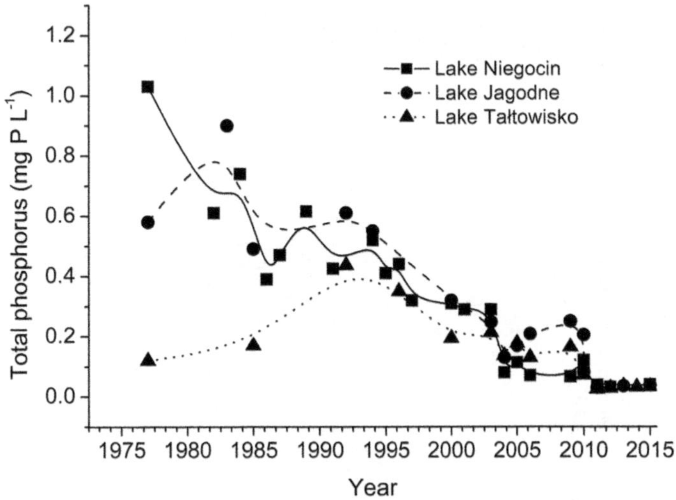

Fig. 6 Changes in total phosphorus concentrations in surface waters of lakes: Niegocin, Jagodne, and Tałtowisko in summer seasons in 1977–2015. Based on published data [19–22], Kufel (unpubl.), and results collected by Department of Microbial Ecology and Environmental Biotechnology, University of Warsaw, in unpublished GMLS database

Lake Jagodne, and its southwestern bay treated sometimes as a separate reservoir known as Lake Szymoneckie, is very diverse, and the whole lake is rather deep (max. depth 37.5 m, mean depth 8.7 m) and deprived of larger areas less than 2 m in depth. Poor afforestation of the larger part (about 30%) of its shoreline and northwestern shores strongly transformed by man activity made it susceptible to degradation processes. Long-term input of phosphorus from Lake Niegocin and supplementation with dissolved organic matter, nitrogen, and other biogenic substances throughout numerous drainage ditches draining low-lying meadows and wetlands covering the southeastern part of lake catchment area caused its rapid hyper-eutrophication. During the last years, water quality in Lake Jagodne has been noticeably improved. However, analysis of physicochemical and biological parameters of its surface waters collected in Table 3 proves that it still remains one of the most eutrophicated and degraded lakes of the GMLS.

The impact of the moving "pollution wave" was particularly evident in Lake Tałtowisko, the deepest gutter lake located in the mainstream of water flow (max. depth 45.5 m, mean depth 14 m). Despite the relatively high phosphorus content in water of this lake, until the early 1970s, it was still mesotrophic, whereas in the mid-1990s, substantial phosphorus input (about 32.4 mg P/m^3/year [1] caused its rapid eutrophication leading to hypertrophy (trophic state index = 92).

The trophic evolution of Lake Tałtowisko was tightly associated with lakes Szymon and Kotek. These small, shallow lakes are another key element of the southern part of the GMLS and work as a kind of natural sewage treatment plants, which protects directly Lake Tałtowisko, and indirectly the rest of the GMLS from

Table 3 Average values (n = 10–20) and the range of variability (in brackets) of basic physicochemical and biological parameters of surface waters of lakes of the southern part of the Great Mazurian Lake System during summer months (July–August) of years 2005–2015

	Lake[a]							
Parameter	N	J	Ta	R	T	M	Ś[b]	B
Trophic state index	56.2	59.4	57.7	60.9	59.5	59.0	58.9	57.3
Secchi disc (m)	1.9	1.3	1.5	1.1	1.3	1.5	1.9	1.6
pH	8.2	8.2	8.2	8.4	8.4	8.2	8.2	8.1
Specific conductivity (μS/cm)	383	380	385	377	376	344	334	334
Total phosphorus (mg P/L)	0.050	0.079	0.037	0.041	0.046	0.042	0.045	0.043
Total nitrogen (mg N/L)	0.912	1.107	1.092	1.140	1.203	0.884	0.857	0.810
DOC (mg C/L)	11.4	12.5	13.9	12.4	12.3	10.9	10.4	10.3
Chlorophyll$_a$ (μg/L)	19.0	38.6	22.2	43.3	39.4	28.5	22.3	25.1
Bacterial number ($\times 10^6$ cells/ml)	9.1	7.8	7.2	10.7	9.8	9.1	7.1	9.1
Primary production[c] (mg C/L/day)	0.904	1.592	2.383	1.938	2.017	1.335	1.804	1.252
Respiration[c] (mg C/L/day)	0.379	0.706	1.292	1.329	0.949	0.479	0.563	0.653

[a]Lake: *N* Niegocin, *J* Jagodne, *Ta* Tałtowisko, *R* Ryńskie, *T* Tałty, *M* Mikołajskie, *Ś* Śniardwy, *B* Bełdany
[b]Sampling site was located about 1 km east of the tributary from Lake Mikołajskie
[c]Mean values (n = 6) for summer seasons 2009–2011

excessive biogenic load flowing from lakes located above. Due to "bottleneck effect," lakes Szymon and Kotek act as a kind of natural trap for phosphorus compounds and organic suspension flowing down Szymoński Channel from Lake Szymoneckie. Capture of phosphorus and mineralization of organic matter in these lakes is favored by rich submerged vegetation and especially location on the navigable route. Permanent mixing of their water body and surface layers of bottom sediments by motorboats and ships of Mazurian Shipping Company increase their oxygenation and thus stimulate phosphorus trapping and organic matter mineralization and affect nitrogen cycling.

It was not a coincidence that during period of economic collapse (tourism and water transport, agriculture) two decades ago, water quality of Lake Niegocin improved very quickly, whereas at the same time, Lake Tałtowisko underwent rapid eutrophication. When, after 2000, shipping and water tourism were reestablished and intensified, the protective action of both of these lakes was restored – the trophic state index (TSI) of Lake Tałtowisko began to rapidly decrease (Fig. 7), and presently parameters of its surface waters are characteristic for typically eutrophic lakes (Table 3).

Lake Ryńskie creates a lake water complex with Lake Tałty, separated by a narrow and shallow neck, which because of the significant rise of the lake bottom in this place does not allow for direct unrestricted flow of waters from Lake Ryńskie to Lake Tałty. Therefore, although whole complex exchanges about 60% of their waters per year, water exchange in Lake Ryńskie alone is probably very slow. At the northern end of the lake lies the touristic town Ryn (about 3,000 inhabitants) with large yacht port and tourist center hosting several thousand tourists during the summer holidays. Over 50% of the direct catchment area of the lake is covered by arable land, about 30% by forests and the rest (about 21%) by urbanized areas.

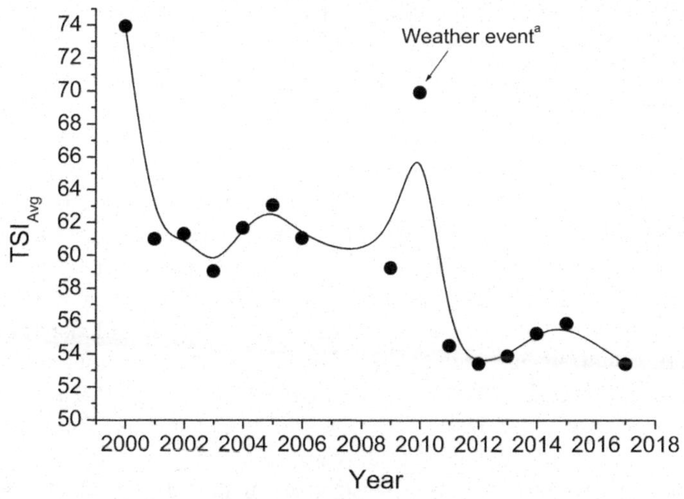

Fig. 7 Changes in the trophic status of Lake Tałtowisko in years 2000–2017 (Weather event[a] – very hot and dry summer period 2010)

The characteristic features of this lake are the large depth (up to 51 m), relatively high and stable TSI, heavy phytoplankton blooms, and physicochemical parameters characteristic for strongly eutrophicated lakes (Table 3). The relatively low production/respiration ratio (1:1) suggests that the lake can be substantially enriched with allochthonous organic matter [7].

In addition to surface runoff, the main source of biogenic substances reaching Lake Ryńskie is an effluent of purified sewage (581 m^3/year) from the municipal sewage treatment plant in Ryn. Monitoring carried out by Polish environmental protection services in 2016 showed a bad ecological potential of the lake. The elements decisive for low assessment were phytoplankton, low water transparency, and anoxic hypolimnetic zone. According to unpublished data (collected by the Department of Microbial Ecology and Environmental Biotechnology, University of Warsaw) summer concentrations of total nitrogen (TN) and total phosphorus (TP) in surface waters of Lake Ryńskie varied from 0.880 to 1.349 mg N/L and from 0.029 to 0.055 mg P/L, whereas in deep anoxic zone, they exceeded 1.351 mg N/L and 0.265 mg P/L, respectively.

Lake Tałty is the one of the deepest lakes of the GMLS (max. depth 44.7 m). In terms of chemical regime and trophic conditions, one could divide it into two different parts: a northern edge situated above Tałcki Channel and connected by narrow and shallow neck with Lake Ryńskie and a southern edge located along the line of water flow through the GMLS. To the end of the twentieth century, the northern part of deep gutter Lake Tałty was supplemented with substantial amounts of biogenic substances by three main sources: system of Jorka River and outputs from Lake Tałtowisko and Lake Ryńskie. During summer months of years 1990–1995 total phosphorus concentrations in surface waters of this lake varied from 0.10 to 0.13 mg P/L, whereas in profundal zone they reached 0.470 mg P/L [7].

The pronounced decline in phosphorus content in the waters of the northern part of Lake Tałty, observed in the beginning of the twenty-first century, was probably an effect of substantial reduction of the import of this element through Jorka River. At the turn of the years 1970–1998, the waters of this river enriched Lake Tałty with significant quantities of P and N carried out from Głębokie Lake which was heavily degraded by rainbow trout aquaculture, quitted in the second half of the 1980s [32]. Phosphorus and nitrogen loads received by the northern part of Lake Tałty from Lake Ryńskie were probably low. The high depth of Ryńskie Lake, high primary production rates [7] binding large quantities of biogenic elements during summer months and relatively slow (probably more than 2 years), water exchange [3, 33] allows to think that the phosphorus retention time in this lake was rather long.

While at present the trophic conditions of the northern part of Lake Tałty seem to be relatively stabilized, its southern part still undergoes substantial but decreasing gradually human impact. The load of biogenic substances leached from Lake Tałtowisko, which reached waters of the northern part of Lake Tałty in the years 1995–2000, was considerably increased by the point input of nitrogen and phosphorus from the inefficiently operating wastewater treatment plant localized in town Mikołajki. According to the data obtained in the municipality of Mikołajki, average capacity of the treatment plant in 2012 was 1,226 m^3/day, and the amount of

N and *P* introduced to the deep waters of the lake was, respectively, 5.4 and 0.11 tons per year. This effect was best seen in depression of the bottom at the southern end of the lake. During summer months (July, August) the greater part of vertical profile (below 6 m depth) was strongly deoxygenated (O_2 conc. < 3–4 mg/L), and the concentrations of DOC and mineral forms of P and N were reaching occasionally even 16 mg C/L, 0.504 mg P/L, and 0.70 mg N/L, respectively [34]. The averaged values of basic physicochemical and biological parameters characterizing surface waters of Tałty Lake during the years 2005–2016 are presented in Table 3.

Lake Mikołajskie elongated from the north to the south and connects Lake Tałty with Lake Śniardwy through shallow narrowing bay. Although it was one of the best-studied lakes in the GMLS, however, the majority of systematic investigations on water chemistry and biology of this lake were carried out in the 1970s and 1980s of the twentieth century [35, 36]. Later studies are relatively few and often occasional [7, 30, 33, 36–38]. Although its northern and northeastern shores along which the city of Mikołajki extends are severed, waters of this lake are still subject to strong fertilization. Two times a year, during mixing periods it is exposed on large amounts of biogenic substances carried out from the bottom depression in the southern edge of the Lake Tałty (Fig. 8). At the peaks of the tourist seasons, when several thousand tourists visit town Mikołajki, substantial amount of pollutants is generated by the yacht port and flows down with storm waters from the streets contaminated more than usual. The shore vegetation is being destroyed by long-term inflow of pollutants generated by ineffective wastewater management, unordered tourism, and agricultural activity in the southeastern edges of its direct catchment area. Additionally, the relatively narrow littoral zone of Lake Mikołajskie constitutes

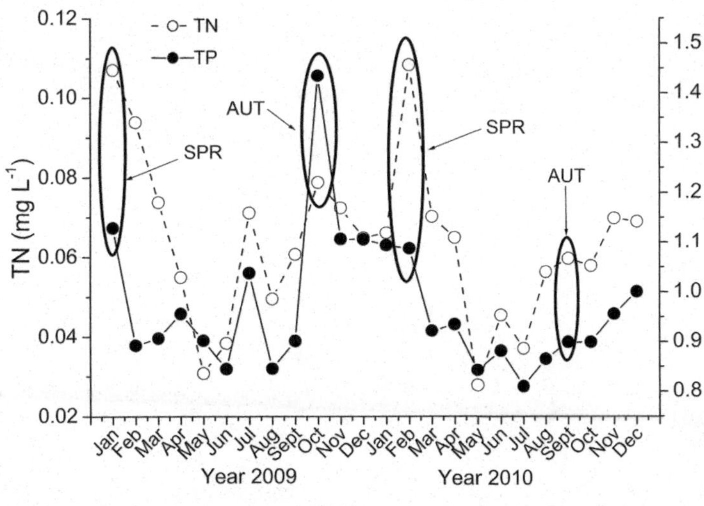

Fig. 8 Delayed effects of translocation of total nitrogen (TN) and total phosphorus (TP) load from the bottom depression in the southern edge of Tałty Lake to surface waters of Lake Mikołajskie. Ellipses marked SPR – translocation of TN and TP after spring mixing periods, AUT – translocation of TN and TP after autumn mixing periods

poor barrier for inflowing wastes. Only southwestern shore of the lake is covered with forest and well protected against inflow of biogenic substances.

All lake waters of the southern part of the Great Mazurian Lake System finally are discharged to Lake Śniardwy. Lake Śniardwy constitutes the largest freshwater lake in Poland (11,340 ha), which is shallow (average 5.8 m depth), and localized in a relatively flat area, which results in high wind water mixing by waves. Large water area and very long diverse shoreline (covered by wetlands, meadows, forests, and agriculture areas) create various water trophic conditions of the different parts of Lake Śniardwy. Generally, the southeastern bay, which receives waters directly from Lake Bełdany and Lake Mikołajskie, is highly eutrophicated contrary to the northern part of Lake Śniardwy (Fig. 9, Table 4).

Selected data of water properties of Lake Śniardwy in summer 2012 (shown in Table 4) indicate that depending on lake area and sampling location (Fig. 9) the lake is generally eutrophic; however, the southeastern part in late summer reached hyper-eutrophic conditions. The clear symptoms of highly eutrophicated conditions (hyper-eutrophy) of this part of Lake Śniardwy were coupled with predominance of unicellular blue-green algae in phytoplankton community. One of the most important factors preventing Lake Śniardwy from heavy eutrophication is high microbial respiratory activity [7] together with sufficient oxygenation due to shallow water body and wind mixing of water column from the surface to bottom sediments.

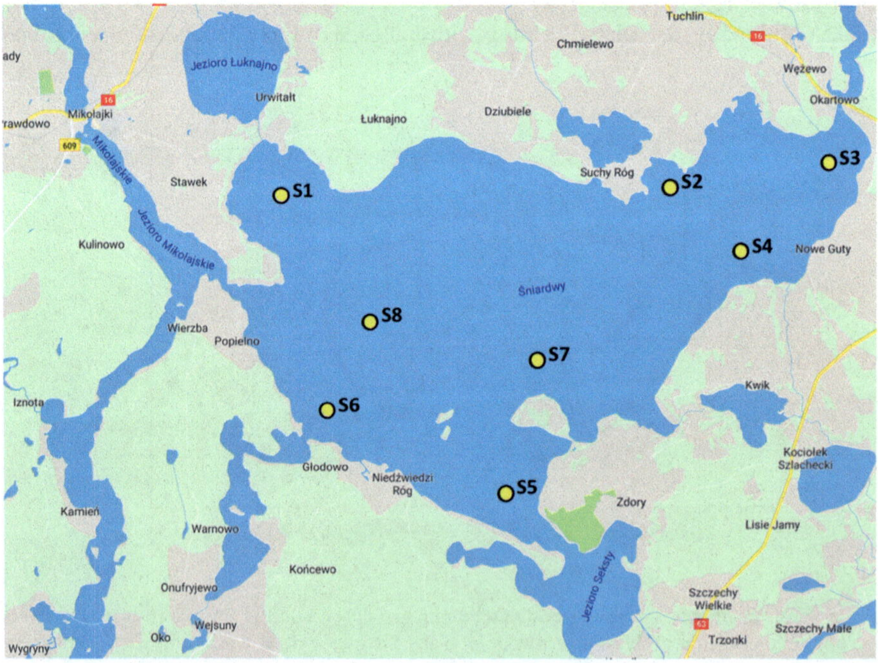

Fig. 9 Lake Śniardwy sampling sites (S1–S8) (July–August 2012)

Table 4 Selected characteristics of the surface water of Lake Śniardwy

Parameter	Range of values of sampling sites S1–S8	Mean and ± standard deviation[a]
Secchi disc (m)	1.12–1.38	1.25 ± 0.11
$Chlorophyll_a$ (µg/L)	21.4–35.8	30.1 ± 6.8
Total P (mg P/L)	0.032–0.045	0.038 ± 0.005
Total N (mg N/L)	0.502–0.846	0.721 ± 0.065
DOC (mg C/L)	12.4–16.2	13.8 ± 1.3
pH	7.9–8.5	13.8 ± 1.3
Bacterial number ($\times 10^6$ cells/mL)	4.3–5.7	4.9 ± 0.2
Bacterial production (µg C/L/day)	30.1–57.6	41.8 ± 0.9
TSI[b]	60.6–65.7	62.6 ± 1.7

Range and mean values in July–August 2012 of the sampling sites shown in Fig. 9
[a]Mean value for surface water calculated from sampling sites S1–S8
[b]Trophic state index (TSI) calculated from $chlorophyll_a$ data according to Carlson [5]

3 Trophic State Evolution of the Great Mazurian Lakes System

Depending on the human impact and usage, and climatic conditions of watershed of the Great Mazurian Lakes System, every particular lake of the system and the total GMLS have significantly changed during the last five decades (Fig. 10).

Very drastic changes of lake water trophic conditions of the GMLS have happened in the 1980s, when its southern part, between Lake Niegocin and Lake Tałty, reached hyper-eutrophic conditions. Since the beginning of the twenty-first century, these hyper-eutrophic lakes have recovered and improved their water quality, and at the present, they show different stages of eutrophication (from moderate eutrophic to high eutrophic state). One of the most important factors responsible for this positive change in water quality were sewage treatment plants built at the end of the twentieth century and changes in regional water and sewage management policy.

Higher rates of eutrophication have been also observed in the northern part of the GMLS (Fig. 10). Unfortunately, the most northern lakes, which were for a long period of time oligotrophic, became mesotrophic and eutrophic during the last decade. This was mostly caused by intensive touristic and recreational activities developed in close vicinity of lakes.

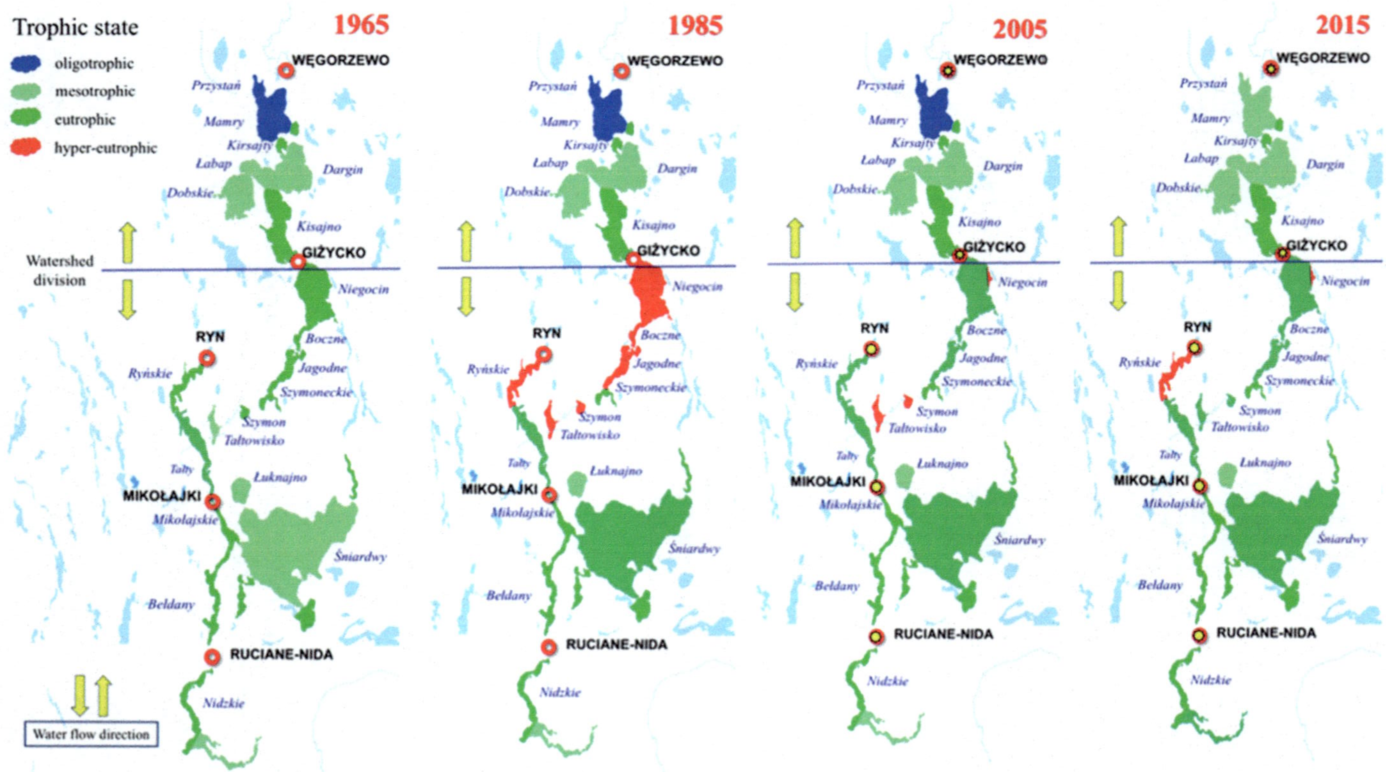

Fig. 10 Trophic conditions of lakes of the Great Mazurian Lake System between 1965 and 2015; yellow circles indicate location of sewage treatment plants

4 Potential Microbial Threats of the Great Mazurian Lake System: Presence of *Legionella* spp. and *Aeromonas* spp. in Lake Water

One of the greatest threats to water quality is accelerated eutrophication, resulting from human activity, like the high intensity of tourism, surface runoffs from fertilized fields, and municipal pollution. Water eutrophication manifests as excessive growth of phytoplankton caused by overabundant nitrogen, phosphorus, and other nutrient supply, which causes deterioration of water quality. In consequence, increased content of organic matter and the amount of microplankton is positively related to the amount of bacterial biomass in eutrophicated water reservoirs [39]. Moreover, in the face of global climate warming, the presence of warming conditions favoring the development of pathogenic microflora and their presence is becoming more likely, at least temporary during periods of warm summers.

There are reports that climate changes related to global warming may affect occurrence and survival of pathogens in the water environments [40]. Manifestations of global warming have been already noticed in Europe in summer 2010 and 2018. We were dealing with heat wave which contributed to the increase of the Great Mazurian Lakes System water temperature above 25–27°C. It was proven that this incident would not have occurred without climate warming [41]. The aforementioned circumstances prompted to assess the impact of trophic state and physicochemical parameters of water of the GMLS on the occurrence and amount of *Legionella* and *Aeromonas* spp.

Both *Legionella* and *Aeromonas* are closely related to environmental water quality, and their presence, especially in large numbers, poses a serious threat to human and animal health [42]. *Legionella* spp. cover more than 61 species, among them 22 are responsible for human diseases. Pathogenic strains of *Legionella* sp. contribute to severe pneumonia called Legionnaires' disease or less-serious Pontiac fever. Inhalation of *Legionella*-contaminated aerosols is the route of infection [43]. This microorganism strictly associated with man-made water systems is often reported in natural water reservoirs [44]. The studies prove that it is able to survive at temperatures from 0 to 63°C and multiplies at temperature ranging from 20 to 45°C [45]. In turn *Aeromonas* sp. initially associated with fishes and cold-blooded animal pathogen is currently described also as emerging human pathogen. Among all *Aeromonas* species, *Aeromonas hydrophila* is at the forefront of causing diseases in human. The main illnesses caused by contagion of *Aeromonas* species are gastrointestinal, skin, soft tissue, and urinary tract infections [46, 47].

Figure 11 presents the results of principal component analysis, aimed to display the impact of various physicochemical and biological parameters on the number of *Legionella* spp. and *Aeromonas* spp. in lakes with different trophic states.

The studies were conducted based on samples collected during summer season in 2016. For the analysis, 16 lakes belonging to the system of the Great Mazurian Lakes System covering the entire geographical location were selected. These lakes differed in the trophic status calculated on the basis of the phosphorus and $chlorophyll_a$ concentration and Secchi disc visibility according with Carlson [5]. Therefore, the studied lakes included water reservoirs from mesotrophic, with

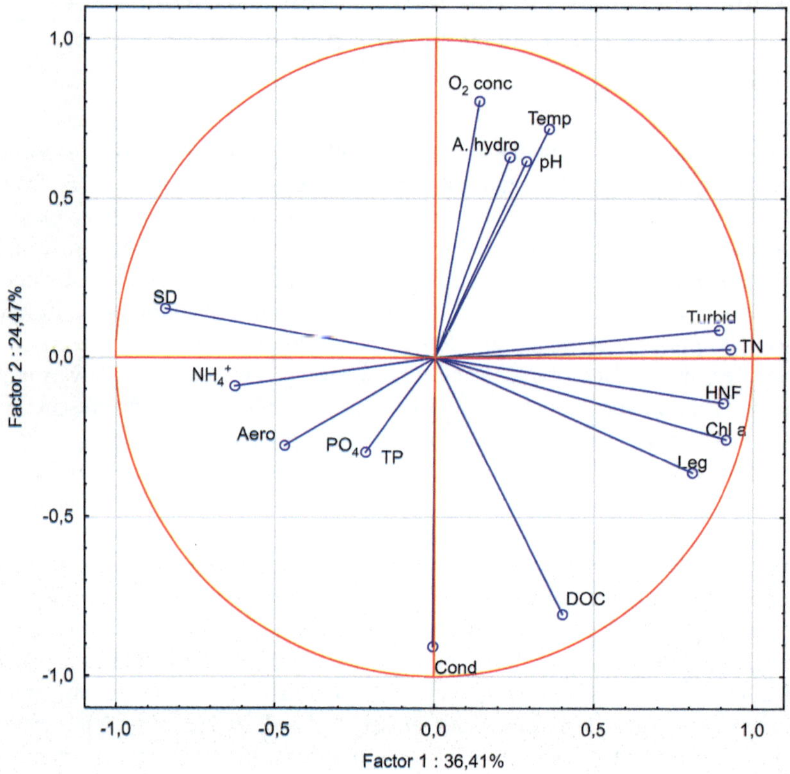

Fig. 11 Principal component analysis (PCA) results showing the impact of physicochemical and biological water parameters of the Great Mazurian Lakes System on *Legionella* spp. and *Aeromonas* spp. abundance. *Cond* conductivity (μs/cm), *DOC* dissolved organic carbon (mg C/L), *Leg Legionella* spp. number (copies/L), *Chl a* chlorophyll *a* concentration (μg/L), *HNF* heterotrophic nanoflagellates (cells/mL), *TN* total nitrogen amount (μg/L), *Turbid* turbidity (NTU), *Temp* temperature (°C), *A. hydro A. hydrophila* number (copies/L), O_2 *conc* oxygen concentration (mg O_2/L), *SD* Secchi disc visibility (m), NH_4^+ ammonium concentration (mg N/L), *TP* total phosphorus concentration (μg P/L), PO_4 orthophosphate concentration (μg P/L), *Aero Aeromonas* spp. number (copies/L)

mean TSI value equals 42.2 in the case of the least eutrophicated lake, to hypereutrophic ones (mean TSI = 70). The principal component analysis indicated that the amount of chlorophyll$_a$, total nitrogen content, water turbidity, and presence of heterotrophic nanoflagellates (HNF) mainly promoted the occurrence of *Legionella* spp. The amount of dissolved organic matter also slightly promoted the presence of *Legionella* spp.

In turn, the presence of *Aeromonas* spp. was promoted by concentrations of total phosphorus and orthophosphates, as well as, ammonium concentration (Fig. 11). Furthermore, the number of *Aeromonas hydrophila* was included in the analysis, and as it turned out, inter alia temperature is a factor promoting its occurrence. All these factors affecting the number of potentially pathogenic bacteria are closely related to eutrophication, and the influence of temperature on *Aeromonas hydrophila* presence can be a sign of the threat posed by climate warming. Particularly noteworthy is the fact that increased amount of nutrients, which is the main reason of eutrophication resulting from excessive anthropogenic pressure, affects the number of *Legionella* spp. and *Aeromonas* spp. Also, chlorophyll$_a$ concentration, which is a measure of the phytoplankton amount and one of the main indicators of water body's trophic state, was positively correlated with the *Legionella* spp. number [48]. Water turbidity, which is a parameter of general amount of suspension in water consisting living microorganisms and dead organic and mineral matter, was also correlated with *Legionella* spp. number. It is important due to the fact that as the turbidity increases, the trophic state of water also rises [49].

5 Antibiotic Resistance of Bacterial Community in the Great Mazurian Lake System

Antibiotics are among one of most often used antimicrobial drugs in modern pharmacology. Because of their irresponsible, excessive use, antibiotic resistance has become an important issue of current microbiology. There are more than 200 different antibiotics registered for use in medicine and veterinary [50]. They belong to many different groups, wherein beta-lactams account for approximately 55–70% of total antibiotic use. Antibiotic consumption, expressed in doses (DDD), increased by 65% (from 21.1 to 34.8 billion DDDs) between 2000 and 2015 [51], and the global consumption of antimicrobials will increase by 67%, from 63.51 $\pm$ 1.560 tons to 105.596 $\pm$ 3.605 tons between 2010 and 2030 [52].

Several antibiotics, such as certain beta-lactams or streptomycins (up to 80%), are naturally produced by microorganisms, and the soil environment is the largest reservoir of bacteria and fungi capable of producing natural antibiotics [53, 54]. Flow from the catchment may play some role in antibiotic input into the aquatic environment. Nevertheless, the largest source of antibiotics and antibiotic-resistant bacteria is human medicine and the food production sector, including animal and plant production and/or aquacultures. The presence of antibiotic resistance genes in natural water is often associated with their transfer from wastewater treatment plants, which receive municipal and hospital wastewater. In this type of sewage, resistant microorganisms are often present and can be transferred into the environment along with the outflow. For example, Korzeniewska et al. [55] have demonstrated the presence of extended-spectrum beta-lactamases (ESBL)-producing *Escherichia coli* from the hospital and the municipal sewage and water treatment area of wastewater treatment plants located in Olsztyn (Poland). The presence of relatively high concentration of antibiotics in Polish sewage treatment plant has also been confirmed [56]. It is no surprise that antibiotic-resistant strain of bacteria is found in many aquatic habitats in Poland as in the Baltic Sea, rivers, and many lakes [57–59].

Though there are many studies concerning antibiotic resistance in the environment, up-to-date information about antibiotic concentration and occurrence of antibiotic-resistant bacteria in GMLS, the biggest lake complex in Northern Poland, are strongly limited. Hereby, we present the preliminary results of an assessment of antibiotic resistance in the bacterial community inhabiting GMLS. Water samples were taken from littoral zone (1–3 m depth) during summer 2016 from the selected lakes of the GMLS.

For the assessment of the expressed antimicrobial drug resistance at the community level, the Biolog PM MicroArrays plate's tests (PM11C i PM12B, Biolog Inc., USA) were used [60]. The analyses were conducted according to the Biolog (USA) manufacturer procedure with modifications. Before measurement, the samples were preincubated for 12 h in 24°C with 0.1× Biolog IF-10b GN/GP Base for GN and GP bacteria (Biolog, Inc.) to avoid potential carbon limitation. For fungi inhibition, 5 fluorocytosine was added (final conc. 0.3 mg/ml). After preincubation, the optical densities of samples were checked; the initial differences were lower than 5%. For respiration intensity detection, the ×1 Biolog Redox Dye Mix D and samples were transferred to the plate wells. The absorbance (ABS) of reduced Redox Dye Mix D was measured every 2 h for 12 h in Synergy H1 (590 nm, BioTek Corporation). Only the wells with maximum concentration of every antibiotic were included in the analysis. Wells without any antimicrobial drugs treated in the same way as wells containing antibiotics or other antimicrobial drugs were used as control wells.

The relative respiration intensity was calculated as the difference between ABS measured in time 0 and time 12 h. The potential antibiotic influence on the bacterial community was calculated as a percentage of respiration in wells containing specific antibiotic or antimicrobial drug compared to respiration in control wells (separately for every lake sample). Table 1 contains groups (first column) and names (second column) of all tested antimicrobial drugs.

We used the following scale to define the relative susceptibility of bacterial communities to tested antimicrobial drugs. When the respiration rates in antibiotic-containing well exceeded 90% of the respective rates in the control well, we assumed that the bacterial community was resistant to tested antibiotics and belongs to class IV. When the respiration rates were lower than 10% of the control, we assumed the bacterial community to be highly susceptible to tested antibiotics and belong to class 1. Between these values we defined the bacterial community as moderately susceptible (class 2, from 10% to 50% of the control respiration rates) and moderately resistant (class 3, from 50% to 90% of the control respiration rates).

Table 5 shows the susceptibility profile of bacterial communities from nine lakes belonging to the GMLS. The highest susceptibility of bacteria in all lakes was recorded with regard to a wide spectrum of macrolides, aminoglycosides, and tetracyclines impairing protein synthesis (like rifampicin, spiramycin, chlortetracycline, minocycline) or DNA replication (like novobiocin). We found the highest resistance of bacteria to beta-lactams, which may be caused by their high occurrence in natural environments. The high resistance of aquatic bacteria to beta-lactams was well documented [61].

Higher antibiotic resistance was demonstrated for the microorganisms present in less eutrophicated, northern lakes (Fig. 12). For example, bacteria communities from Lake Przystań were resistant (class 4) or moderately resistant (class 3) to up to

Table 5 Relative impact of the various antibiotics and other antimicrobial compounds influence on bacterial communities inhabiting lakes of the GMLS

Group	Name	Lake[a]								
		P	M	K	N	J	T_a	T	M_i	Ś
Amino-coumarins	**Novobiocin**	**2**	**1**	**1**	**1**	**1**	**1**	**1**	**1**	**1**
Aminocyclitol antibiotics	Spectinomycin	3	3	3	2	3	3	2	2	2
Aminoglycosides	Amikacin	2	2	2	2	2	2	1	1	2
	Capreomycin	2	2	2	2	1	2	1	1	1
	Gentamicin	2	2	2	3	2	2	1	2	2
	Kanamycin	2	2	2	2	2	2	1	2	2
	Neomycin	**2**	**1**	**2**	**2**	**1**	**1**	**1**	**1**	**2**
	Paromomycin	2	2	2	2	1	2	1	2	1
	Sisomicin	2	2	2	3	2	2	1	2	2
	Tobramycin	2	2	2	2	1	1	1	1	1
Chloramphenicol	Chloramphenicol	2	2	2	2	1	1	1	2	1
Cyclic peptides	Colistin	4	4	4	4	3	3	2	3	2
Fluoroquinolones	Enoxacin	2	2	2	1	1	2	2	2	1
	Lomefloxacin	2	2	2	2	2	3	2	2	1
	Ofloxacin	3	3	2	2	2	3	2	2	2
Glycopeptides	Bleomycin	3	3	3	3	3	4	2	3	3
	Vancomycin	2	2	2	2	2	2	2	2	2
Macrolides	**Spiramycin**	**2**	**1**	**1**	**1**	**1**	**1**	**1**	**1**	**1**
	Erythromycin	2	2	2	2	1	1	2	2	1
	Lincomycin	2	2	2	2	2	2	2	2	2
	Rifampicin[b]	**2**	**2**	**2**	**1**	**2**	**1**	**1**	**1**	**1**
Polypeptides	Polymyxin B	3	4	3	3	3	4	2	4	4
Quinolones	Nalidixic acid	2	2	2	2	2	3	2	2	1
Sulfanilamides	Sulfamethazine	3	4	3	3	3	3	2	2	2
	Sulfadiazine	3	3	3	3	2	3	2	2	2
	Sulfathiazole	3	3	3	3	3	3	2	2	2
	Sulfamethoxazole	3	4	3	3	3	3	3	2	2
Tetracyclines	**Chlortetracycline**	**2**	**2**	**2**	**1**	**1**	**1**	**1**	**1**	**1**
	Demeclocycline	2	2	2	2	2	2	2	2	2
	Tetracycline	2	2	2	2	2	2	2	2	2
	Penimepicycline	3	2	3	2	1	2	2	2	1
	Minocycline[c]	**1**	**1**	**1**	**1**	**1**	**1**	**1**	**1**	**1**
Beta-lactams	Amoxicillin	2	3	3	2	2	2	2	2	2
	Cefazolin	*3*	*3*	*3*	*3*	*4*	*4*	*4*	*4*	*4*
	Ceftriaxone	3	3	3	2	3	2	3	3	3
	Cephalothin	*3*	*4*	*4*	*3*	*4*	*3*	*4*	*4*	*3*
	Cloxacillin	2	2	2	2	2	2	2	2	2
	Nafcillin	2	2	2	2	2	2	2	2	2
	Penicillin	*4*	*4*	*4*	*3*	*3*	*3*	*3*	*3*	*4*
	Oxacillin	2	2	2	2	2	2	2	2	2
	Carbenicillin	*3*	*4*	*4*	*3*	*3*	*3*	*3*	*3*	*3*

(continued)

Table 5 (continued)

Group	Name	Lake[a]								
		P	M	K	N	J	T_a	T	M_i	Ś
Other	Potassium tellurite	3	3	3	3	3	3	2	2	3
	2,4-Diamino-6,7-diisopropyl-pteridine	2	3	3	2	2	3	2	2	3
	D,L-Serine hydroxamate	2	2	2	1	1	1	1	2	2
	Benzethonium chloride	3	2	3	3	2	2	2	2	2
	5-Fluoroorotic acid	2	2	2	2	2	2	2	2	2
	L-Aspartic-B-hydroxamate	2	1	2	2	2	2	2	2	2
	Dodecyltrimethylammonium bromide	3	2	2	2	2	2	2	2	3

Numbers correspond to the susceptibility of bacteria to tested chemicals: 1, highly sensitive (less than 10% of control respiration); 2, moderately sensitive (10–50% of control respiration); 3, moderately resistant (50–90% of control respiration); and 4, resistant (more than 90% of control respiration). The italic characters indicate antibiotics that are characterized by low effectiveness against aquatic bacteria living in the studied lakes (only 3 or 4 for all studied lakes). The bold characters indicate the most effective antibiotics (mostly in class 1)

[a]Lake: *P* Przystań, *M* Mamry, *K* Kisajno, *N* Niegocin, *J* Jagodne, T_a Tałtowisko, *T* Tałty, M_i Mikołajskie, *Ś* Śniardwy

[b]Macrolide analogue

[c]Tetracycline analogue

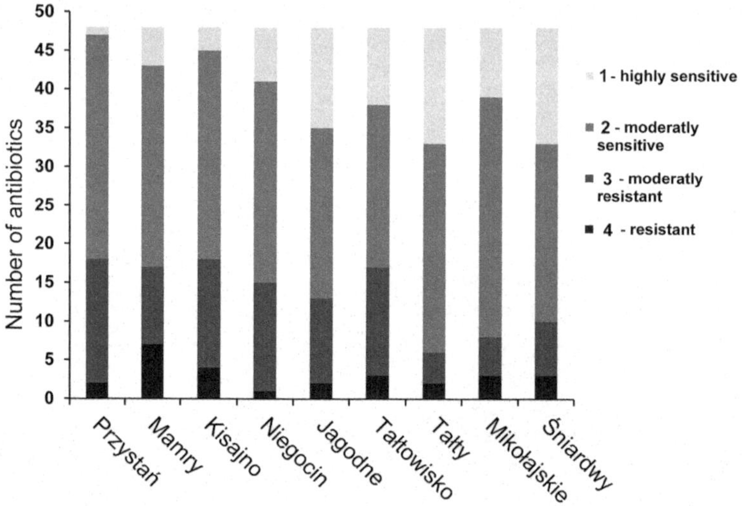

Fig. 12 Summarized numbers of antibiotic and antimicrobial compounds belonging to four classes of effectiveness in microbial respiration inhibition

18 tested chemicals and highly sensitive to only 1 tested antibiotic minocycline. In the southern lakes, generally more eutrophicated, except from Lake Śniardwy, bacteria were highly susceptible (class 1) to 9–15 tested chemicals. However, this result does not necessarily mean that the southern lakes were poorer in potentially antibiotic-resistant bacteria species.

The test used by us is based on the measurement of bacterial respiratory activity. Thus, we did not test the potential antibiotic resistance of bacterial community (the presence of antibiotic resistance genes) but the expressed antibiotic resistance by active microorganisms. The activity of individual bacteria in aquatic ecosystems, also GMLS, is a complex phenomenon and not yet fully explained [61–63]. It is affected by many factors such as the bacterivorous pressure, the competition between bacteria of different species, the quality and quantity of available organic carbon, as well as the physicochemical conditions, including oxygen saturation [64]. Favorable living conditions in a given lake increase the respiratory activity of bacteria, allowing them to cope more easily with the threat of antibiotics. Bacteria living in sub-optimal conditions are exposed to many stress factors, which may handicap the mechanism of active antibiotic detoxification, for instance, energy dependence efflux of tetracyclines. Therefore, due to difficult conditions in eutrophicated environments (strong nutrient competition, potential toxic metabolites of e.g., cyanobacteria, bacterivorous pressure, less carbon source liability), the potential antibiotic resistance may not be fully expressed. This could also explain the lack of the expected influence of sewage treatment plants on bacterial resistance in Lake Tałty and Lake Niegocin (where two sewage treatment plants are located). The relatively long distance (few hundred meters) of the sampling area from the outflow of effluent wastewater and the potentially insufficient survival rate of sewage-derived bacteria in natural lake water may further explain these results. Additionally, using Illumine sequencing we found (data not yet published) that the phylogenetic composition of the bacteria community from the northern and southern parts of the GMLS differs significantly, which may also influence the antibiotic resistance pattern.

Acknowledgments These studies were financially supported by the National Science Centre, Poland, grant OPUS 2015/17/B/NZ9/01552 awarded to R.J. Chróst and grant NN304 080135 awarded to W. Siuda. Field studies were performed in the Research Station in Mikołajki of Nencki Institute of Experimental Biology of Polish Academy of Sciences.

References

1. Siuda W, Kaliński T, Kauppinen ES, Chróst RJ (2014) Eutrofizacja południowej części kompleksu Wielkich Jezior Mazurskich w latach 1977–2011. Technol Wody 35:48–62
2. Skibniewski L, Mikulski Z (1954) Hydrologia Wielkich Jezior Mazurskich. Wiad Służby Hydr Met IV:21–56
3. Mikulski Z (1966) Bilans Wodny Wielkich Jezior Mazurskich. PIHM 19
4. Podział hydrograficzny Polski (Hydrographic Division of Poland) 1980. IMGW, Warszawa

5. Carlson RE (1977) A trophic state index for lakes. Limnol Oceanogr 22:361–369
6. Chróst RJ, Siuda W (2006) Microbial production, utilization, and enzymatic degradation of organic matter in the upper trophogenic water layer in the pelagial zone of lakes along the eutrophication gradient. Limnol Oceanogr 51:749–762
7. Siuda W, Kauppinen ES, Kaliński T, Chróst RJ, Kiersztyn B (2017) The relationship between primary production and respiration in the photic zone of the Great Masurian Lakes (GMLS), in relation to trophic conditions, plankton composition and other ecological factors. Pol J Ecol 65:303–323
8. Siuda W, Kiersztyn B (2015) Urea in lake ecosystem: the origin, concentration and distribution in relation to trophic state of The Great Masurian Lakes (Poland). Pol J Ecol 63:110–123
9. Schindler DW (1975) Whole-lake fertilization experiments with phosphorus, nitrogen, and carbon. Int Ver Theor Angew Limnol Verh 19:3221–3231
10. Odum EP (1971) Fundamental of ecology3rd edn. WB Saunders, Philadelphia
11. Cohn L (1903) Untersuchungen uber das plankton des Löwentin und einigen anderer Seen Masurens. Zeitschrift für Fischerei und deren Hilfwissenschaften 10:201–331
12. Gieysztor M, Odachowska Z (1958) Observations of the themal and chemical properties of Masurian Lakes in the Giżycko Region. Pol Arch Hydrobiol 4:123–152
13. Ławacz W, Planter M, Stasiak K, Tatur A, Wieckowski K (1978) The past, present and future of three Masurian lakes. Pol Arch Hydrobiol 25:233–238
14. Hillbricht-Ilkowska A (2005) Ochrona jezior i krajobrazu pojeziernego - problemy, procesy, Perspektywy. Kosmos 54:285–302
15. Vollenweider RA (1968) The scientific basis of lake eutrophication, with particular reference to phosphorus and nitrogen as eutrophication factors. Tech Rep DAS/DSI/68.27, OECD, Paris, p 159
16. Bartsch AF (1972) Role of phosphorus in Eutrophication. EPA-R3-72-001, National Environmental Research Center Office of Research and Monitoring, US EPA, Corvallis, p 7
17. Reynolds CS (2003) The development of perceptions of aquatic eutrophication and its control. Ecohydrol Hydrobiol 3:149–163
18. Schindler DW (2006) Recent advances in the understanding and management of eutrophication. Limnol Oceanogr 51(part 2):356–363
19. Soszka H, Cydzik D, Kudelska D (1979) The assessment of water quality in Great Masurian Lakes. Inst Kształtowania Środowiska, Warszawa
20. Cydzik D, Kudelska D, Soszka H (1995) Atlas stanu jezior Polski badanych w latach 1989–1993. Państwowa Inspekcja Ochrony Środowiska, Biblioteka Monitoringu Środowiska, Warszawa
21. Kufel I, Kufel L (1993) Monitoring of the Great Masurian Lakes in 1991. Hydrobiological Station Mikołajki Progress Report 1990–1991. Oficyna wydawnicza Instytut Ekologii PAN, Dziekanów Leśny, pp 12–15
22. Kufel I, Kufel L (1999) Spatial and temporal variability of chlorophyll and nutrients in The Great Masurian Lakes. Hydrobiological Station Mikołajki Progress Report 1996–1997. Oficyna wydawnicza Instytut Ekologii PAN, Dziekanów Leśny, pp 10–13
23. Sakamoto M (1966) Primary production by phytoplankton community in some Japanese lakes and its dependence on lake depth. Arch Hydrobiol 62:l–28
24. Downing J, McCauley E (1992) The nitrogen: phosphorus relationship in lakes. Limnol Oceanogr 37:936–945
25. Jeppesen E, Søndegard M, Jensen JP, Havens K, Anneville O, Carvalho L, Coveney MF, Dencke R, Dokulil M, Foy B, Gerdeaux D, Hampton SE, Kangur K, Köhler J, Körner S, Lammens E, Lauridsen TL, Manca M, Miracle R, Moss B, Nöges P, Perrson G, Philips G, Portielie R, Romo S, Schelske CL, Straile D, Tatrai I, Willen E, Winder M (2005) Lake responses to reduced nutrient loading – an analysis of contemporary long-term data from 35 case studies. Freshw Biol 50:1747–1771
26. Pick FR, DRS L (1987) The role of macronutrients (C, N, P) in controlling cyanobacterial dominance in temperate lakes. New Zeal J Mar Fresh Res 21:425–434

27. Schinldler DW (1977) Evolution of phosphorus limitation in lakes. Science 195:260–262
28. Smith VH (1983) Low nitrogen to phosphorus ratios favor dominance by blue-green algae in lake phytoplankton. Science 221:669–671
29. Nöges P, Kangur K, Nöges T, Reinart A, Simola H, Viljanen M (2008) Highlights of large lake research and management in Europe. Hydrobiologia 599:259–276
30. Kauppinen ES (2014) Trophic state of the Great Masurian Lakes system in the past, present and future – causes, mechanisms and effects of changes. Ph.D. dissertation, University of Warsaw
31. Urabe J, Yoshida T, Gurung TB, Sekino T, Tsugeki N, Nozaki K, Maruo M, Nakayama E, Nakanishi M (2005) The production-to-respiration ratio and its implication in Lake Biwa. Ecol Res 20:367–375
32. Chróst RJ, Siuda W, Halemejko GZ (1984) Long-term studies on alkaline phosphatase activity (APA) in a lake with fish-aquaculture in relation to lake eutrophication and phosphorus cycle. Arch Hydrobiol Suppl 70:1–32
33. Chróst RJ, Siuda W (2013) Stan jakości wód oraz zagrożeń eutrofizacyjnych dla jezior w południowej części kompleksu Wielkich Jezior Mazurskich odprowadzajacych wodę do jeziora Śniardwy. Orzysz 2013. http://www.zemuw.pl/pl/files/docs/JM_Jakosc_wod_WJM_2013.pdf
34. Lopata K (2008) Wpływ Miejskiej Oczyszczalni Ścieków w Mikołajkach na wybrane parametry fizyko-chemiczne wód jeziora Tałty i Jeziora Mikołajskiego. M.Sc. thesis, University of Warsaw
35. Pieczyński E, Rybak JI (1990) Wielkie Jeziora Mazurskie. Bibliografia i indeksy. Wydawnictwo SGGW-AR, Warszawa
36. Lewandowski K, Jakubik B (2018) Littoral and sublittoral malacofauna of the eutrophic Lake Mikołajskie (north-eastern Poland). Folia Malacol 26:71–82
37. Ozimek T (2006) The possibility of submerged macrophyte recovery from a propagule bank in the eutrophic Lake Mikołajskie (North Poland). Hydrobiologia 570:127–131
38. Sieńska J, Dunalska J, Łopata M, Parszuto K, Tandyrak R (2016) Trophic state and recreational value of Lake Mikołajskie. Limnol Rev 16:147–153
39. Yang X, Wu X, Hao H, He Z (2008) Mechanisms and assessment of water eutrophication. J Zhejiang Univ Sci B 9:197–209
40. De Toni A, Touron-Bodilis A, Wallet W (2009) Impact of climate change n pathogenic aquatic microorganisms: some examples. Environ Risque Sante 8:311–321
41. Rahmstorf S, Coumou D (2011) Increase of extreme events in a warming world. PNAS 108:17905–17909
42. Percival S, Chalmers R, Embrey M, Hunter P, Sellwood J, Wyn-Jones P (2004) Microbiology of waterborne diseases: microbiological aspects and risks. Elsevier, San Diego
43. Lizana X, López A, Benito S, Augistí G, Ríos M, Piqué N, Marqués AM, Codony F (2017) Viability qPCR, a new tool for Legionella risk management. Int J Hyg Environ Health 220:1318–1324
44. Barna Z, Kàdàr M, Kàlmàn E, Scheirich Szax A, Vargha M (2015) Prevalence of Legionella in premise plumbing in Hungary. Water Res 90:71–78
45. Devos L, Boon N, Verstraete W (2005) Legionella pneumophila in the environment: the occurrence of a fastidious bacterium in oligotrophic conditions. Rev Environ Sci Biotechnol 4:61–74
46. Janda JM, Abbot SL (2010) The genus Aeromonas: taxonomy, pathogenicity and infection. Clin Microbiol Rev 23:35–73
47. Martino ME, Fasolato L, Montemurro F, Novelli E, Cardazzo B (2014) Aeromonas spp.: ubiquitous or specialized bugs? Environ Microbiol 16:1005–1018
48. Petrucio MM, Medeiros AO, Rosa CA, Barbosa FAR (2005) Trophic state and microorganisms community of major sub-basins of the middle Rio Doce basin, Southeast Brazil. Braz Arch Biol Technol 48:625–633
49. Doan PTK, Némery J, Schmid M, Gratiot N (2015) Eutrophication of turbid tropical reservoirs: scenarios of evolution of the reservoir of Coitzo, Mexico. Ecol Inform 29:192–205
50. Kümmerer K (2009) Antibiotics in the aquatic environment - a review - part II. Chemosphere 75 (4):417–434. https://doi.org/10.1016/j.chemosphere.2008.12.006

51. Klein EY, Van Boeckel TP, Martinez EM, Pant S, Gandra S, Levin SA, Laxminarayan R (2018) Global increase and geographic convergence in antibiotic consumption between 2000 and 2015. Proc Natl Acad Sci U S A 115(15):E3463–E3470. https://doi.org/10.1073/pnas.1717295115
52. Van Boeckel TP, Brower C, Gilbert M, Grenfell BT, Levin SA, Robinson TP, Laxminarayan R (2015) Global trends in antimicrobial use in food animals. Proc Nat Acad Sci 112:5649–5654. https://doi.org/10.1073/pnas.1503141112
53. Peláez F (2006) The historical delivery of antibiotics from microbial natural products - can history repeat? Biochem Pharmacol 71(7):981–990. https://doi.org/10.1016/j.bcp.2005.10.010
54. Baltz RH (2008) Renaissance in antibacterial discovery from actinomycetes. Curr Opin Pharmacol 8(5):557–563. https://doi.org/10.1016/j.coph.2008.04.008
55. Korzeniewska E, Korzeniewska A, Harnisz M (2013) Antibiotic resistant Escherichia coli in hospital and municipal sewage and their emission to the environment. Ecotoxicol Environ Saf 91:96–102. https://doi.org/10.1016/j.ecoenv.2013.01.014
56. Ziembińska-Buczyńska A, Felis E, Folkert J, Meresta A, Stawicka D, Gnida A, Surmacz-Górska J (2015) Detection of antibiotic resistance genes in wastewater treatment plant - molecular and classical approach. Arch Environ Prot 41:23–32. https://doi.org/10.1515/aep-2015-0035
57. Harnisz M, Korzeniewska E, Gołaś I (2015) The impact of a freshwater fish farm on the community of tetracycline-resistant bacteria and the structure of tetracycline resistance genes in river water. Chemosphere 128:134–141. https://doi.org/10.1016/j.chemosphere.2015.01.035
58. Giebułtowicz J, Tyski S, Wolinowska R, Grzybowska W, Zaręba T, Drobniewska A, Nałęcz-Jawecki G (2018) Occurrence of antimicrobial agents, drug-resistant bacteria, and genes in the sewage-impacted Vistula River (Poland). Environ Sci Pollut Res Int 25:5788–5807. https://doi.org/10.1007/s11356-017-0861-x
59. Mudryk ZJ, Kosiorek A, Perliński P (2013) In vitro antibiotic resistance of Vibrio-like organisms isolated from seawater and sand of marine recreation beach in the southern Baltic Sea. Hydrobiologia 70(1):141–150. https://doi.org/10.1007/s10750-012-1317-4
60. Chojniak J, Jałowiecki Ł, Dorgeloh E, Hegedusova B, Ejhed H, Magnér J, Płaza G (2015) Application of the BIOLOG system for characterization of Serratia marcescens ss marcescens isolated from onsite wastewater technology (OSWT). Acta Biochim Pol 62:799–805. https://doi.org/10.18388/abp.2015_1138
61. Kiersztyn B, Siuda W, Chróst RJ (2012) Persistence of bacterial proteolytic enzymes in lake ecosystems. FEMS Microbiol Ecol 80:124–134. https://doi.org/10.1111/j.1574-6941.2011.01276.x
62. Shaikh S, Fatima J, Shakil S, SMD R, Kamal MA (2015) Antibiotic resistance and extended spectrum beta-lactamases: types, epidemiology and treatment. Saudi J Biol Sci 22(1):90–101. https://doi.org/10.1016/j.sjbs.2014.08.002
63. Kiersztyn B, Siuda W, Chróst RJ (2017) Coomassie blue G250 for visualization of active bacteria from lake environment and culture. Pol J Microbiol 66:365–373. https://doi.org/10.5604/01.3001.0010.4867
64. Hibbing ME, Fuqua C, Parsek MR, Peterson SB (2010) Bacterial competition: surviving and thriving in the microbial jungle. Nat Rev Microbiol 8(1):15–25. https://doi.org/10.1038/nrmicro2259

Nutrient Balance of North-Eastern Poland Lakes

Jolanta Grochowska, Renata Tandyrak, Renata Augustyniak, Michał Łopata, and Katarzyna Parszuto

Contents

Abstract In order to determine the nutrient balance two groups of the lakes were selected: ten flow lakes, namely, Mielenko, Karczemne, Klasztorne Małe, Klasztorne Duże, Pasłęk, Wymój, Sarąg, Łęguty, Isąg and Suskie (located in Olsztyn Lake District and Kashubian Lake District), and ten non-flow lakes, namely, Czarne, Długie, Podkówka, Starodworskie, Sukiel, Track, Tyrsko, Kepijko, Paskierz and Podąbek (located in Olsztyn Lake District and in Iławskie Lake District).

The annual phosphorus load introduced into flow lakes ranged from 17.2 kg P (Mielenko) to 7,754.5 kg P (Isąg) and nitrogen from 202.6 kg N (Mielenko) to 81,876.5 kg N (Isąg). Nutrient load was mainly introduced with surface water inflow (46–96%). In the group of flow lakes, Mielenko, Karczemne and Suskie were fed primarily with surface runoff from the basin. The surface watercourses flowing into them were periodic, and the amount of water flowing through them was small. Annual load of phosphorus introduced into non-flow lakes were from 4.4 kg N (Czarne) to 169.8 kg N (Track) and nitrogen from 89.4 kg N (Czarne) to 2,311.8 kg N (Track).

In case of non-flow lakes, the main source of supply was usually the runoff from the direct catchment (from 40% to 97%).

It was found that the majority of flow lakes (except Lakes Isąg and Suskie) showed negative retention in relation to nutrients. Non-flow lakes retain the charge of phosphorus and nitrogen, which is introduced from various sources.

J. Grochowska (✉) · R. Tandyrak · R. Augustyniak · M. Łopata · K. Parszuto
Department of Water Protection Engineering, Faculty of Environmental Sciences, University of Warmia and Mazury in Olsztyn, Olsztyn, Poland
e-mail: jgroch@uwm.edu.pl

E. Korzeniewska, M. Harnisz (eds.), *Polish River Basins and Lakes – Part I*, The Handbook of Environmental Chemistry 86,
https://doi.org/10.1007/978-3-030-12123-5_13

Keywords Catchment area · Flow · Nitrogen · Nutrient balance · Phosphorus

1 Introduction

Lake eutrophication is a natural process resulting from the gradual accumulation of nutrients, increased productivity and slow filling in of the basin with accumulated sediments, silt and organic matter from the watershed and long-lasting phenomenon leading to their disappearance [1–3]. Human activities, such as urbanization, industrialization and agricultural use of areas neighbouring water reservoirs, dramatically increase the rates of eutrophication or even degradation of waters. Lakes situated in cities are particularly endangered as they have a role in pollutions receiving from municipal, industrial and storm wastewater. High nutrient load from anthropogenic sources can result in hypoxic or anoxic conditions in stratified lakes and particularly algal blooms and the development of blooms that may include toxic algal species of cyanobacteria [4, 5]. Phosphorus and nitrogen are elements, which are the most responsible for the water eutrophication [6]. When nutrients are introduced into aquatic ecosystems, they participate in a variety of processes known as the circulation of matter [7, 8].

In order to protect lakes from excessive eutrophication, one should precisely determine the area of supplying the lake – the catchment – and recognize all sources from which nutrients reach the lake water. Estimating the amount of nutrients introduced into water is very difficult, especially when the source of emission is dispersed. In the case of preparing the nutrient balance for a given water reservoir, when estimating charges introduced from various sources, factors and indicators are used to determine the approximate or average values of individual components. Preparing a nutrient balance for a specific water reservoir is a statement of their income and loss in an annual period [9]. The difference between the external load (from the catchment) and the load leaving the lake with outflow is called retention [10]. The size of retention is dependent on hydrological and morphometric factors or the physical-chemical processes which favourable immobilization of a particular compound in the sediment [11].

The aim of the work is to present the nutrient balance in selected flow and non-flow lakes located in north-eastern Poland.

2 Material and Methods

Two groups of lakes were selected to determine the balance of nutrients: ten flow lakes (Mielenko, Karczemne, Klasztorne Małe, Klasztorne Duże, Pasłęk, Wymój, Sarąg, Łęguty, Isąg, Suskie) and ten non-flow lakes (Czarne, Długie, Podkówka, Starodworskie, Sukiel, Track, Tyrsko, Kepijko, Paskierz and Podąbek).

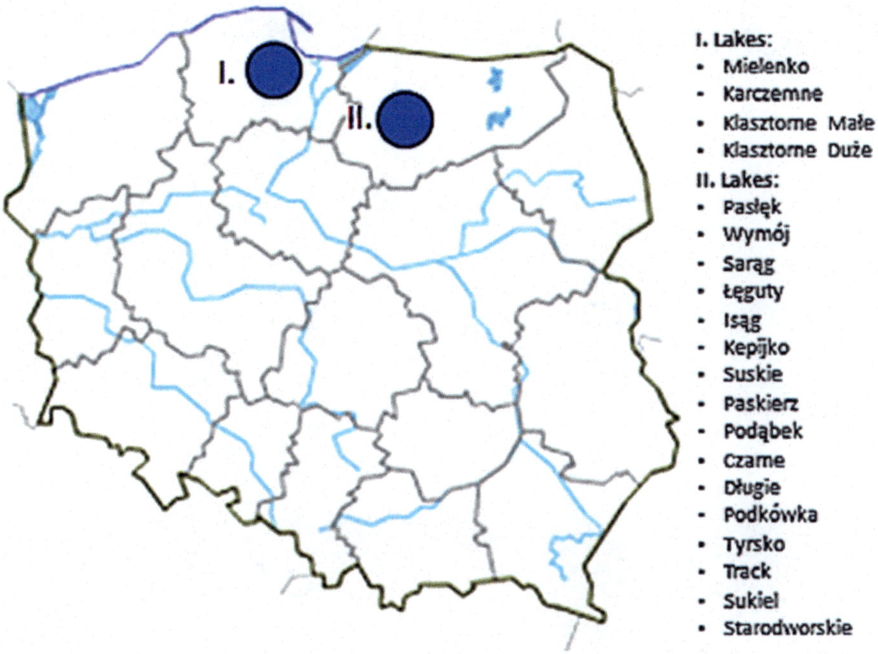

Fig. 1 Location of study lakes

Lakes Pasłęk, Wymój, Sarąg, Łęguty, Isąg, Czarne, Długie, Podkówka, Starodworskie, Sukiel, Track, Tyrsko, Kepijko and Podąbek are located in Olsztyn Lake District (physico-geographical mesoregion in the Masurian Lake District) [12] (Fig. 1). The Olsztyn Lake District stretches along both banks of the upper Łyna, reaching in the west after Pasłęka. The landscape was formed as a result of the last glaciation, whose disappearance phases appear in the form of arches of moraine embankments stretching to the west to Morąg, to the south to Nidzica, and in the east to the Szczytno-Biskupiec line. The height of moraines does not exceed 200 m above sea level.

The ground is mainly covered with boulder clay. In the valleys of glacial gutters and post-lake meadows, there are peat bogs and meadows. Lakes Suskie and Paskierz are situated in Iławskie Lake District (a geographical macroregion in north-eastern Poland. It lies between the Vistula, Osa, Drwęca and Pasłęka. Area – about 4,230 km^2) [12] (Fig. 1). Lakes Mielenko, Karczemne, Klasztorne Małe and Klasztorne Duże are located about 30 km west of Gdańsk, within the administrative boundaries of the city of Kartuzy in the Kashubian Lake District. It is a physical geographical mesoregion belonging to the macroregion of the East Pomeranian Lake District [12] (Fig. 1).

Lakes selected for nutrient balance analysis have a wide range of areas (from 1.5 to 397.5 ha) and depth (from 1.9 to 54.5 m) (Tables 1 and 2).

Table 1 Morphometric data of selected flow lakes (Inland Fisheries Institute in Olsztyn)

Morphometric index	Mielenko	Karczemne	Klasztorne Małe	Klasztorne Duże	Pasłęk	Wymój	Sarąg	Łęguty	Isąg	Suskie
Geographical coordinates	54°19′55″N 18°10′55″E	54°19′42″N 18°11′27″E	54°20′21″N 18°11′35″E	54°20′52″N 18°12′10″E	53°36′21″N 20°19′21″E	53°41′36″N 20°21′03″E	53°41′36″N 20°16′48″E	53°45′05″N 20°09′00″E	53°46′09″N 20°08′02″E	53°42′33″N 19°20′24″E
Location above sea level	204.0	203.7	203.0	202.3	152.9	124.0	112.7	96.3	93.0	100.1
Surface [ha]	7.8	40.4	13.7	57.5	8.5	47.3	183.0	50.9	397.5	62.0
Volume [m^3]	102,900	798,300	1,106,000	2,780,000	325,000	2,413,800	12,627,000	5,234,000	56,189,400	1,377,100
Max. depth [m]	1.9	3.2	20.0	8.5	5.0	16.0	16.5	22.7	54.5	4.1
AD [m]	1.3	1.9	8.1	4.8	3.8	5.1	6.9	8.5	14.2	2.2
Relative depth	0.0068	0.0050	0.0540	0.0112	0.0170	0.0200	0.0120	0.0300	0.0270	0.0052
Depth index	0.7	0.6	0.	0.6	0.7	0.3	0.4	0.4	0.3	0.5
ML [km]	0.460	1.282	0.720	1.320	0.450	1.300	3.200	1.300	4.900	2.251
MW [km]	0.252	0.445	0.250	0.570	0.300	0.500	1.100	0.800	1.100	0.449
Elongation	1.8	2.9	2.9	2.3	1.5	2.6	2.9	1.5	4.5	5.0
LS [km]	1.314	3.163	1.850	4.100	1.100	3.200	9.400	3.700	17.500	5.709
SD	1.3	1.4	1.4	1.5	1.1	1.3	1.9	1.3	2.5	2.0

MD maximum depth, *AD* average depth, *ML* maximum length, *MW* maximum width, *LS* length of shoreline, *SD* shoreline development

Table 2 Morphometric data of selected non-flow lakes (Inland Fisheries Institute in Olsztyn)

Morphometric index	Paskierz	Kepijko	Czarne	Długie	Starodworskie	Sukiel	Podkówka	Tyrsko	Track	Podąbek
Geographical coordinates	53°42′48″N 19°59′10″E	53°41′03″N 20°24′04″E	53°46′09″N 20°27′03″E	53°47′02″N 20°27′08″E	53°44′09″N 20°27′02″E	53°47′03″N 20°26′03″E	53°47′02″N 20°27′01″E	53°48′03″N 20°25′05″E	53°48′03″N 20°33′06″E	53°48′06″N 20°41′06″E
Location above sea level	96.9	119.3	103.7	102.8	110.0	109.0	106.5	105.0	123.9	110.7
Surface [ha]	14.3	7.8	1.5	26.8	6.0	20.8	6.9	18.6	52.8	4.3
Volume [m^3]	364,300	248,100	39,700	1,414,800	540,000	1,365,800	197,369	1,786,100	1,123,000	228,200
MD [m]	10.6	9.0	7.1	17.3	23.3	25.0	6.0	30.4	3.8	17.5
AD [m]	2.6	3.2	2.7	5.3	9.0	6.6	2.8	9.6	2.1	5.3
Relative depth	0.0280	0.0320	0.0580	0.0334	0.0950	0.0550	0.0230	0.0700	0.0052	0.0844
Depth index	0.2	0.4	0.4	0.3	0.4	0.3	0.5	0.3	0.6	0.3
ML [km]	0.750	0.420	0.175	1.670	0.345	0.725	0.365	0.630	1.500	0.270
MW [km]	0.250	0.300	0.110	0.240	0.213	0.400	0.280	0.455	0.575	0.200
Elongation	3.0	1.4	1.6	6.9	1.6	1.8	1.3	1.4	2.6	1.3
LS [km]	1.800	3.240	0.450	4.080	0.900	2.100	1.380	1.620	4.650	0.775
SD	1.3	1.3	1.0	2.2	1.0	1.3	1.5	1.1	1.9	1.1

MD maximum depth, *AD* average depth, *ML* maximum length, *MW* maximum width, *LS* length of shoreline, *SD* shoreline development

Table 3 The catchment areas of analysed lakes

Selected flow lakes	Total catchment area [km^2]	The direct catchment area [km^2]	Selected non-flow lakes	The direct catchment area [km^2]
Mielenko	3.8	0.22	Paskierz	0.30
Karczemne	5.15	0.45	Kepijko	1.95
Klasztorne Małe	7.45	0.13	Czarne	0.13
Klasztorne Duże	12.2	1.03	Długie	1.15
Pasłęk	5.4	0.80	Starodworskie	0.30
Wymój	22.6	0.60	Sukiel	1.02
Sarąg	187.2	10.50	Podkówka	0.26
Łęguty	241.5	1.10	Tyrsko	0.68
Isąg	246.7	4.90	Track	2.48
Suskie	8.5	5.38	Podąbek	0.76

The surface of the catchment of the analysed lakes was determined based on a topographic map at a 1:10,000 scale and field observations in the case of a dubious course of given water divide. Measurements of the surface of the catchment were taken with polar (PL–1) and cylinder (KP90N) planimeters (SOKKIA, Japan). The size of the lake basin is shown in Table 3 and their land-use management in Fig. 2a, b.

In 2017 morphometric measurements of water inflowing and outflowing from the lakes Mielenko, Karczemne, Klasztorne Małe, Klasztorne Duże, Pasłęk, Wymój, Sarąg, Łęguty, Isąg, Kepijko, Paskierz and Podąbek (depth, width) were taken, and water velocity was measured with a Valeport (model 801) electromagnetic flowmeter. The calculations of the amount of nutrients that are annually brought to the lakes with inflow and carried out from the outflow were performed based on their actual concentration in water (TP and TN) and flows measured on the individual stations during yearly field studies. Nutrient loads were calculated with a generally accepted method of time periods. In the water samples, the consecutive parameters were determined: total phosphorus (after mineralization with sulphuric acid and ammonium persulphate by spectrophotometry with ammonium molybdate and tin (II) chloride, Macherey-Nagel NANOCOLOR UV/VIS) and total nitrogen (carbon and nitrogen analyser type IL 550 TOC-TN, HACH Inc.).

Data for other lakes were obtained from the Lossow [13] study and Dunalska [14].

The balances of nutrients (phosphorus and nitrogen) were calculated as follows:

$$\mathrm{LDO} + \mathrm{LDC} + \mathrm{LP} + \mathrm{LR} + \mathrm{LF} + \mathrm{LAT} + \mathrm{LB} = \mathrm{LOD} + R$$

where expressed in kg $year^{-1}$:

LDO – load discharged with surface water inflow
LAS – load discharged from the direct catchment-area sources

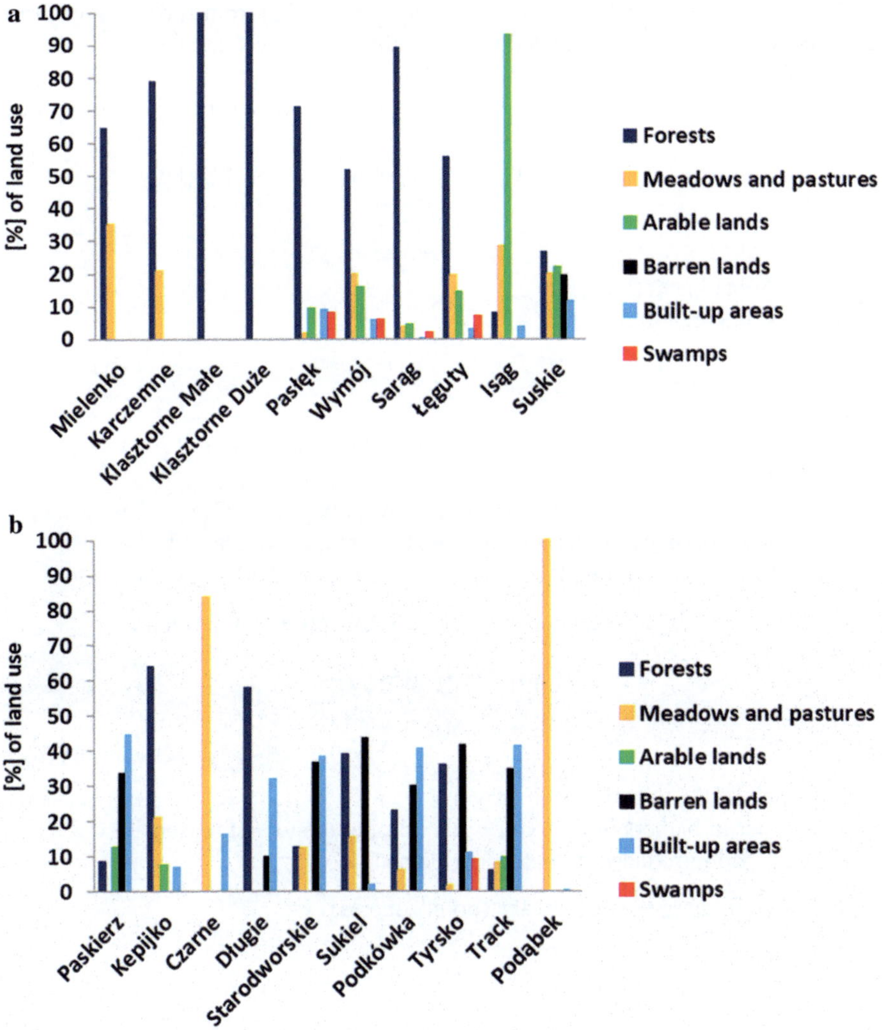

Fig. 2 (**a**) The land use of direct catchment (flow lakes). (**b**) The land use of direct catchment (non-flow lakes)

LP – load discharged from the point sources
LD – load discharged from dispersed sources
LA – load discharged by anglers with groundbait
LAT – load discharged from the atmosphere delivered with precipitation
LB – load discharged from bathers
LOD – load flowing out along with the outflow water
R – retention of given element in the lake ecosystem

The magnitude of a lake load with nutrients originating from surface flows from the direct catchment was calculated with a method that is recommended and applied by OECD; this method consists of calculating the loads with flow coefficients [15] which depend on land management way and the denivelation surface of a given area.

$$\mathrm{LAS} = A_{\mathrm{go}}\, W_{\mathrm{go}} + A_{\mathrm{uz}}\, W_{\mathrm{uz}} + A_l\, W_l + A_n\, W_n + A_z\, W_z \ \left(\mathrm{kg\ y^{-1}}\right)$$

where:

A_{go} – arable land area (ha)
A_{uz} – grassland area (ha)
A_l – forest area (ha)
A_n – barren land area (ha)
A_z – urban area (ha)
W_{go} – the flow coefficient of TP or TN from arable lands (kg $\mathrm{ha^{-1}\ y^{-1}}$)
W_{uz} – the flow coefficient of TP or TN from grassland (kg $\mathrm{ha^{-1}\ y^{-1}}$)
W_l – the flow coefficient of TP or TN from forests (kg $\mathrm{ha^{-1}\ y^{-1}}$)
W_n – the flow coefficient of TP or TN from barren lands (kg $\mathrm{ha^{-1}\ y^{-1}}$)
W_z – the flow coefficient of TP or TN from urban areas (kg $\mathrm{ha^{-1}\ y^{-1}}$)

The load taken by the bathers (LB) was determined by the formula:

$$\mathrm{LB} = \mathrm{NB} \cdot \mathrm{WB} \cdot D$$

where:

D – duration of the tourist season (days)
NB – number of bathers in reservoir waters during the tourist season
WB – unit load introduced by bathers (g $\mathrm{person^{-1}}$)

The load discharged with sewage (point sources)

$$\mathrm{LP} = L_{\mathrm{DO}} \cdot V \cdot D$$

where:

L_{DO} – TP or TN concentration in sewage (kg $\mathrm{m^{-3}}$)
D – number of days in which sewage is dumped into the lake
V – sewage volume ($\mathrm{m^3}$)

The load discharged from dispersed sources

$$\mathrm{LD} = (M \cdot W \cdot 365 + T \cdot W \cdot D) \cdot (1 - R_G)$$

where:

M – number of residences
T – number of tourists

D – duration of the tourist season
W – unit load introduced by person (g person^{-1})
R_G – infiltration in soil

The load discharged by anglers

$$\mathrm{LA} = \mathrm{NA} \cdot D \cdot Z$$

where:

NA – number of anglers
D – duration of the fishing season
Z – unit load introduced with groundbait (g P, g N)

The load of nutrients introduced to a given lake with precipitation was determined based on the coefficients of pollution deposition per surface unit (a lake in this case) that is annually published by VIEP (Regional Inspectorate for Environmental Protection).

3 Results and Discussion

The annual phosphorus load introduced into flow lakes ranged from 17.2 kg P (Mielenko) to 7,754.5 kg P (Isąg) and nitrogen from 202.6 kg N (Mielenko) to 81,876.5 kg N (Isąg) (Table 4). Nutrients were mainly introduced with surface water inflow (46–96%).

In the group of flow lakes, only Mielenko, Karczemne and Suskie were fed primarily with surface runoff from the basin, because the surface watercourses flowing into them were periodic and the amount of water flowing through them was small. Maximum loads were recorded during maximum flows associated with rain-snow and snowmelt supply. In the lakes forming the river and lake system (Mielenko, Karczemne, Klasztorne Małe, Klasztorne Duże–Klasztorna Struga lakes complex and Pasłęk, Wymój, Sarąg, Łęguty, Isąg–upper Pasłęka lakes complex), the load grew with the increase of the basin. In the case of lakes Wymój, Sarąg, Łęguty and Isąg, the load of phosphorus and nitrogen carrying by Pasłęka water fluctuated around 90% of the total external load, whereas in the Klasztorne Małe and Klasztorne Duże, the load of phosphorus and nitrogen supplied by Klasztorna Struga ranged between 47% and 95%. The high load of nutrients introduced with river water was dependent on the flow rate and, above all, was the result of the highly agricultural character of the Pasłęka drainage basin and the introduction of domestic and industrial sewage to Klasztorna Struga. According to Somorowski and Mosiej [16] and Somorowski [17], migration of substances from farmland occurs as a result of complex physical, chemical and biological processes in soil, which – apart from the surface runoff – are caused by water and wind erosion. The non-point load enters waters via the surface runoff during the thaws and rains, mainly in the form of

Table 4 The external load of selected flow lakes

Sources of P and N	Nutrient load kg $year^{-1}$ (% share in total load)	Mielenko	Karczemne	Klasztorne Małe	Klasztorne Duże	Pasłęk	Wymój	Sarąg	Łęguty	Isąg	Suskie
LDO	P	1.4 (8.1%)	54.2 (33.5%)	311.9 (60.4%)	930.0 (95.2%)	51.3 (73.1%)	542.4 (94.5%)	1,654.1 (84.3%)	7,148.3 (93.6%)	7,439.5 (95.9%)	9.3 (4.6%)
	N	12.4 (6.1%)	402.2 (30.2%)	732.9 (46.6%)	4,468.2 (86.2%)	1,176.0 (72.7%)	4,821.2 (87.3%)	11,702.1 (85.6%)	69,177.4 (86.5%)	75,238.3 (91.9%)	305.4 (7.4%)
LP	P	2.4 (14.0%)	68.1 (42.1%)	185.7 (35.9%)	0.2 (<0.1%)	–	–	–	–	–	–
	N	32.3 (16.0%)	454.1 (34.0%)	663.0 (42.2%)	3.6 (0.1%)	–	–	–	–	–	–
LAS	P	3.0 (17.3%)	4.5 (2.8%)	1.3 (0.3%)	1.3 (0.1%)	15.9 (22.6%)	15.0 (2.6%)	243.7 (12.4%)	468.6 (6.1%)	175.9 (2.3%)	150.7 (74.4%)
	N	72.0 (35.5%)	108.4 (8.1%)	26.0 (1.6%)	206.0 (4.0%)	381.9 (23.6)	361.2 (6.5%)	676.2 (4.9%)	10,318.0 (12.9%)	3,815.3 (4.6%)	3,150.8 (76.6%)
LD	P	–	–	–	–	–	–	–	–	–	–
	N	–	–	–	–	–	–	–	–	–	–
LB	P	–	–	–	–	–	0.1 (<0.1%)	0.4 (<0.1%)	0.1 (<0.1%)	0.4 (<0.1%)	0.8 (0.4%)
	N	–	–	–	–	–	2.2 (<0.1%)	3.7 (<0.1%)	2.6 (<0.1%)	8.6 (<0.1%)	17.5 (0.4%)
LA	P	7.7 (44.8%)	20.8 (12.9%)	13.1 (2.5%)	25.1 (2.6%)	–	–	–	–	–	27.4 (13.5%)
	N	30.7 (15.2%)	83.2 (6.2%)	52.6 (3.4%)	100.7 (1.9%)	–	–	–	–	–	109.5 (2.7%)
LAT	P	2.7 (15.8%)	14.1 (8.7%)	4.8 (0.9%)	20.1 (2.1%)	3.0 (4.3%)	16.5 (2.9%)	63.9 (3.2%)	21.2 (0.3%)	138.7 (1.8%)	14.4 (7.1%)
	N	55.2 (27.2%)	286.0 (21.5%)	97.0 (6.2%)	407.1 (7.8%)	60.2 (3.7%)	334.9 (6.1%)	1,295.6 (9.5%)	431.2 (0.5%)	2,814.3 (3.4%)	530.1 (12.9%)
Total	P	17.2 (100%)	161.7 (100%)	516.8 (100%)	976.7 (100%)	70.2 (100%)	574.0 (100%)	1,962.1 (100%)	7,638.2 (100%)	7,754.5 (100%)	202.6 (100%)
	N	202.6 (100%)	1,333.9 (100%)	1,571.5 (100%)	5,186.5 (100%)	1,618.1 (100%)	5,519.5 (100%)	13,677.6 (100%)	79,929.2 (100%)	81,876.5 (100%)	4,113.3 (100%)

LDO surface water inflow, *LP* point sources, *LAS* area sources, *LD* dispersed sources, *LB* bathers, *LA* anglers, *LAT* atmospheric sources organic and mineral suspension. Nitrogen compounds are leached from soil while phosphorus compounds migrate mainly through erosion of the soil material. A close relationship has been found between the amount of fertilizers introduced to a land and phosphorus and nitrogen load runoff to surface water [18–21]. Increased runoff of nutrients takes place in plants with shallow root systems and poorly developed side roots (e.g. rapeseed). This type of plants is requiring strong mineral and organic fertilization.

Annual phosphorus load introduced into non-flow lakes ranged from 4.4 kg N (Czarne) to 169.8 kg N (Track) and nitrogen load from 89.4 kg N (Czarne) to 2,311.8 kg N (Track). In the case of this lake type, the main source of supply was usually the surface runoff from the direct catchment (from 40% to 97%). The size of the external load depended on the size of the catchment area and on the way of land use. The lakes surrounded by arable lands were burdened with a particularly high nitrogen load (Podąbek). In the nutrient balance of lakes Sukiel, Podkówka and Tyrsko, important load of phosphorus and nitrogen has been introduced from the atmospheric precipitation (60% of total charge) (Table 5). Pool of nutrients reaching the other analysed lakes with atmospheric precipitation did not matter in the annual balance sheet. Kopáček [22] and Lo and Chu [23] emphasized that due to the seasonality, atmospheric precipitation can be a serious source of nutrients, but only during their occurrence.

Lakes Mielenko, Karczemne and Klasztorne Małe received a high load of nutrient from point sources – mouths of rainwater collectors (from 14% to 42% of the total phosphorus load and from 16% to 42% of the total nitrogen load). Lakes Mielenko, Karczemne, Klasztorne Małe, Klasztorne Duże and Suskie were fed with phosphorus and nitrogen load introduced by anglers with groundbait. Nutrient load reaching from this source was varied from 7.7 kg P to 27.4 kg P and from 30.7 kg N to 109.5 kg N (Table 4). According to Wołos [24], anglers introduce 3.27 g of phosphorus and 18.15 g of nitrogen into the lake and with the fish biomass remove about they extract about 7.35 g of phosphorus and 31.08 g of nitrogen.

Matter circulation in a lake is not a fully closed cycle. It is not a full cycle because part of the matter is precipitated and deposited in bottom sediment and does not participate in further circulation. It is also not a closed cycle, because the lake constantly is receiving matter from the outside. The part of the matter is carried out by the river water outflowing from the lake [25].

The estimated retention or removal (negative retention) values of phosphorus and nitrogen in the analysed flow and non-flow lakes are summarized in Figs. 3a, b and 4a, b. It was found that the majority of flow lakes (except Lakes Isąg and Suskie) showed negative retention in relation to nutrients. The highest loads of phosphorus and nitrogen were discharged from Sarąg and Łęguty lakes (over 5 tons of phosphorus and 50 tons of nitrogen) (Fig. 3a, b).

The relative retention of both phosphorus and nitrogen depends strongly on the duration of the hydraulic retention time in the lake.

The longer retention time the higher the loss percentage of the added phosphorus or nitrogen during its passage through the lake.

Table 5 The external load of non-flow lakes

Sources of P and N	Nutrient load kg $year^{-1}$ (% share in total load)	Paskierz	Kepijko	Czarne	Długie	Starodworskie	Sukiel	Podkówka	Tyrsko	Track	Podąbek
LP	P	–	–	–	–	–	–	–	–	–	–
	N	–	–	–	–	–	–	–	–	–	–
LAS	P	15.1 (83.4%)	80.8 (78.6%)	4.10 (93.1%)	40.6 (88.3%)	12.4 (91.2%)	3.4 (45.0%)	3.4 (70.8%)	8.3 (36.7%)	159.2 (93.8%)	30.6 (97.1%)
	N	168.8 (59.0%)	1,886.8 (93.2%)	77.4 (86.6%)	398.6 (65%)	141.8 (74.7%)	93.0 (35.9%)	94.8 (63.2%)	188.9 (49.3%)	1,889.4 (81.8%)	763.6 (95.7%)
LD	P		17.5 (17.0%)	–	–	–	–	–	10.4 (46.0%)	–	–
	N		43.8 (2.2%)	–	–	–	–	–	41.7 (10.9%)	–	–
LB	P	0.1 (0.5%)	–	–	–	–	–	–	–	–	–
	N	2.9 (1.0%)	–	–	–	–	–	–	–	–	–
LA	P	–	–	–	–	–	–	–	–	–	–
	N	–	–	–	–	–	–	–	–	–	–
LAT	P	2.9 (16.1%)	4.5 (4.4%)	0.3 (6.9%)	5.4 (11.7%)	1.2 (8.8%)	7.6 (55.0%)	1.4 (29.2%)	3.7 (16.4%)	10.6 (6.2%)	0.9 (2.9%)
	N	114.4 (40.0%)	93.6 (4.6%)	12.0 (13.4%)	214.4 (35%)	48.0 (25.3%)	166.4 (64.1%)	55.2 (36.8%)	148.8 (38.8%)	422.4 (18.1%)	34.4 (4.3%)
Total	P	18.1 (100%)	102.8 (100%)	4.4 (100%)	46.0 (100%)	13.6 (100%)	7.6 (100%)	4.8 (100%)	22.6 (100%)	169.8 (100%)	31.5 (100%)
	N	286.1 (100%)	2,024.2 (100%)	89.4 (100%)	613.0 (100%)	189.8 (100%)	166.4 (100%)	150.0 (100%)	383.4 (100%)	2,311.8 (100%)	798.0 (100%)

LP point sources, *LAS* area sources, *LD* dispersed sources, *LB* bathers, *LA* anglers, *LAT* atmospheric sources

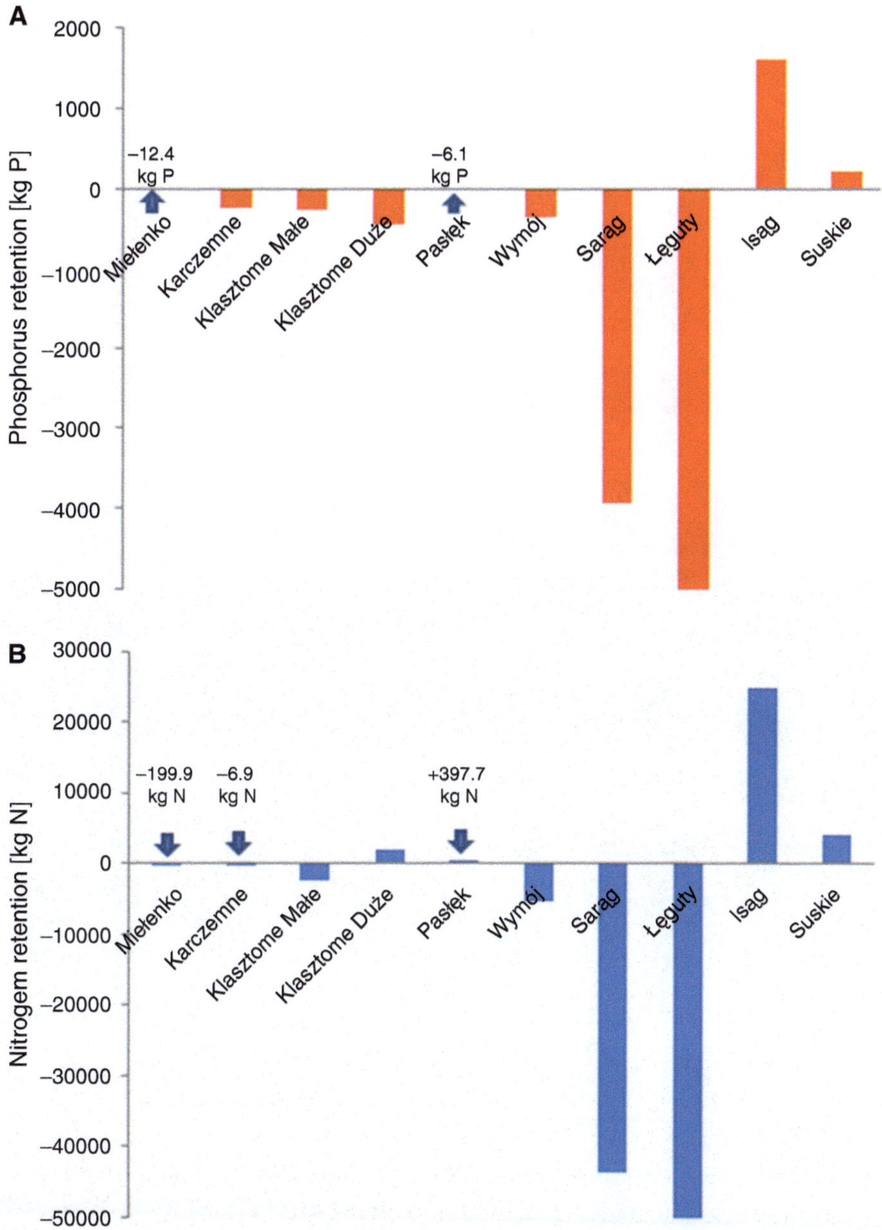

Fig. 3 (**a**) Phosphorus retention in flow lakes. (**b**) Nitrogen retention in flow lakes

The whole groups of lakes without flow stored nutrients were a "trap" for them. Nitrogen retention in lakes does not only occur as incorporation in sediment organic matter but also, and largely so, via denitrification, where nitrate is exploited

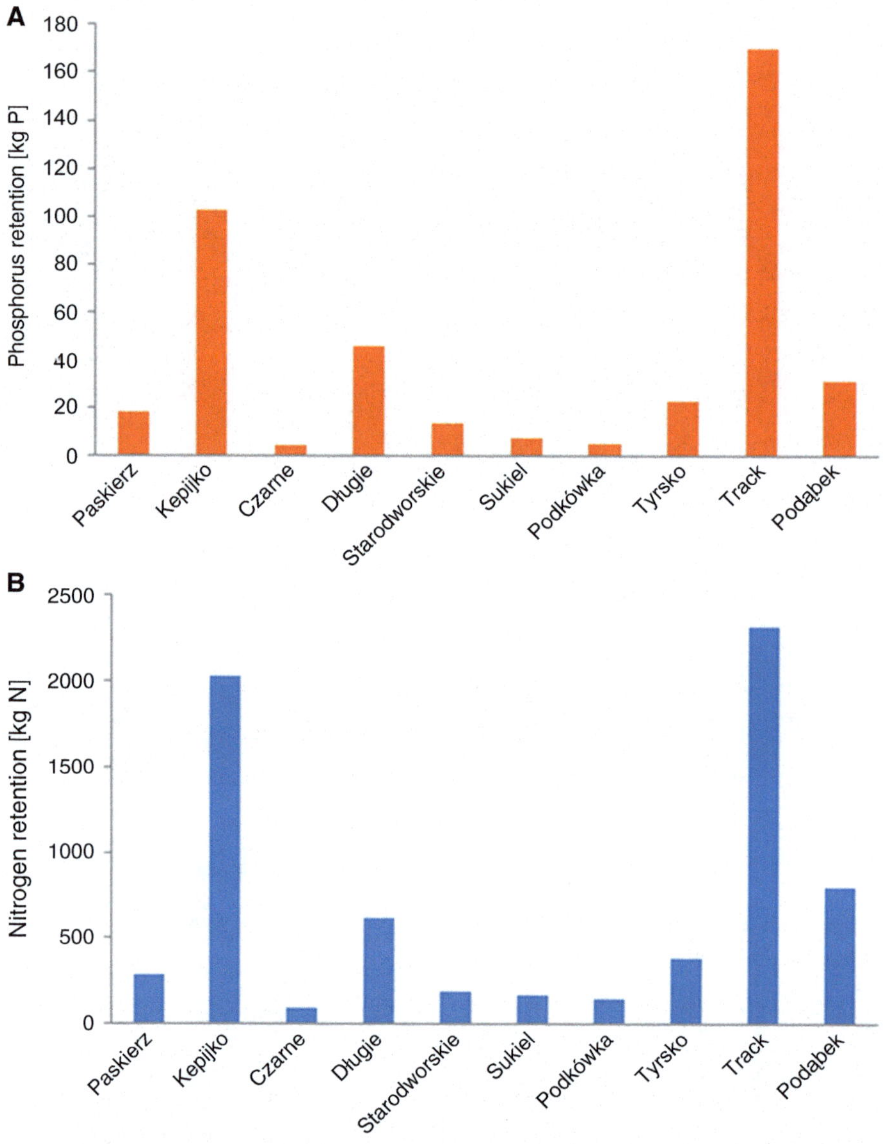

Fig. 4 (**a**) Phosphorus retention in non-flow lakes. (**b**) Nitrogen retention in non-flow lakes

for the bacterial turnover of organic matter [26]. Thereby nitrate converts to ammonium or to free nitrogen (N_2) that may diffuse into the water phase and the atmosphere and thus is lost from the system. Average 77% of nitrogen retention can be ascribed to denitrification [27]. In eutrophic lake estimated ca 90% of the nitrogen is to be removed via denitrification, while 10% remains permanently buried in the sediment [27]. As for phosphorus, the connection between retention time and phosphorus retention has been modelled by Vollenweider [28]. In turn, Grochowska [29] developed models of nitrogen and phosphorus retention in flow lakes. The models of P and N retention indicate that the retention of nutrients in flow lakes is primarily affected by the size of the volume of flow at the inflow and outflow of river to lake. Important factor to be considered in determining the value of retention is P and N release from sediment.

4 Conclusion

In the group of flow lakes, nutrient load was introduced mainly with surface water inflow (46–96%). Only Mielenko, Karczemne and Suskie were fed primarily with surface runoff from the basin, because the surface watercourses flowing into them were periodic and the amount of water flowing through them was small. The size of the load introduced to non-flow lakes depended on the size of the catchment and on the way it was managed. The lakes surrounded by arable lands were burdened with a particularly high nitrogen load (Podąbek). Undoubtedly, the nutrient balance in the lake depends on the amount of water exchanging in the reservoir during the year. The whole groups of examined non-flow lakes stored nutrient were a "trap" for them.

It is noteworthy that in most of the analysed flow lakes, they showed negative retention in relation to phosphorus and nitrogen, but the total load introduced from different sources into these lakes was many times higher than that introduced into non-flow lakes. Such a high external load, even with negative retention, is intensified by the eutrophication process.

References

1. Schauser I, Lewandowski J, Hupfer M (2003) Decision support for the selection of an appropriate in-lake, measure to influence the phosphorus retention in sediments. Water Res 37:801–812
2. Waters S, Webster-Brown JG (2016) The use of a mass balance phosphorus budget for informing nutrient management in shallow coastal lakes. J Hydro Environ Res 10:32–49
3. Grochowska J, Brzozowska R, Łopata M (2017) How durable is the improvement of environmental conditions in a lake after the termination of restoration treatments. Ecol Eng 104:23–29
4. Smith VH, Schindler DW (2009) Eutrophication science: where do we go from here? Trends Ecol Evol 24:201–207
5. Schindler DW, Hecky RE, McCullough GK (2012) The rapid eutrophication of Lake Winnipeg: greening under global change. J Great Lakes Res 38:6–13
6. Kajak Z (2001) Hydrobiology – limnology. Inland water ecosystems. PWN, Warsaw
7. Chapman P, Thornton JA (1986) Nutrients in aquatic ecosystems: an introduction to similarities between freshwater and marine ecosystems. J Limnol Soc South Afr 12(1/2):2–5
8. Pathak H, Pathak D (2012) Eutrophication: impact of excess nutrient status in Lake water ecosystem. J Environ Anal Toxicol 2(5):1–5
9. Małecki A (2007) Nitrogen and phosphorus balance in partial catchments of the Sławskie Lake. Environ Eng 13:269–279

10. Dillon PJ, Rigler FH (1974) A test of simple nutrient budget model predicting the phosphorus concentration in lake water. J Fish Res Board Can 3:1771–1778
11. Moses SA, Janaki L, Joseph S et al (2010) Influence of lake morphology on water quality. Environ Monit Assess 182(1–4):443–454. https://doi.org/10.1007/s10661-011-1888-y
12. Kondracki J (1994) Polish geography. Physico-geographical mesoregions. Warsaw, Poland
13. Lossow K, Gawrońska H, Mientki Cz et al. (2005) Olsztyn Lakes. Trophic state, threats. Olsztyn, Poland
14. Dunalska J, Brzozowska R, Grochowska J (2013) Program for the protection of the Suskie lake with the recultivation project. Olsztyn, Poland
15. Giercuszkiewicz-Bajtlik M (1990) Forecasting changes in standing water quality. Warsaw, Poland
16. Somorowski C, Mosiej J (1994) Some aspect of reducing nutrients leaching from agricultural areas in Poland. Proceedings of International workshop: measures to reduce the leakage of nutrients from agriculture in the Baltic Sea Area. Royal Academy of Science, Stockholm
17. Somorowski C (1998) The balance-balance criteria of shaping the habitats in the agricultural landscape. SGGW, Warsaw
18. Terelak H, Motowicka-Terelak T, Sadurski W (1999) Leaching of chemical components from arable land to drainage water. Protection of resources and quality of surface and underground waters. Economics and Environment, Białystok
19. Cymes I, Szymczyk S (2005) Impact of the land use, drainage and natural factors on the concentration of sodium, calcium and magnesium in groundwater and their drainage by a drainage network from heavy soils. Ecol Eng 13:44–49
20. Grabińska B, Koc J, Skwierawski A et al (2005) Concentrations and outflows of total phosphorus with river waters from the catchment of diversified use. Ecol Eng 13:87–92
21. Skorbiłowicz M, Ofman P (2015) Changes in specific loads of mineral components outflowing from catchment area of river Supraśl in 2001–2009. J Ecol Eng 16(1):135–140
22. Kopáček J, Hejzlar J, Procházková L (1998) Contribution of direct atmospheric deposition to nitrogen and phosphorus load to reservoirs. Int Rev Hydrobiol 83:339–346
23. Lo SL, Chu HU (2006) Evaluation of atmospheric deposition of nitrogen to the Feitsui Reservoir in Taipei. Water Sci Technol 53(2):337–344
24. Wołos A, Teodorowicz M, Grabowska K (1992) Effect of ground-baiting on anglers' catches and nutrient budget of water bodies as exemplified by Polish lakes. Aquac Fish Manag 23:499–509
25. Bajkiewicz-Grabowska E (2002) Matter circulation in lake – river systems. UW, Warsaw
26. Wetzel R (2001) Limnology. Lake and river ecosystems. Academic, New York
27. Jensen JP, Jeppesen E, Kristensen P et al (1992) Nitrogen loss and denitrification as studied in relation to reduction in nitrogen loading in a shallow, hypertrophic lake (Lake Sobygard, Denmark). Int Revue Ges Hydrobiol 77:29–42
28. Vollenweider RA (1976) Advances in defining critical loading level for phosphorus in lake eutrophication. Mem Inst Ital Idrobiol 33:53–83
29. Grochowska J (2015) Circle of selected macroelements and biogenic compounds in the lake-river system for example of upper Pasłęka. Environmental Engineering Committee, Lublin

Internal Phosphorus Loading in Eutrophic Lakes in Western Poland

Katarzyna Kowalczewska-Madura, Renata Dondajewska, and Ryszard Gołdyn

Contents

Abstract We hypothesized that the variability of P internal loading in lakes and reservoirs depends on the trophic state of analysed waterbodies, as well as spatial (different water depths) and temporal (season) aspect in particular waterbodies. Additionally, year-to-year changes of P loading in lakes restored with variable methods (sustainable restoration, P inactivation with iron compounds and magnesium chloride, nitrate treatment and effective microorganisms (EM) application) were expected. To verify these assumptions, we have analysed the process of internal P loading in 40 waterbodies situated in Western Poland, based on the ex situ experiments on intact sediment cores, collected from different water depths in various seasons. Additionally, basic sediment characteristics (TP and its fraction content, organic matter, SRP and TP in pore water and water above sediments) were studied. The most intensive P release from sediments into water column was noted in summer, especially at greater depths and in heavily eutrophicated lakes. Internal P loading in restored lakes usually decreased, apart from lakes in which EM were used.

K. Kowalczewska-Madura (✉) · R. Dondajewska · R. Gołdyn
Department of Water Protection, Faculty of Biology, Adam Mickiewicz University, Poznań, Poland
e-mail: madura@amu.edu.pl

E. Korzeniewska, M. Harnisz (eds.), *Polish River Basins and Lakes – Part I*,
The Handbook of Environmental Chemistry 86,
https://doi.org/10.1007/978-3-030-12123-5_14

Keywords Bottom sediments · Eutrophic lakes · Internal loading · Phosphorus

1 Introduction

Freshwater reservoirs are complex ecosystems with functioning strongly dependent on nutrient concentrations. One of them – phosphorus (P) – often plays a limiting role for primary production. It comes from many external sources such as watercourses feeding the lake, surface runoff from catchment areas with variable cover, point sources as well as precipitation and recreation. Bottom sediments are crucial source of P for water column due to long-term P accumulation in various compounds; thus sediments are considered as 'black box'. Ca. 80–90% of P in lake ecosystem is buried in a 10-cm-thick surface sediment layer. This ability to bind and release P determines the role of sediments in lake as a trap or a source of this element. P retention in sediments is a consequence of two processes with opposite directed flux rate, i.e. sedimentation of different particles as a downward flux and the release of dissolved P via the interstitial water as upward flux. The domination of P release over its accumulation is called 'internal P loading' [1–5]. P release from sediment is influenced by a number of factors being biological (microorganism activity, organic matter decomposition), chemical (oxygen concentration, redox potential, pH, nitrates or Fe-P ratio) as well as physical (temperature, wind causing resuspension) [6, 7]. Variable factors may play the most significant role in a particular lake due to local conditions [8].

The exchange of phosphorus between sediments and water is one of the most important pathways of P circulation in lake ecosystem. The intensity and duration on internal P loading may exert a crucial impact on P concentrations in lake water as well as on lake water quality [9–11]. Therefore, P release shall be introduced into a broad program of lake research prior the restoration, to assess the impact of sediments on lake productivity and to propose proper methods of reclamation. The diagnosis of processes responsible for P loading enables the use of treatment dedicated to efficient P release diminish.

We hypothesized that the variability of P internal loading in lakes and reservoirs depends on the trophic state of analysed waterbodies, as well as spatial (different water depths) and temporal (season) aspect in particular waterbodies. Additionally, year-to-year changes of P loading in lakes restored with variable methods were expected. Furthermore, selected sediment characteristics related to P release/accumulation were of concern, i.e. P and organic matter content in sediment, P content in interstitial and over-bottom water and P fractions in sediment.

2 Study Site

Internal P loading was studied in 40 lakes and man-made reservoirs located in Western Poland, most of them in a radius of 45 km from Poznań, in Wielkopolski Lake District (Fig. 1). Waterbodies varied in size, maximum and mean depth as well as water volume. Nine of them had surface area over 100 ha, with the greatest 742.5 ha, and twelve were characterized by area 30–100 ha, while the most were small (under 15 ha). In case of maximum water depth, the majority of lakes were shallow (under 10 m), and only six of them had depth over 15 m. Mean water depth varied from 0.5 to 8.6 m, but usually it did not exceed 5 m (Table 1).

Nine of the studied waterbodies were under restoration treatment. Hypolimnion aeration, P inactivation (with iron compounds and/or magnesium chloride) and biomanipulation were conducted in lakes Durowskie, Góreckie and Swarzędzkie [17–19], whilst in Jelonek and Winiary, restoration program included P inactivation in sediments with a Proteus device [20], biomanipulation, barley straw and macrophyte mowing. Lakes mentioned above were included into Group 1, i.e. waterbodies restored with combined physical, chemical and biological methods, but only to the extent that is necessary for the gradual reconstruction of the ecosystem – so-called sustainable restoration [21]. P inactivation with iron compounds and magnesium chloride was introduced in Rusałka reservoir throughout the analysed period [22] and in Uzarzewskie Lake in years 2006–2007 [23, 24]; thus these waterbodies were turned into Group 2. Uzarzewskie Lake has been treated with nitrate doses from natural springs to hypolimnion since 2008 [24, 25]; therefore the results from this period were introduced to Group 3. Finally, lakes Konin and Wielgie were restored by Effective Microorganisms (EM) application, forming Group 4. In part of these lakes, internal P loading was analysed year to year for 3–10 years, whilst for the rest, it was for 1 year only.

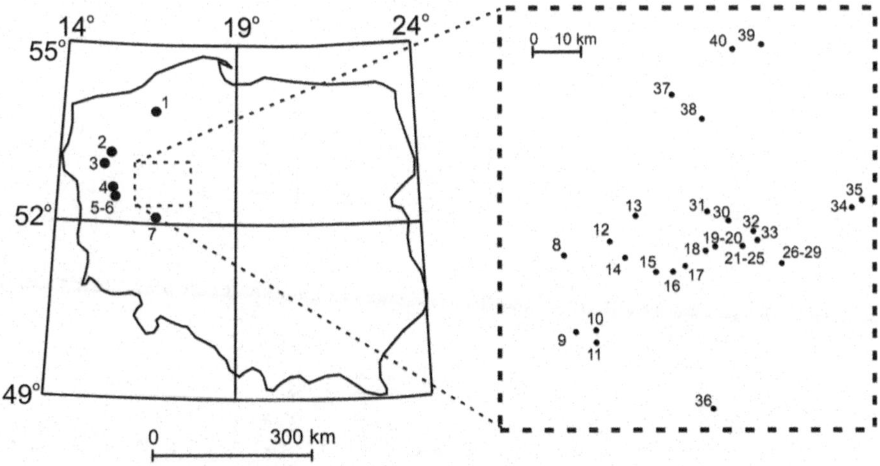

Fig. 1 The location of lakes presented in this paper (numbering convergent with the Table 1)

Table 1 Morphometric features of lakes presented in this paper (after, [12], unless stated otherwise with asterisks)

N°	Lake	Lake district	Coordinates	Surface area (ha)	Water volume (10^3 m^3)	Max depth (m)	Mean depth (m)
1	Bielsko	Południowopomorski	53°51′N, 16°51′E	247.9	15,977.1	23.0	6.2
2	Piaseczno	Południowopomorski	53°07′N, 16°00′E	58.7	4,519.2	25.9	7.6
3	Wielgie	Południowopomorski	52°58′N, 15°45′E	136.9	3,077.6	6.8	2.2
4	Konin	Lubuski	52°23′N, 15°52′E	79.5	874.5	3.0	1.1
5	Nowowiejskie	Lubuski	52°11′N, 15°53′E	29.0	319.0	2.1	1.1
6	Błędno	Lubuski	52°14′N, 15°54′E	742.5	26,178.7	9.6	3.5
7	Zbęchy	Leszczyńskie	52°00′N, 16°54′E	108.9	4,636.1	8.5	4.3
8	Lusowskie	Wielkopolski	52°25′N, 16°40′E	121.9	10,479.9	19.5	8.6
9	Lipno	Wielkopolski	52°17′N, 16°43′E	9.0**	500.0**	9.5**	5.6**
10	Jarosławieckie	Wielkopolski	52°17′N, 16°47′E	11.8	259.6	4.5	2.2
11	Góreckie	Wielkopolski	52°16′N, 16°47′E	97.4	6,136.2	15.5	6.3
12	Strzeszyńskie	Wielkopolski	52°27′N, 16°49′E	34.9	2,847.1	17.8	8.2
13	Glinowieckie	Wielkopolski	52°30′N, 16°54′E	18.5***	–	5.0***	–
14	Rusałka	Wielkopolski	52°25′N, 16°52′E	36.7	701.4	9.0	1.9
15	Maltański	Wielkopolski	52°24′N, 16°58′E	67.5****	2,000.0****	3.7****	2.8****
16	Antoninek	Wielkopolski	52°24′N, 17°02′E	7.2****	30.0****	0.8****	0.5****
17	Swarzędzkie	Wielkopolski	52°24′N, 17°04′E	93.7	2,038.5	7.2	2.6
18	Uzarzewskie	Wielkopolski	52°26′N, 17°08′E	10.6	360.4	7.5	3.4
19	Prawe (Zachodnie)	Wielkopolski	52°27′N, 17°09′E	3.1*****	–	6.2*****	–
20	Lewe	Wielkopolski	52°27′N, 17°10′E	2.2*****	–	6.6*****	–
21	Góra	Wielkopolski	52°27′N, 17°12′E	37.8	505.6	3.0	1.3
22	Brzostek	Wielkopolski	52°27′N, 17°15′E	6.0******	270.0******	7.8******	4.5******
23	Kazanie	Wielkopolski	52°27′N, 17°18′E	1.8***	–	6.0	–
24	Wójtostwo	Wielkopolski	52°27′N, 17°16′E	9.2	248.4	5.7	2.7
25	Dębiniec	Wielkopolski	52°28′N, 17°13′E	15.0	510.0	7.4	3.4
26	Baba	Wielkopolski	52°25′N, 17°22′E	2.2***	–	6.0*****	–

27	Cyganek	Wielkopolski	52°25′N, 17°22′E	1.7***	–	3.4*****	–
28	Uli	Wielkopolski	52°25′N, 17°22′E	6.4***	–	6.5*****	–
29	Ósemka	Wielkopolski	52°25′N, 17°22′E	1.5***	–	1.3*****	–
30	Wronczyńskie Wielkie	Wielkopolski	52°30′N, 17°11′E	39.9	932.3	4.7	2.3
31	Tuczno	Wielkopolski	52°31′N, 17°08′E	12.7	254.0	4.0	2.0
32	Biezdruchowo	Wielkopolski	52°29′N, 17°16′E	48.8	2,815.5	17.7	5.7
33	Dobra	Wielkopolski	52°28′N, 17°17′E	12.0	768.0	15.0	6.4
34	Jelonek	Wielkopolski	52°32′N, 17°35′E	14.4	172.8	2.4	1.2
35	Winiary	Wielkopolski	52°32′N, 17°36′E	14.4	302.4	4.2	2.1
36	Raczyński	Wielkopolski	52°08′N, 17°09′E	84.4	2,342.9	5.8	2.7
37	Rogoźno	Wielkopolski	52°44′N, 17°00′E	125.8	3,808.5	5.8	3.0
38	Budziszewskie	Wielkopolski	52°41′N, 17°17′E	163.0	7,842.9	14.0	4.8
39	Łeknińskie	Wielkopolski	52°50′N, 17°17′E	85.2	1,376.5	2.8	1.6
44	Durowskie	Wielkopolski	52°49′N, 17°11′E	143.0	11,322.9	14.6	4.6

** [13]
*** [14]
**** [15]
***** Own research
****** [16]

3 Materials and Methods

Internal P loading research were conducted in the years 2005–2017 in ex situ experiments on undisturbed sediment cores. Samples were collected in three seasons (Sp, spring; Su, summer; A, autumn) from stations differing in water depth, oxygen content by the bottom and macrophyte cover. In case of depth spectrum, four zones were distinguished: zone I (Z I) included water depth from 0 to 2 m, zone II (Z II) 2–5 m, zone III (Z III) 5–10 m and zone IV (Z IV) over 10 m. According to Carlson [26] criteria, trophic state was determined. Studied lakes were classified as mesoeutrophic (M-E), eutrophic (E) and hypertrophic (H).

Sediment cores were collected with modified Kajak sediment sampler to transparent tube made from PMMA (polymethyl methacrylate). Each transparent tube, 6 cm in diameter, containing the collected sample of intact sediment core (ca. 15 cm in length) and the overlying water, was sealed with rubber stoppers. The cores were incubated under laboratory conditions at temperature and oxygen concentrations that corresponded to the ambient values determined during field research. Samples of water from above the sediment cores were collected at definite intervals (1–3 days) over a period of 2–3 weeks from every tube (Fig. 2). Analyses of total phosphorus content were done spectrophotometrically with ascorbic acid as the reducer [27]. Water temperature, dissolved oxygen concentration, pH and conductivity were measured before water sampling in tubes with a WTW Multi 350i metre. P concentrations were calculated into mgP/m^2 d, based on water volume and sediment surface. Positive values indicate P release from sediment whilst negative its accumulation.

Selected sediment characteristics were analysed as well in samples collected from each station: total P and its fractions, according to Psenner et al. [28]; organic matter (OM, after combustion in 550°C); TP and orthophosphates (SRP) in pore water (after centrifugation for 1 h at 3,000 rpm); and in the water collected above sediment. Statistical calculations were made using STATISTICA version 10.0 software.

4 Results

4.1 Internal P Loading and Sediment Characteristics in Trophic, Spatial and Temporal Aspect

Ex situ experiments revealed explicit variability of *internal P loading* in relation to trophic status, season as well as water depth. In most cases the highest P release from sediments was noted in summer (Fig. 3). In the shallowest zone (Z I), the greatest prevalence of P loading on its accumulation was observed in hypertrophic lakes, where it reached 18.9 mgP/m^2 d in spring. Littoral zone of mesoeutrophic lake was characterized by low P release – 0.6 mgP/m^2 d on average – and P binding in sediments was stated in autumn, zone II, with water depth 2–5 m, notable P loading

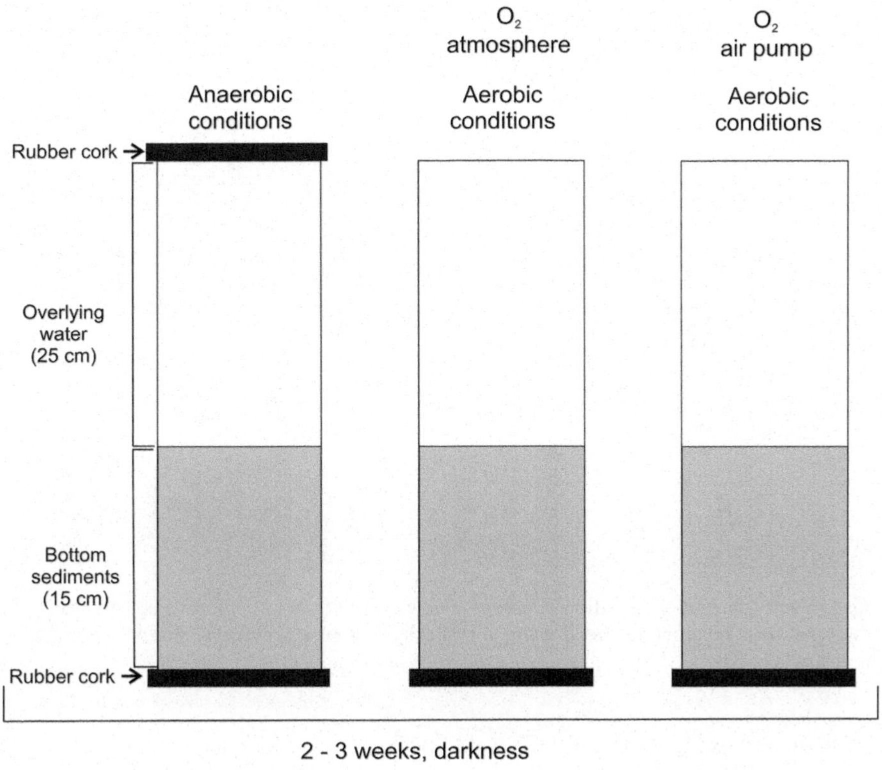

Fig. 2 Scheme of laboratory ex situ experiment

occurred in summer, both in eutrophic (6.1 mgP/m^2 d) and hypertrophic (6.6 mgP/m^2 d) lakes. Higher amounts of P were released into water columns in the deepest zones Z III (5–10 m) and Z IV (>10 m), especially in case of hypertrophic lakes in summer – mean ca. 16.0 mgP/m^2 d. To sum up, internal P loading increased with water depth, particularly in zones Z I to Z III, whilst in Z IV it was similar to Z III or slightly lower. More intense P release was observed in lakes with higher trophic state. The greatest values characterized hypertrophic waterbodies in zones I and III, ca. 13 mgP/m^2 d (10.41 mgP/m^2 d on average), lower in eutrophic lakes (6.57 mgP/m^2 d on average) and the lowest in mesoeutrophic lakes (1.48 mgP/m^2 d on average) (Fig. 6).

No significant variability in *sediment P content* was noted in particular season. The lowest concentrations were observed in the littoral zone (Z I) of mesoeutrophic lakes – up to 0.8 mgP/g DW in summer. The same zone in hypertrophic lakes was characterized by P content exceeding 1.2 mgP/g DW. In other zones TP values were slightly higher: in Z II and Z IV maximum was noted in hypertrophic water bodies (1.5–1.7 mgP/g DW) whilst in Z III eutrophic ones (1.53 mgP/g DW). Mesoeutrophic lakes had clearly lower P content in Z III, whilst in Z IV it was

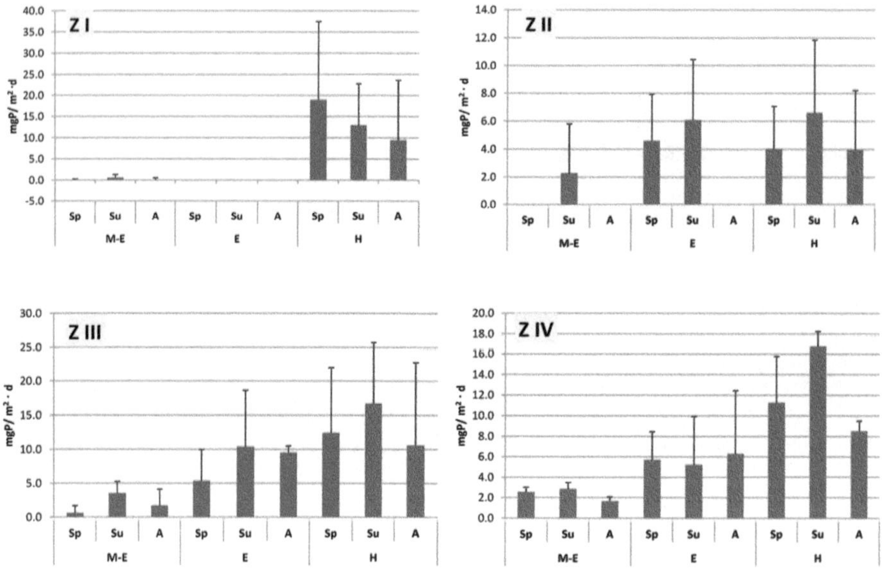

Fig. 3 The accumulation (negative values) or release (positive values) of P from bottom sediments of waterbodies with different trophic states in particular seasons at different depth zones (see Sect. 3 for abbreviations)

similar to values noted in eutrophic waterbodies; however, in general, the higher the trophic status, the greater the P concentration in sediments (0.92 mgP/g DW for mesoeutrophic, 1.34 mgP/g DW for eutrophic and 1.33 mgP/g DW for hypertrophic, on average (Figs. 4 and 6)).

The higher amounts of *organic matter* in the shallowest zone were observed in hypertrophic lakes – up to 31.2% on average in spring. With increasing depth of strongly eutrophicated waterbodies, the percentage of OM decreased, reaching 25.2% in Z II and 17.4% in Z III, wherein higher values were noted in spring. Nevertheless, in Z IV of studied lakes, organic matter amount rose again to 21.8%. In mesoeutrophic lakes higher contribution of OM was stated in summer – from mean 10.1% in Z I to 43.6% in Z II. Eutrophic lakes were characterized by higher amounts of OM in Z II and Z III – ca. 26–28% with maximum in summer (Fig. 5). To conclude, organic matter content was lowest in mesoeutrophic lakes (18.02% on average) and similar in eutrophic and hypertrophic ones (23.7% and 24.5%, respectively, on average (Fig. 6)).

Both *SRP and TP concentrations in pore water* of sediments increased with water depth. Highest values were usually noted in summer. It reached up to 0.25 mgP/L (SRP) and 0.47 mgP/L (TP) in Z I of mesoeutrophic lakes and up to 1.91 mgP/L and 2.63 mgP/L, respectively, in Z IV. In case of eutrophic waterbodies, noted values fluctuated from mean 1.42 mgP/L (SRP) and 1.82 mgP/L (TP) in Z II up to 3.61 mgP/L (SRP) and 4.05 mgP/L (TP) in Z IV. The highest concentrations of P compounds were observed in hypertrophic lakes – from mean 1.67 mgP/L (SRP) and

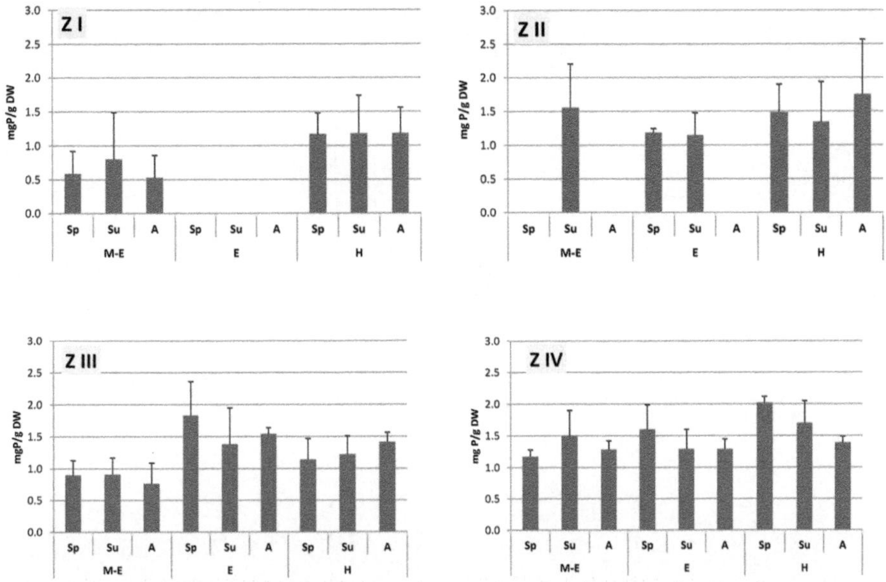

Fig. 4 P content in sediments in waterbodies with different trophic states in particular seasons at different depth zones (see Sect. 3 for abbreviations)

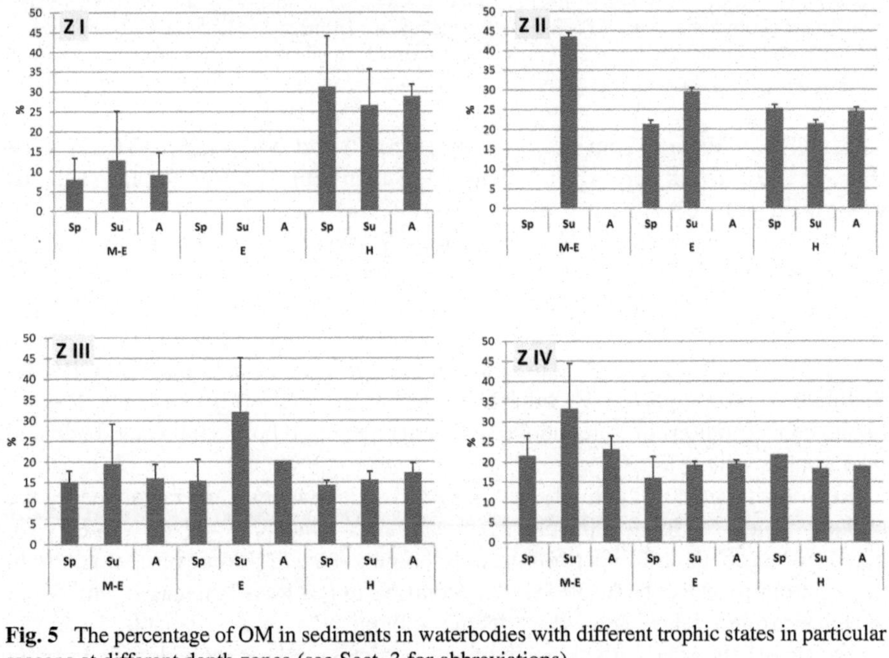

Fig. 5 The percentage of OM in sediments in waterbodies with different trophic states in particular seasons at different depth zones (see Sect. 3 for abbreviations)

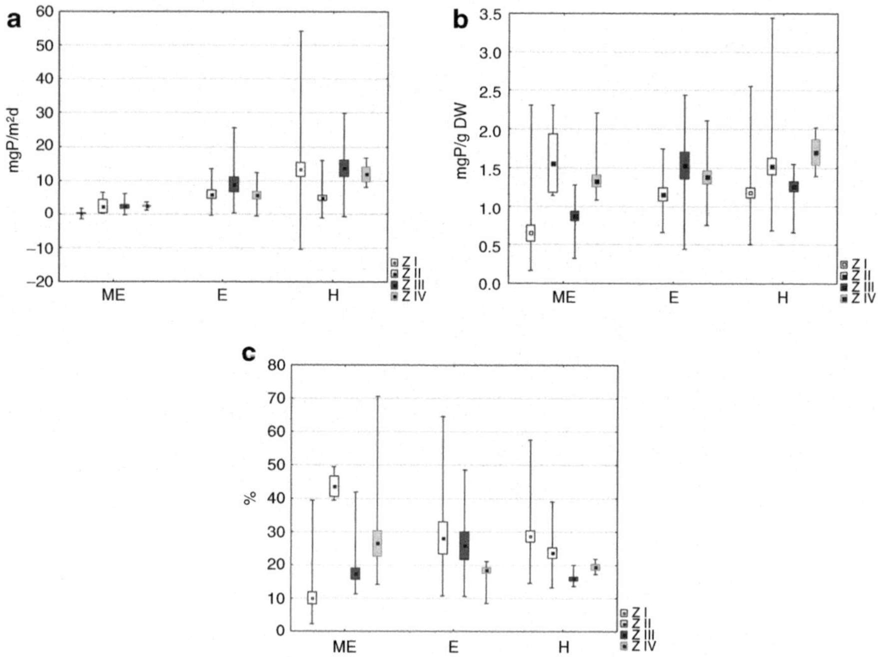

Fig. 6 Mean values of selected sediment characteristics in waterbodies with different trophic states at different depth zones (see Sect. 3 for abbreviations): (**a**) internal P loading, (**b**) P content in sediments, (**c**) organic matter in sediments (box, mean ± standard deviation; whiskers, min and max)

2.16 mgP/L (TP) in Z I up to 3.61 mgP/L (SRP) and 4.05 mgP/L (TP) in Z IV (Fig. 7). P content in pore water increased together both with lake trophic state and water depth (Fig. 8).

The greater the water depth, the higher the *SRP and TP concentrations in water above bottom*. Maximum values were usually noted in summer or autumn. SRP content varied from 0.05 mgP/L in Z I to 0.29 mgP/L in Z IV whilst TP from 0.11 mgP/L to 0.40 mgP/L, respectively, in mesoeutrophic lakes. In case of eutrophic waterbodies, this range increased to 0.04 mgP/L (Z I)–0.88 mgP/L (Z IV) for SRP and to 0.05 mgP/L–0.91 mgP/L for TP, respectively (Fig. 9). Thus, P compound concentrations in water above sediments increased both with water depth and lake trophic state (Fig. 10).

Res-P *fraction*, i.e. P permanently buried in sediments, was characterized by the highest mean contribution in total P in sediments. Distinctly greater values were noted in eutrophic and hypertrophic waterbodies, over 53% on average, whilst in mesoeutrophic lakes, it was 39.8%. The contribution of Res-P increased with depth in hypertrophic lakes (Fig. 11). Three biologically most available fractions, i.e. NH_4Cl-P (loosely sorbed on sediment particles), BD-P (bound with Fe) and NaOH-P (bound with Al), had a share in total P in sediment fluctuating from 14 to 16%. In case of NH_4Cl-P and BD-P, lower values were noted in hypertrophic lakes

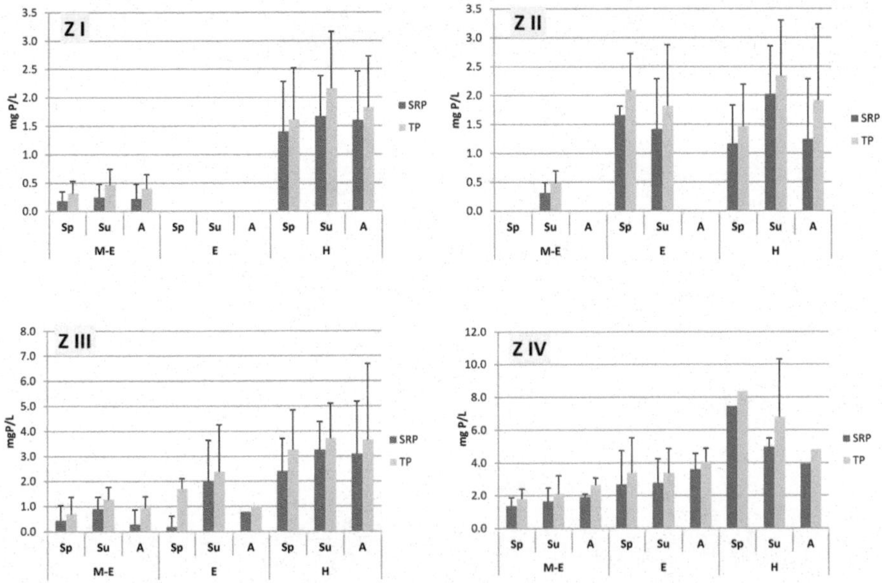

Fig. 7 The concentration of SRP and TP in interstitial waters of sediments in particular seasons in waterbodies with different trophic states at different depth zones (see Sect. 3 for abbreviations)

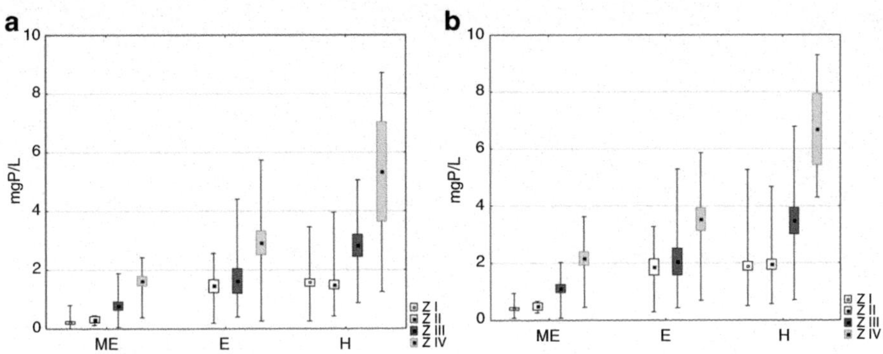

Fig. 8 The concentration of SRP (**a**) and TP (**b**) in interstitial waters of sediments in waterbodies with different trophic states at different depth zones (box, mean ± standard deviation; whiskers, min and max)

(under 5%) and higher in eutrophic ones (ca. 6.2%). In mesotrophic lakes, on the other hand, greater contribution was observed for NaOH-NRP (P bound in organic compounds) and HCl-P (bound with Ca) – 23.4% and 20.7%, respectively. In comparison, it was under 20% in eutrophic and hypertrophic lakes.

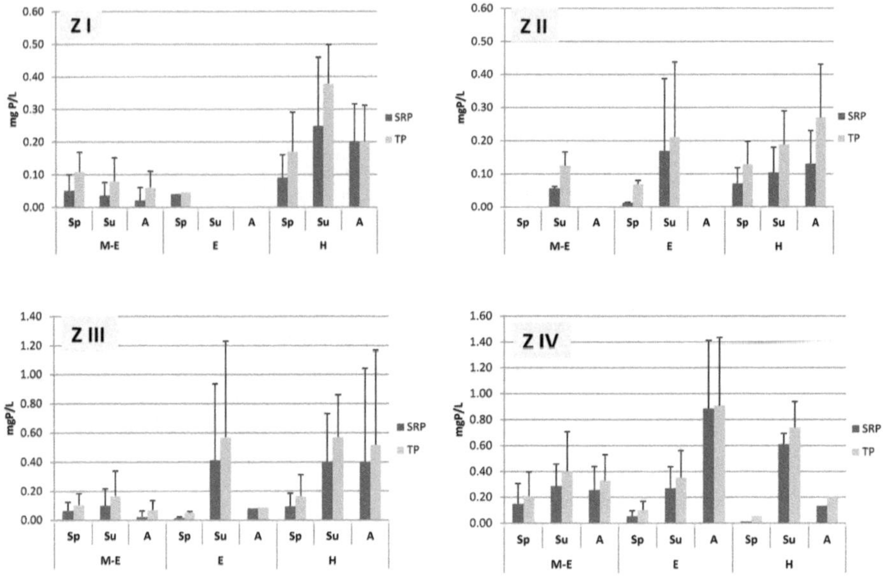

Fig. 9 The concentration of SRP and TP in water above sediments in particular seasons in waterbodies with different trophic states at different depth zones (see Sect. 3 for abbreviations)

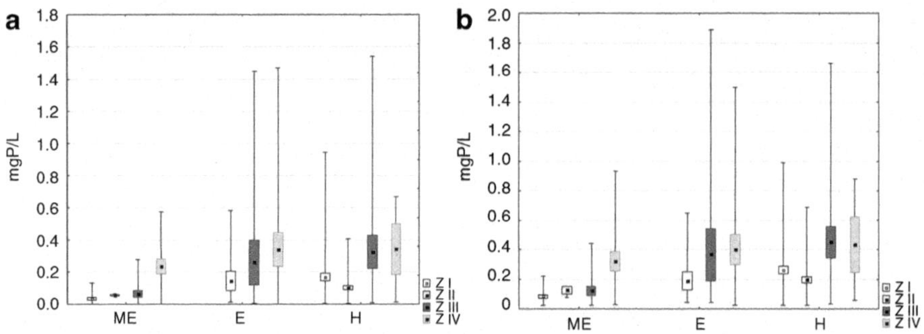

Fig. 10 The concentration of SRP (**a**) and TP (**b**) in water above sediments in waterbodies with different trophic states at different depth zones (box, mean ± standard deviation; whiskers, min and max)

4.2 Internal P Loading and Sediment Characteristics in Lakes Under Restoration

Four depth zones were represented in lakes introduced to *Group I*. In the shallowest one, internal P loading did not exceed 7.5 mgP/m^2 d. Lake Jelonek, analysed in the 9th year from the beginning of restoration only, stood out with greater values in summer (up to 18.6 mgP/m^2 d). Mean P loading varied from 1.8 to 4.6 mgP/m^2 d. In case of the lakes studied for 5 subsequent years, it remained stable; however, the

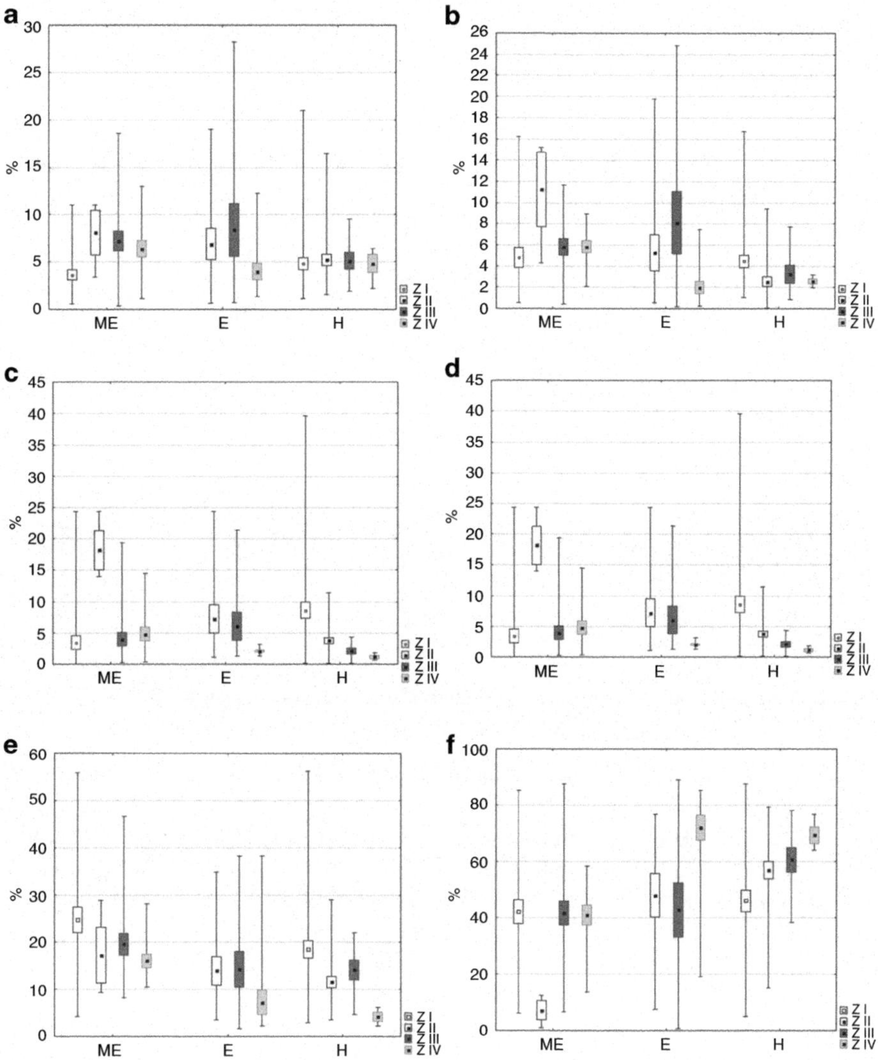

Fig. 11 The percentage of P fractions in sediments of analysed waterbodies (**a** NH_4Cl-P, **b** BD-P, **c** NaOH-P, **d** NaOH-NRP, **e** HCl-P, **f** Res-P) (box, mean ± standard deviation; whiskers, min and max)

range of values decreased (Fig. 12). Zones Z II and Z III were characterized by gradual diminish in internal P loading. The change of mean values was from ca. 7.5 mgP/m^2 d to ca. 2 mgP/m^2 d in Z II and from 17.3 mgP/m^2 d to less than 4.0 mgP/m^2 d in Z III. Slighter changes were noted in Z IV, where P loads were less than 4 mgP/m^2 d in the first and last 3 years, increasing in the period from the 4th to 6th year of restoration to 20.8 mgP/m^2 d maximally (Fig. 12).

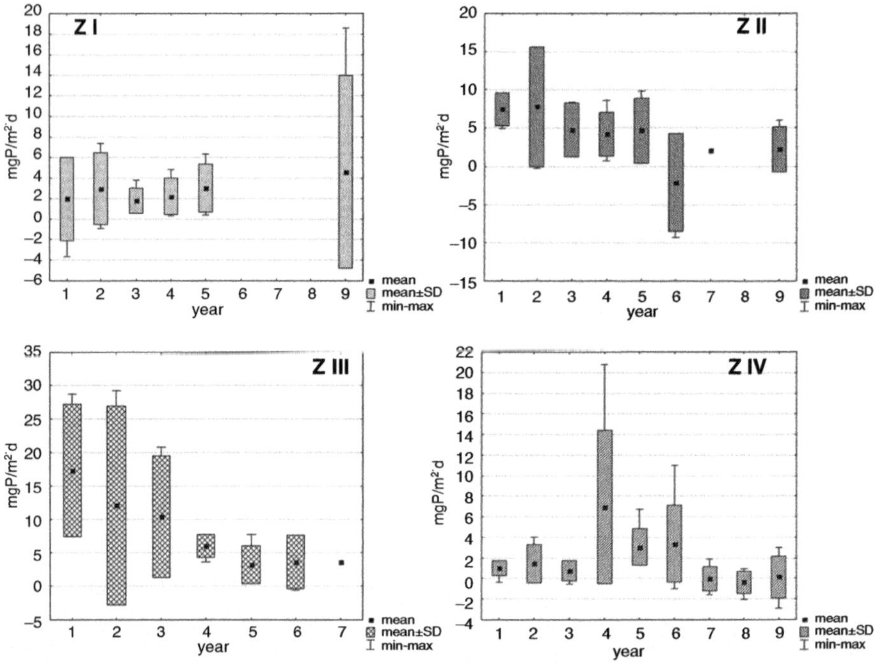

Fig. 12 Internal P loading in Group I restored lakes at different depth zones

TP content in sediments has gradually risen in Z I and Z II – mean values were ca. 1.0 mgP/g DW at the beginning of treatment, increasing up to 1.7–1.8 mgP/g DW in the latter years (Fig. 13). Similar tendency was noted in Z III, where TP content grew from less than 0.9 mgP/g DW to 1.30 mgP/g DW on average. However, a fall was observed in the 6th year of restoration to 0.88 mgP/g DW. Such a tendency was not noted in the deepest lake parts. The amount of P in sediments varied from 0.88 to 1.42 mgP/g DW on average. Lowest values, both in case of mean and minimum, were stated in the middle of the research period (Fig. 13).

Pore water TP concentrations in Z I diminished from 1.44 mgP/L initially to 0.80 mgP/L in the 2nd year of restoration and then increased again back to 1.46 mgP/L. Once more Lake Jelonek stood out from other lakes with the greatest range of values 0.45–2.56 mgP/L (Fig. 14). A decreasing trend was observed in Z II and Z III, but the zones differ in concentrations. Mean values in Z II varied from 0.27 to 2.62 mgP/L and from 2.69 to 5.16 mgP/L in Z III. Highest pore water TP was observed in the first 2 years, decreasing in the subsequent ones. The deepest zone was characterized by mean TP in interstitial water, 2.08–2.93 mgP/L. It increased in years 2–5 of the restoration period in comparison to initial year but fell again in years 6–9, both mean and minimum concentrations (Fig. 14).

Two depth zones were represented in lakes introduced to *Group II*, and in both, internal P loading decreased as a result of restoration. Mean values changed from 6.1–9.8 mgP/m^2 d in the first 2 years to 1.8–4.1 mgP/m^2 d in the last 3 years in Z I. A

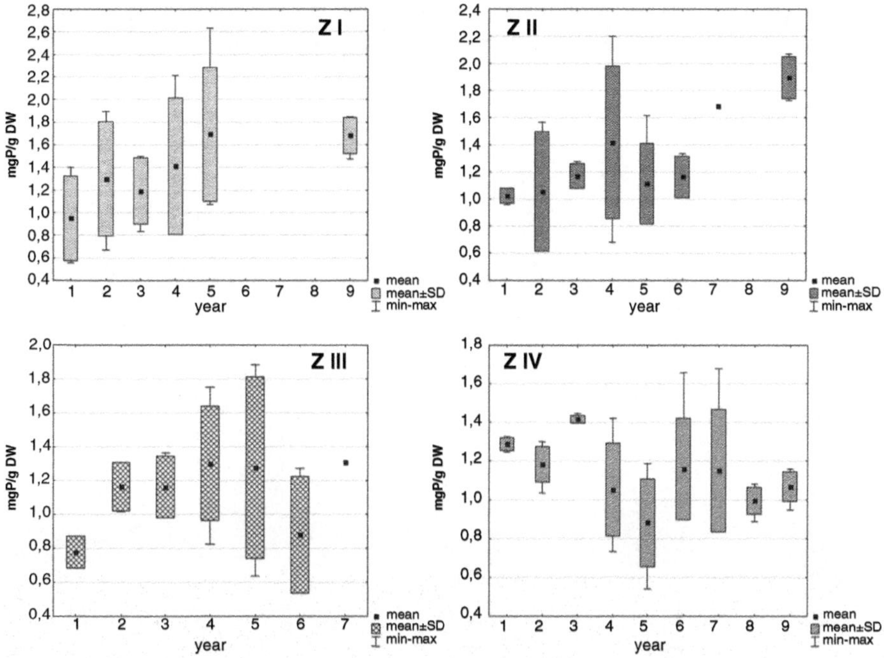

Fig. 13 TP content in sediments in Group I restored lakes at different depth zones

slight accumulation of P in sediments was noted in the 9th year (−0.24 mgP/m^2 d on average). Z III was characterized by initial mean P loading of 32.0 mgP/m^2 d, decreasing to 19.6 mgP/m^2 d in the 2nd year of restoration and gradually to less than 10.0 mgP/m^2 d in following years (Fig. 15).

TP content in sediments of Z I was quite stable, varying in case of annual mean values from 0.96 to 1.11 mgP/g DW. Lower variability was noted in the 10th and 12th year of treatment, higher in the 11th. In Z II the amount of P increased significantly between the first (0.98 mgP/g DW) and second (1.44 mgP/g DW) year of treatment and remained on the level of ca. 1.5 mgP/g DW in the following years, except a fall in the last year to 1.15 mgP/g DW (Fig. 16).

Opposite tendencies of pore water TP concentrations were stated in depth zones. In Z I mean values were decreasing from 2.4 to 0.38 mgP/L whilst in Z III rising from 4.6 to 9.38 mgP/L (Fig. 17).

Two depth zones were represented in lakes from *Group III*, differing in temporal changes of internal P loading. Mean annual values varied in the range of 3.65–13.85 mgP/m^2 d in Z I. Maximum loading reached even 35.0 mgP/m^2 d in summer, whilst in in spring and autumn, an accumulation was noted periodically. Deeper lake part was much more susceptible to restoration treatment, and the amount of P released from sediment decreased gradually from 19.38 mgP/m^2 d in the 1st year to −1.63 mgP/m^2 d in the last one (Fig. 18).

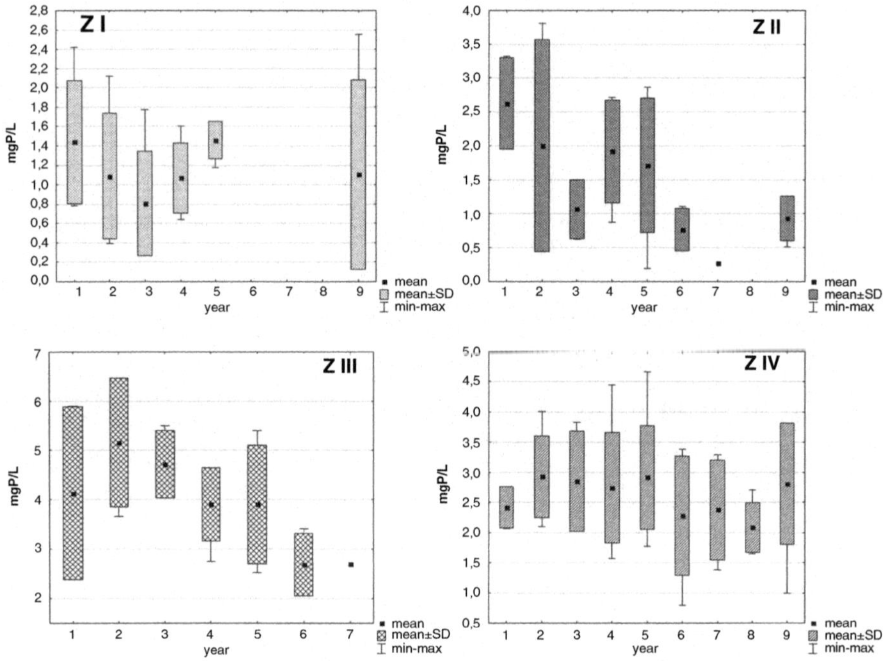

Fig. 14 Pore water TP concentrations in sediments in Group I restored lakes at different depth zones

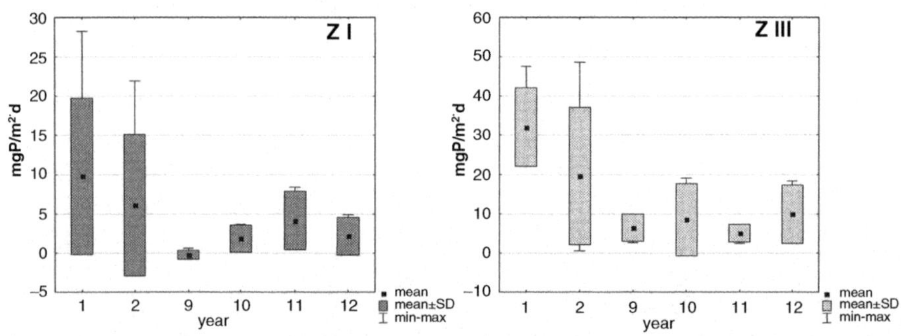

Fig. 15 Internal P loading in Group II restored lakes at different depth zones

Mean TP content in sediments of littoral zone was high at the beginning of restoration (1.63 mgP/g DW), falling to less than 1.3 mgP/g DW in years 3–7, then increased, reaching finally over 1.21 mgP/g DW. Similar trends were observed in Z III, but at the end of research period, the amount of P bounded in sediments was higher than prior the restoration, reaching 3.1 mgP/g DW maximally (Fig. 19).

Pore water TP concentrations varied in time in Z I, fluctuating from 0.74 mgP/L in the 7th year to 3.58 mgP/L in the 3rd year. Z III was characterized by much higher

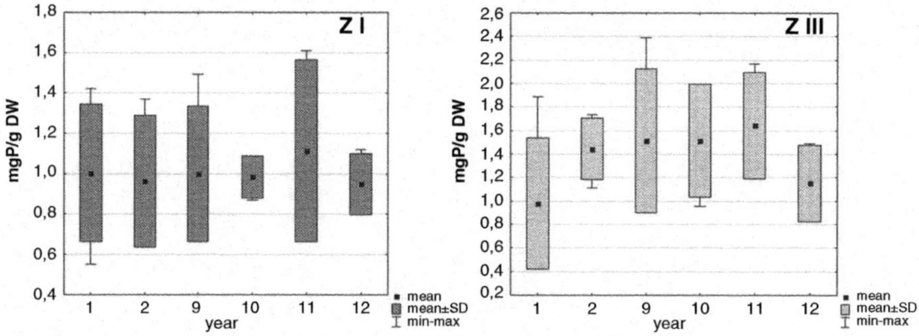

Fig. 16 TP content in sediments in Group II restored lakes at different depth zones

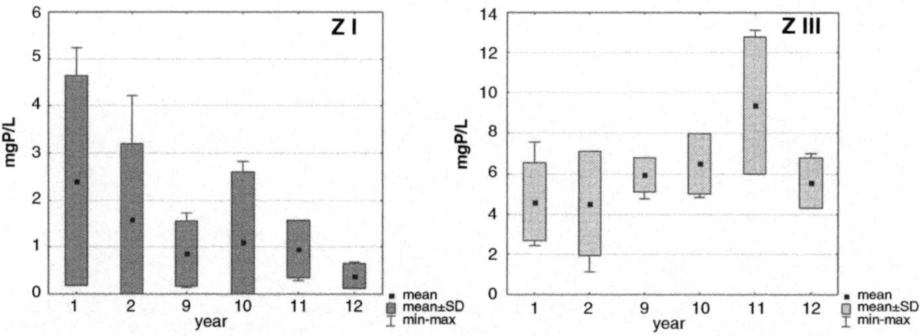

Fig. 17 Pore water TP concentrations in sediments in Group II restored lakes at different depth zones

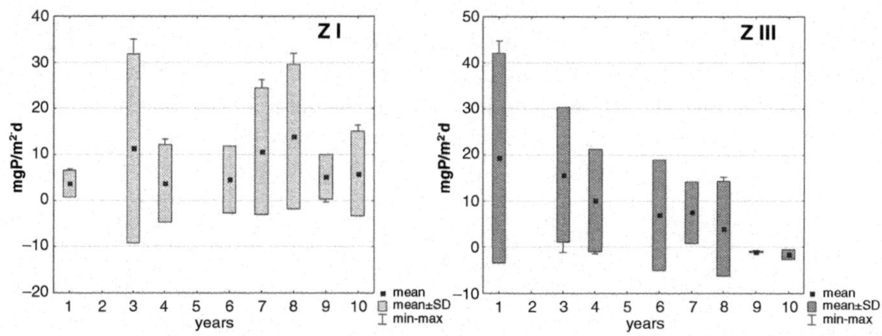

Fig. 18 Internal P loading in Group III restored lakes at different depth zones

values in years 1–4 (4.98–7.30 mgP/L), falling down since the 6th year (2.12–3.47 mgP/L, Fig. 20).

Lakes introduced to *Group IV* were shallow, with depth varying from 2 to 5 m; thus only one depth zone was represented. An increase of mean values was observed in the 2nd year of treatment in case of internal P loading and pore water TP

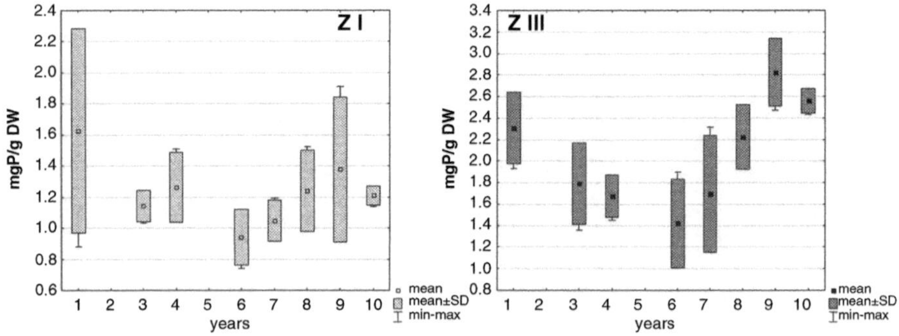

Fig. 19 TP content in sediments in Group III restored lakes at different depth zones

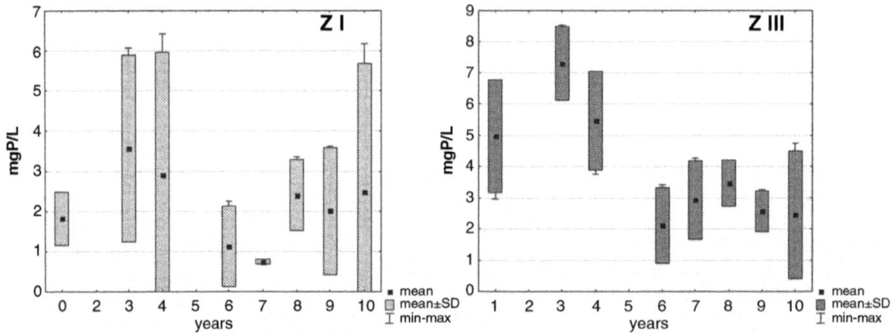

Fig. 20 Pore water TP concentrations in sediments in Group III restored lakes at different depth zones

concentrations. Amounts of P released from sediments grew fivefold from 1.22 to 6.60 mgP/m^2 d and decreased again to 2.60 mgP/m^2 d in the 3rd year. Pore water TP content rose from 1.07 to 1.89 mgP/L, but in the last year, a fivefold fall was observed to 0.37 mgP/L. TP content in sediments was very similar in the first 2 years (1.24 mgP/g DW), increasing to 1.59 mgP/g DW in the last year of restoration (Fig. 21).

5 Discussion

Trophic state of lake waters as well as season and water depth strongly influenced the internal P loading. Mesoeutrophic lakes were characterized by the lowest P amounts released from sediments, especially in littoral zone, where P binding was observed as well [29–31] as a result of good oxygen conditions in sediment-water interface and thus high redox potential. These two factors determine the direction of P exchange in the lake ecosystem [7]. In other trophic types of lakes, P loading from sediments

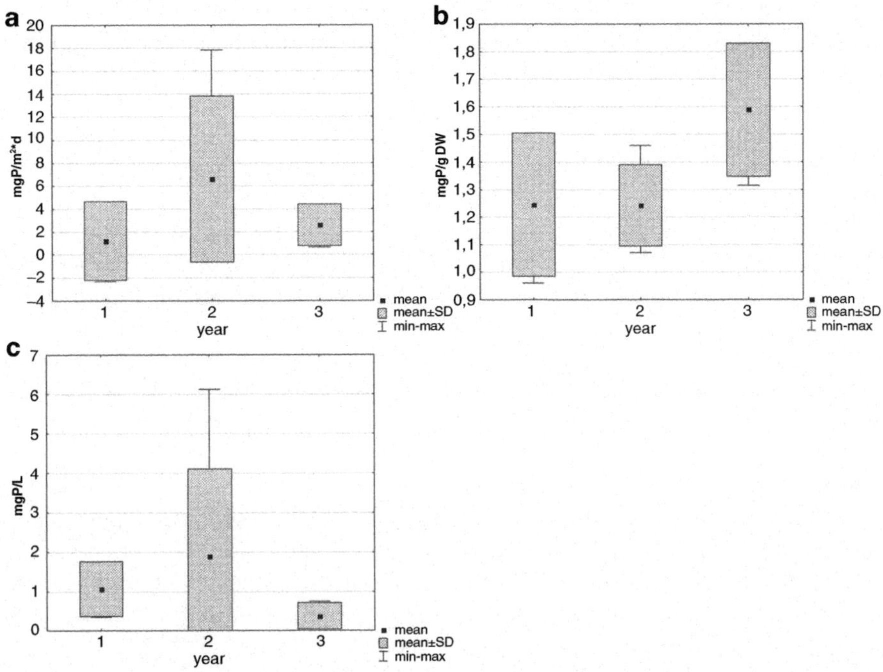

Fig. 21 Internal P loading (**a**), TP content (**b**) and pore water TP concentrations (**c**) in sediments in Group IV restored lakes at Z II

dominated, with greatest intensity in hypertrophic waterbodies, both in profundal, suffering from oxygen depletion, and well-oxygenated littoral [29, 32–36]. Statistically significant differences of the intensity of P loading between depth zones were confirmed (Kruskal-Wallis test, $p < 0.05$), apart from Z II. The increase of P release with lake trophic state resulted from oxygen deficits in sediment-water interface, typical in summer in eutrophic lakes. Anaerobic conditions are one of the most important causes of a significant P release from bottom sediments [37]. Under aerobic ones, iron is in the form of insoluble compounds and adsorbs P, whilst during anoxia, iron is reduced and its compounds are soluble, releasing P to the water column. Anaerobic conditions are accompanied by a decrease in redox potential and an accumulation of organic matter in sediments, subjected to decomposition. Total oxygen depletion results in redox potential −100 to −200 mV, being conducive to P release from sediments [7].

Internal P loading was usually increased with water depth. Statistically significant difference in P loads in depth zones was noted for mesoeutrophic and hypertrophic lakes (Kruskal-Wallis test, $p < 0.05$). It was determined by the dissimilarities in P circulation in deep and shallow lakes. In the first ones, hypolimnion acts as a temporary sink for settled P as the release of the compound is much greater and faster than biotic uptake as well as the chemical precipitation reactions. Thus, profundal zone is a place of seasonal dissolved P accumulation. Shallow waterbodies are characterized by the lack of thermocline; therefore P release dominated over its binding [38]. Additionally, P release in shallow lakes is accelerated by instant changes in oxygen content by the bottom, short-term thermal stratification or wind-driven circulation from throughout the water column [6, 39, 40]. This is of paramount importance for shallow waterbodies, where the external nutrient loads determine the overall water quality, yet the sediments play a crucial role for the in-lake P cycling [41].

The hand-in-hand relation between internal P loading and water depth was as well observed in restored lakes due to mentioned reasons. It was explicit in Groups II and III, where summer oxygen deficits strongly affected the increase in P loads from sediments in Z III. In case of Group I lakes, Z IV stood out with lower P release in comparison with more shallow lake zones. This results from trophic status of two lakes taken into consideration in this case. One of them – Lake Durowskie – is of eutrophic character [17, 36, 42], whilst the second, Lake Góreckie, is hypertrophic [43]. Higher P loads in years 4–6 were determined by results from the latter lake, which was analysed for 3 years only. Nevertheless, a reduction of P loads under the influence of combined restoration measures was noted, both in mean and maximum values.

The intensity of P release was affected by season as well. Both oxygen depletion by the sediment in deeper lakes and increased water temperature were observed in summer [35, 44]. The latter factor strongly accelerates the organic matter decomposition, thus increasing P loads from sediments [45, 46], being especially important in shallow eutrophic lakes at maximum pelagic primary productivity during large parts of the summer and controlling the P concentration in the lake water [41, 47]. Additionally, gas bubble emission boosts in summer, increasing the resuspension of sediments together with pore water rich with P [4, 6]. In case of deep lakes, thermocline limits the P circulation from hypolimnion to epilimnion, but the process of P release occurs at the greatest depth due to oxygen deficits. It was confirmed, i.e. in mesotrophic Lake Strzeszyńskie, where the dominance of P release over its binding was largest in summer [31]. Eutrophic and hypertrophic lakes were characterized by the highest internal P loads in Z III (5–10 m) than Z IV (>10 m) due to abundance of fresh organic matter, reaching the profundal zone of the shallower area, particularly in summer. This matter of phytoplanktonic origin was mineralized by bacteria present in sediment in anaerobic conditions [39, 48]. It was already stated that microorganisms have a significant impact on P circulation due to organic matter decomposition [38].

Biological processes like primary production and plankton sedimentation are of great importance for internal P loading in lake littoral, influencing as well the possibility to reduce the amount of P released within restoration treatment. Z I (0–2 m) was represented in all three groups of restored water bodies, and in one only there was a decrease in internal P loads, namely, Group II, i.e. lakes restored with P inactivation with iron sulphide and magnesium chloride. The method of P inactivation aims at increasing sorption complex of sediments, thus reducing the amount of P released to the water column. Magnesium chloride reacts to both phosphates and ammonium nitrogen, forming struvite (magnesium ammonium phosphate), which in the form of crystals sediments to the lake bottom. Additionally, positive effect was accelerated by submerged macrophytes, covering the bottom of Lake Rusałka in the first half of the year. Macrophytes, which may occur in great number in shallow lakes, have both negative and positive impacts on phosphorus cycle. Oxygen released from the roots may increase redox-sensitive phosphorus sorption; however when the macrophyte cover is dense, the phosphorus release may increase due to a diminish in oxygen concentration in water [6, 49, 50]. Increasing macrophyte cover might be then responsible for an increase of internal P loads in Z I of Group II in years 10–12 in comparison to year 9. Nevertheless, the addition of P-binding agents during restoration resulted in clear fall of P loading from sediments between years 1–2 and 9–12, especially since the deeper zone in Group II lakes (Z III) was subjected to a decrease in internal P release as well, what was related to macrophyte cover to a much lesser extent. As it was mentioned above, in Z I of restored lakes from Groups I and III, internal P loads remained quite stable despite the restoration treatment, however on a lower level than deeper lake parts. Nonetheless, P released from the sediment in the littoral zone passes directly to the trophogenic zone, causing water blooms, especially in summer [24].

Z II and Z III in Groups I and III revealed a steady decreasing trend in P loading from sediments due to restoration results expressed by the gradual sustainable rebuilding of the lake ecosystem. The application of three reclamation methods, supporting each other, resulted in a decrease of primary production and organic matter sedimentation as well as the reduction of organic matter readily decomposed by bacteria [48]. In case of Group I lakes, it was stimulated by hypolimnetic oxygenation by wind-driven aerators. Although it did not provide high amounts of oxygen, the organic matter decomposition was not accelerated [17]; however the content of oxygen was enough to enable denitrification as well as anammox reaction to occur and thus to increase the redox potential [51]. Rising P internal loads in the 2nd year of restoration with the use of aerator might be a consequence of sediment resuspension immediately after the installation, but in the following years, this phenomenon was not observed. In case of Uzarzewskie Lake, representing Group III with hypolimnetic nitrate treatment [25], the addition of nitrates to water above the bottom resulted in an increase of the redox potential in the sediment-water interface as well, causing internal P loading reduction [24]. Additionally, this method eliminates the hydrogen sulphide in the water overlying the sediments and reduces nitrogen concentration, due to the denitrification process. Hydrogen sulphide was observed in all deep lakes hypolimnion during summer as a product of bacterial sulphate decomposition that can block Fe^{+2} as an insoluble sulphide. Furthermore, P mobilization and SO_4^{2-} reduction in anoxic sediments generate a pH increase that would inhibit P sorption [52].

A completely different restoration method applied on two lakes introduced to Group IV, i.e. Effective Microorganisms (EM), resulted in boosted internal P loading. The use of a composition of photosynthetic bacteria, lactic acid bacteria, actinomycetes, yeasts and fermenting fungi [53] resulted in an acceleration of organic matter decomposition and P release to the water column in the 2nd year of restoration. In the following vegetation season, sediment P loads decreased but remained higher than prior the treatment, indicating the lack of success in case of lake sediments.

Internal loading of P in lakes is a complex process which is mainly determined by the amount and species of settled P as well as their subsequent diagenetic transformation in the sediment [38]. TP content was distinctly lower in mesoeutrophic lakes and higher in eutrophic and hypertrophic waterbodies; however statistically it was confirmed only in Z I (littoral) and Z III (5–10 m) (Kruskal-Wallis test, $p < 0.05$), whereas the differences between all depth zones were statistically important for mesoeutrophic and hypertrophic lakes (Kruskal-Wallis test, $p < 0.05$).

In most groups of restored waterbodies and depth zones, TP content increased together with falling internal P release, proving the augmentation of P sorption complex and successful binding of P both with Fe (from iron compounds) and magnesium (from magnesium chloride, binding both phosphates and ammonium nitrogen as struvite). Not only was the amount of binding agent increased, but the conditions for P binding were created by increased redox potential by additional restoration method (Group I), the main method itself in a case of a lake already supplied with Fe (Group III) or the macrophyte cover exert a positive impact on sediments (Group II). In Group IV lakes, TP content decreased together with internal P loading in the second year of treatment whilst distinctly increased in the third year as the P loads fell down.

P fractions in sediments are more informative in the aspect of P internal loads than total P amount in sediments [4]. Res-P, i.e. biologically inaccessible P fraction of insoluble inorganic and organic compounds, was dominant in studied lakes. As this P is permanently buried in sediments, its high contribution limits the role of sediments as an internal P source [4]. No statistically significant differences were stated for Res-P between lake trophic states in case of Z I, whereas in all eutrophic and hypertrophic waterbodies, there was a significant difference between depth zones (Kruskal-Wallis test, $p < 0.05$). Greatest amounts of this fraction were already noted, i.e. in lakes Swarzędzkie, Góra and Uzarzewskie, situated in the River Cybina watercourse [33, 54]. Res-P was followed by NaOH-NRP fraction, i.e. P in organic matter, which dominated in mesoeutrophic lakes, what was determined by the presence of macrophytes and charophytes. This relation was confirmed in Lake Strzeszyńskie [30]. Statistically important differences in NaOH-NRP content in lakes of various trophic states were observed only in Z IV (Kruskal-Wallis test, $p < 0.05$), but no differences were stated between water depths. P in compounds with calcium is represented in sediments as biologically inaccessible HCl-P fraction, with the highest contribution in mesoeutrophic lakes (especially littoral) and the lowest in eutrophic lakes (deep zones). Calcium-bound P is not sensitive to redox potential, and once deposited in bottom sediments, it may be stored there for a long time [55]. Statistically significant differences for this P fraction were stated between trophic lake types in Z IV only whereas in eutrophic and hypertrophic lakes between all depth zones (Kruskal-Wallis test, $p < 0.05$). Fractions of paramount importance for internal P loading were the most mobile, i.e. NH_4Cl-P, BD-P and NaOH-P, representing P bound to iron and aluminium and loosely sorbed on sediment particles. Total amount of those three fractions diminished together with lake trophy due to P loading increasing in the same direction. Eutrophic lakes were characterized by higher contribution of NH_4Cl-P whilst mesotrophic waterbodies by BD-P, where it decreased with water depth. Lower pH is stated as more conducive for insoluble Fe-P bound formation [4] in mesotrophic lakes, whilst oxygen depletion promotes iron reduction and the release of adsorbed phosphorus into the water column in eutrophic and hypertrophic ones [4, 56].

Organic matter content in sediment depends on numerous factors, i.e. primary production and allochthonic matter from catchment [4]. Our results indicated lower amounts in mesoeutrophic lakes, whilst in eutrophic and hypertrophic ones, its content decreased together with depth. This finding was consistent with higher productivity of waterbodies with greater trophy, where sediments are supplied with fresh matter from sedimenting phytoplankton. Both organic matter sources, allochthonic and autochthonic, increased the amount of organic compounds in littoral sediments; thus a relation between internal P loading and organic matter content was stated for Z I only ($p < 0.05$). In all depth zones, statistically important differences were stated among trophic lake types as well as depth zones (Kruskal-Wallis test, $p < 0.05$).

The pore waters of the sediment are important for the SRP transport in sediment-water interface [57, 58]. SRP in this medium are a direct link with the water above and a solid-liquid boundary between water and sediment [8]. In our research both SRP and TP in pore water increased together with lake trophy and depth. It was already noticed in lakes Swarzędzkie, Strzeszyńskie and Uzarzewskie [30, 31, 59]. Usually highest P content was observed in summer, as well as internal P loading, and this relation was confirmed in long-term studies in Lake Swarzędzkie [59] and statistically significant in analysed lakes ($p > 0.05$). In all depth zones, statistically important differences were noted between lake trophic type and depth zones (Kruskal-Wallis test, $p < 0.05$).

The relations between pore TP concentrations and internal P loading in restored lakes proved that P content in the interstitial water allows to predict the P pool potentially available to be released from the bottom sediments [7]. In most groups and depth zones, the lower the P loads, the lower the TP in pore water. Statistical significance of this relation was revealed: in Z II Group I ($b = 0.48$, $p < 0.05$, $n = 27$), in Z III Group I ($b = 0.42$, $p < 0.05$, $n = 23$), in Z IV Group I ($b = 0.36$, $p < 0.01$, $n = 56$), in Z I Group II ($b = 0.79$, $p < 0.01$, $n = 25$), in Z I Group III ($b = 0.60$, $p < 0.01$, $n = 23$), in Z III Group III ($b = 0.52$, $p < 0.05$, $n = 23$) and in Z II Group IV ($b = 0.80$, $p < 0.01$, $n = 13$).

SRP and TP concentrations in water above bottom increased together with lake trophy and water depth. Maximum values were noted in summer due to an accumulation of released from sediments P in hypolimnion, isolated from shallower waters. This relation was already confirmed in mesoeutrophic Lake Strzeszyńskie and hypertrophic Lake Uzarzewskie [24, 31, 60]. A correlation between internal P loading and P concentration in water above sediments was noted only for Z III ($p < 0.05$).

6 Conclusion

Internal P loading is of great importance for lake water quality of freshwater lakes and reservoirs. Analysis based on results of multiannual research on numerous waterbodies in Western Poland indicated distinct variability of this process in relation to lake trophic status, water depth and season. P accumulation outweighed P loading periodically in mesoeutrophic lakes, whilst in eutrophic and hypertrophic, P release was more common. Most intensive P loading was observed during summer due to oxygen depletion, emission of gas bubbles and higher water temperature in case of shallower zones. Additionally, greater content of TP and organic matter was found in sediments of heavily eutrophicated lakes due to intensive sedimentation of organic particles. The higher the trophic status and water depth, the greater the SRP and TP concentrations noted in pore and above bottom waters, and these sediment characteristics were correlated with internal P loads. To diminish them, and to improve water quality as well, restoration treatment has been conducted in part of analysed lakes. A decrease in P release and, at the same time, higher TP content in sediments were found in lakes from Groups I to III, i.e. restored with different methods, mainly P inactivation. This resulted from increased sorption complex and formation of conducive conditions for P binding with used agents. Such results have not been observed in case of lakes restored with Effective Microorganisms.

Acknowledgements This study was supported by (1) the Ministry of Science and Higher Education (Poland), grant no. N 305 108 31/3632 and partly by grant no. NN305 372838; (2) KBN, grant no. 3T09D 01527; and (3) the Fund for the statutory activities of the Department of Water Protection.

References

1. Boström B, Andersen JM, Fleischer S, Jansson M (1988) Exchange of phosphorus across the sediments-water interface. Hydrobiologia 170:229–244
2. Golterman HL (1995) The role of iron hydroxide-phosphate-sulphide system in the phosphate exchange between sediments and overlying water. Hydrobiologia 297:43–54
3. Wiśniewski RJ (1995) Rola zasilania wewnętrznego w eutrofizacji zbiorników zaporowych [The role of internal loading in dam reservoirs eutrophication]. In: Zalewski M (ed) Procesy biologiczne w ochronie i rekultywacji nizinnych zbiorników zaporowych [Biological processes in protection and restoration of lowland dam reservoirs] Biblioteka Monitoringu Środowiska, WIOŚ Łódź, pp 61–70. Polish
4. Kentzer A (2001) Fosfor i jego biologicznie dostępne frakcje w osadach jezior o różnej trofii [Phosphorus and its bioavailable fractions in the sediments of lakes of different trophy] Nicolaus Copernicus University Press, Toruń, Dissertations. 111 p. Polish
5. Wang H, Appan A, Gulliver JS (2003) Modelling of phosphorus dynamics in aquatic sediments: I-model development. Water Res 37(16):3928–3938
6. Søndergaard M, Jensen JP, Jeppesen E (2001) Retention and internal loading of phosphorus in shallow, eutrophic lakes. Sci World 1:427–442

7. Søndergaard M, Wolter KD, Ripl W (2002) Chemical treatment of water and sediments with special reference to lakes. Handbook of ecological restoration, vol 1. Cambridge University Press, Cambridge
8. Boström B, Ahlgren I, Bell R (1982) Internal nutrient loading in a eutrophic lake, reflected in seasonal variations of some sediment parameters. Verh Internat Verein Limnol 22:3335–3339
9. Jeppesen E, Kristensen P, Jensen JP, Søndergaard M, Mortensen E, Lauridsen T (1991) Recovery resilience following a reduction in external phosphorus loading of shallow, eutrophic Danish lakes: duration, regulating factors and methods for overcoming resilience. Mem Ist Ital Idrobiol 48:127–148
10. Gonsiorczyk T, Casper P, Koschel R (1997) Variations of phosphorus release from sediments in stratified lakes. Water Air Soil Pollut 99:427–434
11. Søndergaard M, Jensen JP, Jeppesen E (1999) Internal phosphorus loading in shallow Danish lakes. Hydrobiologia 408/409:145–152
12. Jańczak J (1996) Atlas jezior Polski. [The atlas of Polish lakes] Bogucki Publishing, Poznań. Polish
13. Szyper H, Romanowicz W, Gołdyn R (2000) Zagrożenie jezior Wielkopolskiego Parku Narodowego przez czynniki zewnętrzne [The threats to Wielkopolski National Park lakes by external factors] In: Burchardt L (ed) Ekosystemy wodne Wielkopolskiego Parku Narodowego [Water ecosystems of Wielkopolski National Park]. Poznań. Polish
14. Choiński A (2006) Katalog jezior Polski [The catalog of Polish lakes]. UAM Scientific Publishing, Poznań. Polish
15. Gołdyn R, Grabia J (1998) Program ochrony wód rzeki Cybiny [Protection program of waters in the Cybina River]. Department of Environmental Protection, Poznań Town's Authority, Poznań. Polish
16. Pułyk M, Tybiszewska E (1995) Raport o stanie środowiska w województwie poznańskim w roku 1994. [The report on environment state in poznań province in 1994] Biblioteka Monitoringu Środowiska, Poznań. Polish
17. Gołdyn R, Messyasz B, Domek P, Windhorst W, Hugenschmidt C, Nicoara M, Plavan G (2013) The response of Lake Durowskie ecosystem to restoration measures. Carpath J Earth Environ Sci 8(3):43–48
18. Rosińska J, Kozak A, Dondajewska R, Gołdyn R (2017) Cyanobacterial blooms before and during the restoration process of a shallow urban lake. J Environ Manage 198:340–347
19. Sobczyński T, Joniak T, Pronin E (2012) Assessment of multi-directional experiment to restore Lake Góreckie (Western Poland) with particular focus on oxygen and light conditions: first results. Pol J Environ Stud 21(4):1025–1031
20. Wiśniewski R, Ślusarczyk J, Kaliszewski T, Szulczewski A, Nowacki P (2010) "Proteus", a new device for application of coagulants directly to sediment during its controlled resuspension. Verh Internat Verein Limnol 30(9):1421–1424
21. Gołdyn R, Podsiadłowski S, Dondajewska R, Kozak A (2014) The sustainable restoration of lakes – towards the challenges of the Water Framework Directive. Ecohydrol Hydrobiol 14(1):67–74
22. Kowalczewska-Madura K, Dondajewska R, Gołdyn R (2011) Seasonal changes of phosphorus release from the bottom sediments of Rusałka Lake during the restoration process. Ecol Chem Eng A 18(2):219–224
23. Kowalczewska-Madura K, Dondajewska R, Gołdyn R (2008) Influence of iron treatment on phosphorus internal loading from bottom sediments of the restored lake. Limnol Rev 8(4):177–182
24. Kowalczewska-Madura K, Dondajewska R, Gołdyn R, Podsiadłowski S (2017) The influence of restoration measures on phosphorus internal loading from the sediments of a hypereutrophic lake. Environ Sci Pollut Res 24(16):14417–14429. https://doi.org/10.1007/s11356-017-8997-2
25. Dondajewska R, Kozak A, Kowalczewska-Madura K, Budzyńska A, Podsiadłowski S, Tomkowiak A (2018) The response of a shallow hypertrophic lake to innovative restoration measures – Uzarzewskie Lake case study. Ecol Eng 121:72–82. https://doi.org/10.1016/j.ecoleng.2017.07.010
26. Carlson RE (1977) A trophic state index for lakes. Limnol Oceanogr 22(2):361–369

27. ISO 2004. 6878:2004 Water quality – determination of phosphorus – ammonium molybdate spectrometric method
28. Psenner R, Boström B, Dinka M, Pettersson K, Pucsko R, Sager M (1988) Fractionation of phosphorus in suspended matter and sediment. Arch Hydrobiol Beih Ergebn Limnol 30:83–112
29. Kowalczewska-Madura K, Dondajewska R, Gołdyn R (2009) Possibilities of phosphorus release/accumulation in the bottom sediments of chosen Wielkopolska and Pomeranian Lakes. In: Marszelewski W (ed) Anthropogenic and natural transformations of lakes, vol 3. PTLim, Toruń, pp 129–134
30. Kowalczewska-Madura K, Gołdyn R, Dera M (2015) Spatial and seasonal changes of phosphorus internal loading in two lakes with different trophy. Ecol Eng 74:187–195
31. Kowalczewska-Madura K, Gołdyn R, Bogucka J, Strzelczyk K (2019) Impact of environmental variables on spatial and seasonal internal phosphorus loading in a mesoeutrophic lake. Int J Sediment Res 34:14–26. https://doi.org/10.1016/j.ijsrc.2018.08.008
32. Dondajewska R (2008) Internal phosphorus loading form bottom sediments of a shallow preliminary reservoir. Oceanol Hydrobiol Stud 37(2):89–97
33. Kowalczewska-Madura K, Dondajewska R, Gołdyn R (2010) a. Internal phosphorus loading in selected lakes of the Cybina River valley. Oceanol Hydrobiol Stud 39(3):35–45
34. Kowalczewska-Madura K, Gołdyn R, Dondajewska R (2010) b. The bottom sediments of Lake Uzarzewskie – a phosphorus source or sink? Oceanol Hydrobiol Stud 39(3):81–91
35. Kowalczewska-Madura K, Gołdyn R, Dondajewska R (2010) c. Phosphorus release from the bottom sediments of Lake Rusałka (Poznań, Poland). Oceanol Hydrobiol Stud 39(4):135–144
36. Kowalczewska-Madura K, Dondajewska R, Gołdyn R, Kozak A, Messyasz B (2018) Internal phosphorus loading from the bottom sediments of a dimictic lake during its sustainable restoration. Water Air Soil Pollut 229(280). https://doi.org/10.1007/s11270-018-3937-4
37. Ishikawa M, Nishimura H (1989) Mathematical model of phosphate release rate from sediments considering the effect of dissolved oxygen in overlying water. Water Res 23:351–359
38. Hupfer M, Lewandowski J (2008) Oxygen controls the phosphorus release from lake sediments – a long-lasting paradigm in limnology. Int Rev Hydrobiol 93(4-5):415–432
39. Kleeberg A, Dudel GE (1997) Changes in extent of phosphorus release in a shallow lake (Lake Grosser Muggelsee; Germany, Berlin) due to climatic factors and load. Mar Geol 139:61–75
40. Golterman HL (2004) The chemistry of phosphate and nitrogen compounds in sediments. Kluwer Academic Publishers, Dordrecht
41. Søndergaard M, Bjerring R, Jeppensen E (2012) Persistent internal phosphorus loading during summer in shallow eutrophic lakes. Hydrobiologia 710(1):95–107
42. Messyasz B, Gąbka M, Barycki J, Nowicki G, Lamentowicz Ł, Goździcka-Józefiak A, Rybak A, Dondajewska R, Burchardt L (2015) Phytoplankton, culturable bacteria and their relationships along environmental gradients in a stratified eutrophic lake. Carpath J Earth Environ Sci 10(1):41–52
43. Sobczyński T, Joniak T (2013) The variability and stability of water chemistry in a deep temperate lake: results of long-term study of eutrophication. Pol J Environ Stud 22(1):227–237
44. Kowalczewska-Madura K, Gołdyn R (2009) The internal loading of phosphorus from the sediments of Swarzędzkie Lake (Western Poland). Pol J Environ Stud 18(4):635–643
45. Kelderman P (1984) Sediment – water exchange in Lake Grevelingen under different environmental conditions. Neth J Sea Res 18:286–311
46. Forsberg C (1989) Importance of sediments in understanding nutrient cycling in lakes. Hydrobiologia 176/177:263–277
47. Ramm K, Scheps V (1997) Phosphorus balance of a polytrophic shallow lake with the consideration of phosphorus release. Hydrobiologia 342/343:43–53
48. Katsev S, Dittrich M (2013) Modelling of decadal scale phosphorus retention in lake sediment under varying redox conditions. Ecol Model 251:246–259
49. Frodge JD, Thomas GL, Pauley GB (1991) Sediments phosphorus loading beneath dense canopies of aquatic macrophytes. Lake Reserv Mange 7:61–71

50. Stephen D, Moss B, Philips G (1997) Do rooted macrophytes increase sediment phosphorus release. Hydrobiologia 342:27–34
51. Kozak A, Rosińska J, Gołdyn R (2018) Changes in phytoplankton structure due to prematurely limited restoration treatments. Pol J Environ Stud 27(3):1097–1103
52. Clavero V, Garcia-Sanchez MJ, Niell FX, Fernandez JA (1997) Influence of sulphate enrichment on the carbon dioxide and phosphate fluxes across the sediment-water interface. Hydrobiologia 345:59–69
53. Higa T (1995) What is EM technology. College of Agriculture, University of Ryukyus, Okinawa, Japan
54. Kowalczewska-Madura K, Gołdyn R (2005) Phosphorus content of bottom sediments in the Swarzędzkie Lake. Verh Internat Verein Limnol 29:1063–1068
55. Kufel L, Biardzka E, Strzałek M (2013) Calcium carbonate incrustation and phosphorus fractions in five charophyte species. Aquat Bot 109:54–57
56. Nürnberg GK (2009) Assessing internal phosphorus load – problems to be solved. Lake Reserv Manage 25:419–432
57. Katsev S, Tsandev I, L'Heureux I, Rancourt DG (2006) Factors controlling long-term phosphorus efflux from lake sediments: exploratory reactive-transport modeling. Chem Geol 234:127–147
58. Ignatieva NV (1996) Distribution and release of sedimentary phosphorus in Lake Ladoga. Hydrobiologia 322:129–136
59. Kowalczewska-Madura K, Gołdyn R (2012) Spatial and seasonal variability of pore water phosphorus concentration in shallow Lake Swarzędzkie, Poland. Environ Monit Assess 184:1509–1516
60. Kowalczewska-Madura K, Dondajewska R, Gołdyn R (2007) Changes of phosphorus concentration in bottom sediments and in overlying water of two eutrophicated lakes in Wielkopolska Region. Limnol Rev 7(4):201–207

The Effect of a Dam Reservoir on Water Trophic Status and Forms of River Transport of Nutrients

Stanisław Chmiel, Joanna Sposób, Katarzyna Męsiak-Wójcik, Zdzisław Michalczyk, and Sławomir Głowacki

Contents

Abstract Hydrochemical research conducted in the dam reservoirs often show strong eutrophication of their waters. This usually results from high supply of the mineral forms of nitrogen and phosphorus from the river waters. The reduction of the supply of nutrients to the reservoir below the level causing water quality deterioration can be limited in the case of the agricultural land use in the catchment. In order to reduce nutrients input and to improve the ecological state in the one of the reservoirs in the Eastern Poland, it has been proposed solution: a change of the functioning of the reservoir from dammed to lateral and construction of an additional preliminary reservoir above the existing one.

Keywords Biogenes · Dam reservoir · Rivers · Water quality

S. Chmiel (✉) · J. Sposób · K. Męsiak-Wójcik · Z. Michalczyk
Department of Hydrology and Climatology, Faculty of Earth Sciences and Spatial Management, Maria Curie-Skłodowska University, Lublin, Poland
e-mail: stanislaw.chmiel@poczta.umcs.lublin.pl

S. Głowacki
Regional Water Management Board, Polish Waters, Lublin, Poland

E. Korzeniewska, M. Harnisz (eds.), *Polish River Basins and Lakes – Part I*,
The Handbook of Environmental Chemistry 86,
https://doi.org/10.1007/978-3-030-12123-5_15

1 Introduction

Rivers constitute open systems exchanging energy and matter with the surrounding environment. Along the course of water flow, the ecological status of the river is characterised by relative homoeostasis the continuum of which [1] is disturbed in the case of construction of a dam reservoir [2]. A change in the conditions of water flow affects the physical, chemical, and biological processes occurring in the reservoir and usually causes deterioration of water quality [3]. The scope of such changes largely depends on the parameters of the reservoir, and particularly on its area, depth, time of retention, and water exchange, as well as quality of the waters inflowing to the water body [2, 3].

Physical and chemical indicators strongly degrading the quality of surface waters include biogenic substances. They contribute to the acceleration of water eutrophication, and deterioration of the ecological state of the water body, and its useful values. In the case of weak/bad water quality in water bodies, costly methods of their protection and reclamation have frequently been implemented. They rarely provided expected results [2, 4].

This confirms the thesis that efficient improvement of water cleanliness should first of all involve the identification of the primary pollutants reaching the water body and their limiting/liquidation. The last stage should involve the implementation of reclamation methods in reference to the water body. A threat of eutrophication and risk of appearance of cyanobacterial blooms constitute the basic elements to be considered while designing, using, and reclaiming a water body [5, 6]. Such an approach to the protection of water bodies constitutes the basis of the area of knowledge described as ecohydrology, developing in recent years [7]. One of the methods of limiting deterioration of water in dam reservoirs is construction of preliminary reservoirs [8, 9].

2 Study Area

The objective of this paper is the assessment of the effect of the Zemborzyce dam reservoir on forms of biogene migration and trophic status of the waters of the Bystrzyca River (Fig. 1), as well as the determination of solutions which may contribute to an improvement of water quality in the reservoir and river below it.

The Zemborzyce Reservoir was established in 1974 in the Bystrzyca River valley. It is a dam reservoir. It has an area of 282 ha, length of 3 km, maximum width of 1.3 km, maximum depth near the dam of 4 m, and volume of 6.3 million m^3. The area of the Bystrzyca River catchment reaching the dam profile amounts to 725 km^2. The land use structure of the catchment is dominated by arable land: cultivated land occupies approximately 71% of the area, meadows and pastures 4%, and orchards 3%. Forests constitute almost 11% of the area and built-up areas approximately 10%. The eastern part of the direct catchment is occupied by forests. The western part is dominated by built-up areas and land under agricultural use [10].

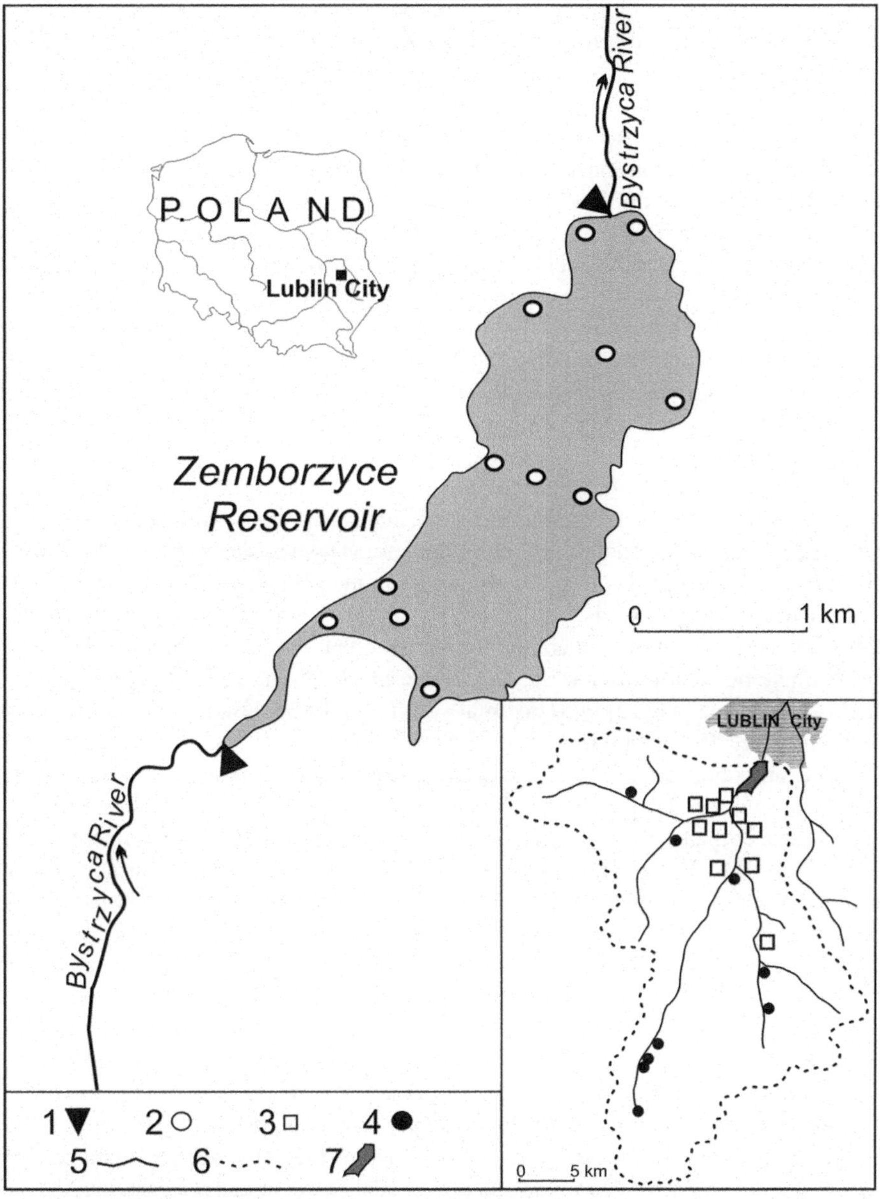

Fig. 1 Location of water sampling sites for analyses in the Bystrzyca River catchment and Zemborzyce Reservoir (water sampling sites: 1, above and below the reservoir; 2, in the reservoir; 3, from overland flow; 4, from springs; 5, main rivers; 6, watershed of the Bystrzyca River catchment; 7, reservoir)

The water body is fed by the waters of the Bystrzyca River. In the multiannual 1960–2010, minimum discharges below the reservoir amounted to 0.5 m^3/s, average 3.0 m^3/s, and maximum discharges exceeded 50 m^3/s. Therefore, water exchange in the reservoir during low discharges occurred after 146 days, during average discharges after 25 days, and during maximum discharges even within 1 day. Based on hydrometeorological data from the period 2008–2012, components of the water balance of the Bystrzyca River until the dam profile had the following values: precipitation 625.0 mm, total runoff 129.1 mm, and surface runoff 10.9 mm. The balance of the catchment in the profile above the reservoir should be supplemented by losses to evaporation from its area which was calculated in accordance with the Iwanow formula (for the period with no ice cover) to amount to 2.0 mm. The river particularly transported waters from underground drainage which constituted approximately 90% of total runoff. In the multiannual, the contribution of groundwater recharge in total runoff amounted to approximately 25% [10].

During groundwater alimentation of the river, its discharges showed average and low values. Water in the river channel originated from drainage of Upper Cretaceous and Palaeogene rocks, opokas and marly opokas, as well as sandy-gravel Quaternary formations filling river valleys. Groundwaters in the area were abundant in nitrogen and phosphorus compounds, particularly of agricultural origin [6]. Surface alimentation of the river was considerable during snowmelts and after abundant rainfalls, resulting in the highest discharges. The river was also discharged with municipal and industrial sewage [11, 12] with the total volume in the analysed period constituting approximately 2% of total runoff.

An example of a dam reservoir where considerable deterioration of water quality occurred is the Zemborzyce Reservoir on the Bystrzyca River (Fig. 1). Its ecological state has been bad for many years (reports on the state of the environment in the years 2008–2012), particularly as a result of "water blooms" – a consequence of the occurrence of cyanobacteria, chlorophytes, and diatoms. The discussion on the manner of improvement of water quality in the reservoir was a subject of three conferences [13–15]. The effect of the first two meetings was the implementation of hydrobiological and biomanipulation methods of improvement of the cleanliness of the water body. They were expected to improve water quality through the implementation of biological barriers in the form of rosettes with emergent vegetation for capturing biogenes, stocking with predatory fish and harvesting herbivorous fish, planting vegetation limiting direct supply of biogenes from agricultural fields to the water body, and improvement of the state of the water-sewage management in the catchment. All of the implemented methods, however, did not contribute to the improvement of the ecological state of the water body.

The objective of this paper is the assessment of the effect of the Zemborzyce dam reservoir on forms of biogene migration and trophic status of the waters of the Bystrzyca River, as well as the determination of solutions which may contribute to an improvement of water quality in the reservoir and river below it.

3 Study Methods

Hydrochemical research on the waters of the Zemborzyce Reservoir was conducted in the period 2008–2012. The research works involved field measurements of the basic physical and chemical water properties, water sampling, and laboratory analyses (Table 1). Approximately a dozen measurements of water transparency by means of a Secchi disc were also performed in the field. Water sampling sites were located on the inflow and outflow of the reservoir and periodically within the basin (Fig. 1). Measurements and water sampling from the profile above and below the reservoir were performed at least once per quarter. Measurements inside the water body were performed three times in the summer season of 2010 and 2012 and in the winter season of 2012. Water samples for identification of the basic anions and cations were filtered through a membrane filter 0.45 μm by means of a vacuum set by Nalgene. After drying, the filters were used for TSS determinations. Water for TN and TP analyses (without filtering) was preserved with sulphuric acid in glass containers of 40 mL.

The level of dissolved mineral/dissolved inorganic carbon (DIC) was calculated as the sum of concentrations of $C\text{-}HCO_3^-$ and $C\text{-}CO_{2(aq)}$. The concentration of dissolved mineral/dissolved inorganic nitrogen (DIN) was calculated as the sum of ions $N\text{-}NO_3^-$, $N\text{-}NO_2^-$, and $N\text{-}NH_4^+$.

In the analytic process, calibration solutions by Merck were used for the calibration of devices and certified reference material Environment MISSIPPI-03 and QC NUTRIENTS and complex and simple RTC for the verification of the accuracy of the performed analyses. Results of determinations of the basic ions were also verified by means of the method of water ionic balance. The assessment of differences in mean values of the studied indices on the inflow and outflow from the reservoir involved the application of a t-Student test for dependent samples from package

Table 1 Methods of research on physical and chemical indices and equipment

Index	Method/norm	Equipment
Temperature (T)	Electrometric	Hach HQ40d
Water reaction (pH)	Electrometric	
Dissolved oxygen(O_2)	Optical	
Carbon dioxide (CO_2)	Electrometric	OxyGuard CO_2 Analyser
Chlorophyll (CHLa)	Fluorometric	AquaFluor by Turner Designs
NO_3^-, NO_2^- Cl^-, SO_4^{2-}	[20]	Ion Chromatograph MIC 3 by Metrohm
NH_4^+, Na^+, K^+, Mg^{2+}	[21]	
Ca^{2+}	[22]	Digital Burette Solarus
Alkalinity	[23]	Digital Burette Solarus
Total suspended solids (TSS)	[24]	Analytical Balance Radwag AS 60/220/C/2
Orthophosphates (PO_4^{3-})	[25]	Spectrophotometer UV/VIS AquaMate Plus by Thermo
Total nitrogen (TN)	[26]	
Total phosphorus (TP)	[27]	
Total organic carbon (TOC)	Deconvolution	Spectrophotometer Pastel UV

Statistica 6.1. Indices of water saturation with calcite (SI_{cal}) were calculated by means of hydrogeochemical programme AQQA.

The trophic status of the analysed waters was assessed based on Trophic State Indices (TSI) [16, 17]. TSI were calculated for total phosphorus concentration $TSI_{TP} = 14 \cdot \ln(TP) + 4.15$, total chlorophyll "a" concentration $TSI_{CHLa} = 9.81 \cdot \ln(CHLa) + 30.6$, and total nitrogen concentration $TSI_{TN} = 14.43 \cdot \ln(TN) + 54.45$. The level of TSI signifying oligotrophy takes values below 40, mesotrophy 40–50, eutrophy 50–70, and hypertrophy above 70.

The load of N, P, and C provided to the reservoir, their runoff, and retention were also calculated. The calculations applied the values of the median of biogene concentration in the analysed water samples collected from the Bystrzyca River in the period 2008–2012 and mean water discharge (SQ) from groundwater alimentation and overland flow. The level of biogenes supplied directly to the reservoir from atmospheric precipitation was calculated based on data on the chemistry of precipitation waters collected in a daily cycle in Lublin. Concentrations of N, P, and C in river waters originating from groundwater alimentation were determined based on analyses of water samples collected from nine springs draining the main Cretaceous aquifer in the Bystrzyca River catchment. Water samples from the springs were collected at least once a year. The content of biogenes in overland flow was determined based on 76 water samples collected from episodic overland flow on slopes. Forty water samples were collected from overland flow from snowmelt and 36 from that resulting from rainfall. The difference between total load of biogenes reaching the reservoir and total load from groundwater alimentation, overland flow, and atmospheric precipitation was interpreted as the load of biogenes discharged to the river together with sewage.

Loads of nitrogen and phosphorus reaching the Zemborzyce Reservoir were also compared with the values of acceptable and dangerous load of the reservoir with biogenes according to the static (no inflow-outflow) and hydrodynamic model (inflow-outflow) by Vollenweider [18, 19]. Acceptable load is the amount of nitrogen or phosphorus supplied to the water which should not cause algal blooms. Dangerous load of nitrogen or phosphorus is their amount which can cause algal blooms.

4 Physical and Chemical Water Properties of the Bystrzyca River

The physical and chemical properties of water samples collected from the Bystrzyca River above and below the Zemborzyce Reservoir usually showed high diversity (Table 2). In the case of 17 analysed indices, mean values above and below the reservoir were statistically significantly different ($p < 0.05$). In the case of five indices, Cl^-, SO_4^{2-}, Na^+, Mg^{2+}, and NH_4^+, the calculations showed no statistically significant differences between mean values. Therefore, for the majority of the

Table 2 Characteristic values of physical and chemical indices analysed in the period 2008–2012 (above/below the reservoir)

Index	Unit	Number of samples	Minimum	Maximum	Median	Mean	Difference of means ($p < \mathbf{0.05}$)	Standard deviation	Variability coefficient
Temp.	°C	56	1.6	20.1	9.3	9.3	**−2.94**	4.96	0.53
		51	0.5	28.6	10.2	11.1		8.01	0.72
pH	–	56	6.93	8.20	7.96	7.92	**−2.52**	0.25	0.03
		51	6.74	8.76	8.20	8.10		0.46	0.06
O_2	mg/L	44	5.6	12.9	9.4	9.3	**−4.34**	1.83	0.20
		38	5.9	17.4	10.8	10.9		2.69	0.25
CO_2	mg/L	22	1	34	6	5	**2.13**	8.6	0.96
		22	<1	29	2	4		7.5	1.50
N-NH_4^+	mg/L	49	0.01	0.45	0.07	0.10	−0.77	0.09	0.93
		44	<0.01	0.79	0.06	0.13		0.16	1.26
N-NO_2^-	mg/L	49	<0.01	0.09	0.02	0.03	**6.81**	0.02	0.82
		44	<0.01	0.07	<0.01	0.01		0.02	1.83
N-NO_3^-	mg/L	49	1.26	6.61	1.99	2.14	**11.8**	0.83	0.39
		44	0.01	3.35	0.27	0.77		0.89	1.16
DIN	mg/L	49	1.35	6.75	2.41	2.26	**13.7**	0.86	0.38
		44	0.01	3.44	0.35	0.78		0.90	1.16
TN	mg/L	34	1.79	7.87	3.12	3.23	**7.64**	1.05	0.32
		33	1.15	4.93	1.95	2.28		0.80	0.39
PO_4^{3-}	mg/L	44	0.12	1.39	0.54	0.46	**10.2**	0.21	0.46
		38	<0.01	0.50	0.07	0.10		0.12	1.39
TP	mg/L	34	0.07	0.81	0.25	0.26	**4.43**	0.13	0.55
		33	0.04	0.33	0.12	0.14		0.05	0.34
CHLa	mg/L	34	1.0	31.0	12.0	14.2	**−7.76**	9.0	0.64
		33	12.0	360	103	115		78.9	0.68
TSS	mg/L	44	2.1	83.0	16.3	14.0	**4.03**	12.4	0.76
		38	1.9	24.7	9.0	7.1		6.3	0.58

(continued)

Table 2 (continued)

Index	Unit	Number of samples	Minimum	Maximum	Median	Mean	Difference of means ($p < $ **0.05**)	Standard deviation	Variability coefficient
TOC	mg/L	44	1.5	10.8	3.2	4.0	**−3.04**	2.1	0.54
		38	1.6	11.7	5.1	5.6		2.6	0.49
ALK	mg/L	49	2.65	5.21	4.58	4.57	**10.0**	0.28	0.06
		44	2.36	5.13	3.61	3.67		0.80	0.27
Cl^-	mg/L	49	8.7	18.4	15.4	14.8	−1.21	2.11	0.15
		44	11.4	17.5	15.1	15.1		0.97	0.06
SO_4^{2-}	mg/L	49	15.4	32.2	22.6	22.9	0.61	4.38	0.19
		44	19.2	33.1	22.7	23.3		3.75	0.16
Na^+	mg/L	49	4.46	9.82	6.84	6.86	0.17	1.09	0.16
		44	5.41	8.44	6.95	6.96		0.53	0.08
K^+	mg/L	49	2.17	4.97	2.85	3.05	**3.36**	0.62	0.20
		44	2.13	4.02	2.78	2.84		0.39	0.14
Mg^{2+}	mg/L	49	5.23	8.78	7.21	7.04	0.94	0.78	0.11
		44	4.92	9.06	6.91	7.05		1.01	0.14
Ca^{2+}	mg/L	49	51.0	114	92.6	94.3	**17.2**	11.5	0.11
		44	53.9	94.0	71.1	71.8		8.28	0.11

Bolded values mark correlation coefficients statistically significant for $p < 0.05$

analysed indices, the reservoir had a considerable impact on the development of physical and chemical properties.

In hydrochemical terms, the analysed water samples were characterised by the dominance of pairs of ions HCO_3-Ca. Electrolytic conductivity in the waters of the Bystrzyca River above the reservoir was usually at a level of 500 μS/cm and below the reservoir 330–450 μS/cm. Water reaction changed in a range from approximately pH 7 to almost pH 9 and suggested a seasonal rhythm. Higher values were usually recorded in the summer half-year. In the near-shore zone of the reservoir, during abundant development of phytoplankton, water reaction values of even more than pH 9 were recorded (Fig. 2). Significant changes caused by the impact of the reservoir also occurred in the case of dissolved oxygen concentration. In the reservoir, the highest contents of oxygen and water saturation, similarly as in the case of water reaction, occurred in the period of intensified development of phytoplankton.

In the waters of the Zemborzyce Reservoir, chlorophyll "a" showed a high concentration. Particularly in the summer half-year, it frequently exceeded 100 μg/L. In the winter half-year, it usually showed a lower concentration, from a dozen to several tens of μg/L. The level, however, was several times higher than that in the waters of the Bystrzyca River above the reservoir. Periodically, during intensive development of phytoplankton, "water blooms" were observed in the reservoir, and maximum contents of chlorophyll "a" exceeded 1,000 μg/L [28]. High fertility of the waters of the reservoir was the cause of low water transparency. In the summer season, the values usually varied from 0.3 to 0.6 m and in the winter season from 0.4 to 0.7 m. Water temperature was an important factor determining phytoplankton development. The low depth of the reservoir facilitated the process of its warming in the vegetative season. Water temperature in the reservoir exceeding 20°C, favourable for the development of cyanobacteria, was frequently recorded already at the end of April. The maximum temperature of surface water in the summer season exceeded 35°C, and the minimum temperature under the ice cover was below 1°C. Periodical thermal water stratification occurred in the reservoir, usually not higher than 3°C. On the outflow from the reservoir, water temperature in the Bystrzyca River was usually below 20°C. The transparency of water inflowing to the reservoir during groundwater alimentation amounted to 0.5–1 m. During overland flow, when the waters were abundant in suspension, it decreased to less than 0.2 m.

On the inflow to the reservoir, the concentration of suspension in the period of groundwater alimentation of the river amounted to approximately 10 mg/L. In periods of high water stages, it was maintained at a level of approximately 30 mg/L, and the maximum values exceeded 100 mg/L. The reduction of suspension in the reservoir during high water stages was almost fivefold. In the winter half-year, on the outflow from the reservoir, the concentration of suspension reached up to several mg/L, and in the summer half-year, it was at a level of approximately 10 mg/L and was often higher than on the inflow. The reservoir also affected the character of suspension. On the inflow, its mineral form was dominant and on

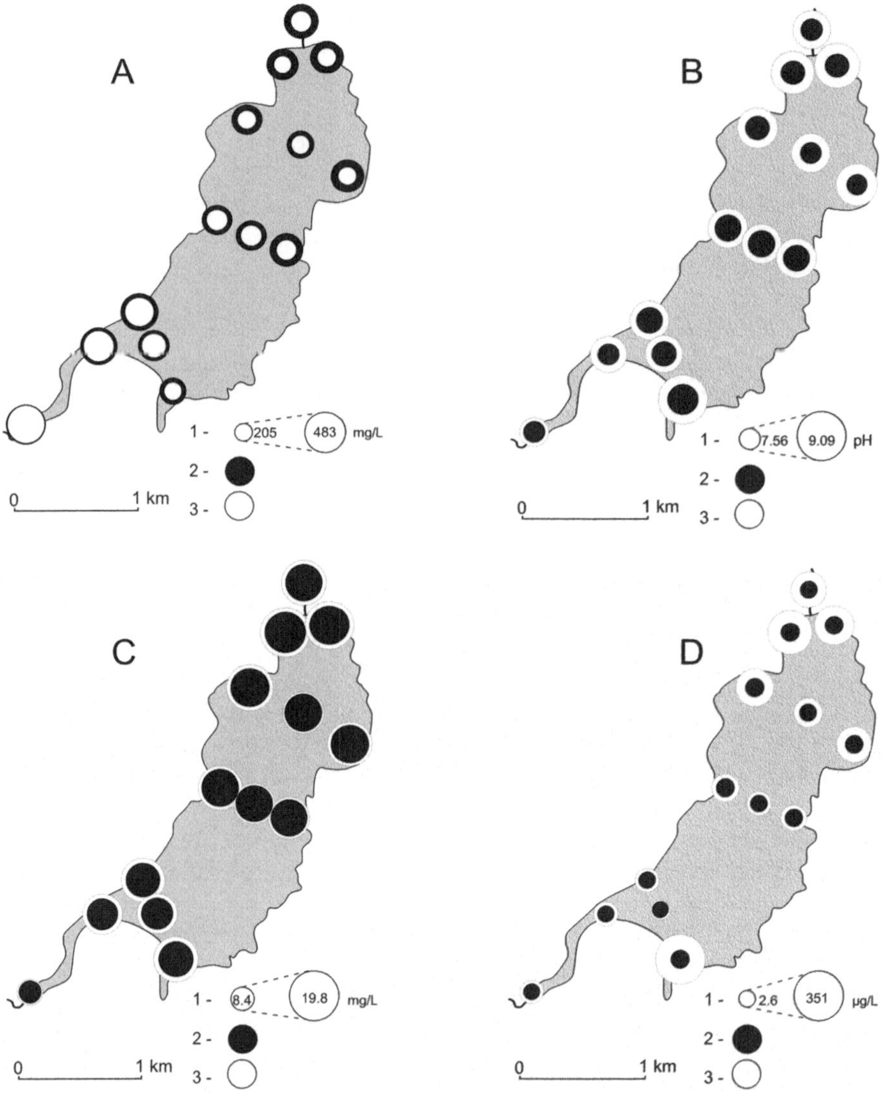

Fig. 2 Content of selected physical-chemical properties in Zemborzyce Reservoir waters in summer and winter seasons in 2012. (**a**) Total dissolved solids, (**b**) water reaction, (**c**) dissolved oxygen, (**d**) chlorophyll "a". 1, concentration of the indices corresponds to circle diameter; 2, winter 2012; 3, summer 2012

the outflow, organic form. Over the period of functioning of the reservoir (40 years), suspension sedimentation led to a decrease in its volume by approximately 20% [29].

5 Effect of the Reservoir on the Content of Biogenes and Trophic Status of Water

Waters of the Bystrzyca River above the reservoir were characterised by considerable abundance of biogenes, particularly their mineral forms (DIN and DIP). Mineral nitrogen in the profile of the Bystrzyca River above the reservoir constituted approximately ¾ of total nitrogen. Below the reservoir, the concentration of mineral nitrogen showed considerably lower values, approximately 20% of total nitrogen. Similar correlations occurred in the case of phosphorus. The dominant role in the outflow was therefore played by organic forms of N and P, suggesting intensive development of biological life in the reservoir and, as a consequence, low water quality.

The trophic indices of waters inflowing to the reservoir, TSI_{TP} [16] and TSI_{TN} [17], were characteristic of waters varying from eutrophy to hypertrophy (Table 3). In the case of TSI_{CHLa} [16], the values were characteristic of mesotrophic to eutrophic waters. Below the reservoir, the values of TSI_{TP}, TSI_{TN}, and TSI_{CHLa} usually exceeded 70. They suggested the hypertrophic status of the waters, i.e. their excessive fertility. Differences in the assessment of the trophic status of waters above the reservoir between indices TSI_{TP}, TSI_{TN}, and a TSI_{CHLa} suggest a limited possibility of application of TN and TP values in the assessment of the trophic status of river waters. The indices, of synthetic character, included mineral as well as organic forms of nitrogen and phosphorus. In the case of DIN and DIP, they were forms easily absorbed by plants in the processes of photosynthesis. Therefore, they did not determine the trophic status of waters, but the possibility of production of organic compounds from inorganic matter. They pointed to the "trophic potential" and not "trophic status" of the waters. In the studied object, after including DIP and DIN supplied to the reservoir in the biological chain, organic forms of N and P with similar weights were obtained. The observed indicator of such processes was greenish water colour and a decrease in water transparency. Analytically, the process was identified as an increase in chlorophyll "a" concentration. Therefore, dissolved mineral forms of nitrogen and phosphorus (DIN, DIP) can be used for predicting the trophic status of river waters resulting from the construction of a dam reservoir, e.g. in documents such as environmental impact assessments.

Research on the concentration of biogenes in different phases of water circulation (Table 4) was used for the calculation of their load introduced to the Zemborzyce Reservoir through precipitation waters, overland flow, groundwater alimentation,

Table 3 Trophic State Index (TSI) [16, 17] of water above/below the Zemborzyce Reservoir in the years 2008–2012

Value	Minimum	Maximum	Median	Mean
TSI_{TP}	65.4	100.7	83.8	84.3
	57.3	87.8	73.2	75.4
TSI_{Chla}	30.6	64.3	54.9	56.6
	54.9	88.3	76.1	77.1
TSI_{TN}	62.9	84.2	70.9	71.4
	56.5	77.5	64.1	66.3

Table 4 Characteristic values of nitrogen, phosphorus, and carbon concentration in atmospheric precipitation, overland flow, and groundwaters (springs) of the Bystrzyca River catchment (minimum-maximum value/median from years 2008 to 2012)

Index	Precipitation (384 samples)	Overland flow (76 samples)	Groundwaters (9 springs – 69 samples)
TN mg/L	0.2–12.5/1.5	0.3–31.4/6.3	0.4–4.5/2.1
DIN mg/L	0.17–9.93/1.08	0.15–14.1/3.38	0.4–4.2/1.88
TP mg/L	<0.01–0.11/0.03	0.10–8.42/0.76	0.05–0.30/0.15
DIP mg/L	<0.01–0.05/0.02	0.05–4.81/0.42	0.05–0.28/0.13
TC mg/L	<1–10/3	8–60/20	51–78/66
DIC mg/L	<1–6/1	6–50/11	51–77/65
TOC mg/L	<1.0–7.5/1.6	1.4–15.2/8.6	<1.0–2.2/<1.0

Table 5 Comparison of the actual and acceptable and dangerous load of phosphorus and nitrogen inflowing into the Zemborzyce Reservoir in the years 2008–2012 according to Vollenweider's static and hydraulic model [18, 19]

		Acceptable		Dangerous	
		Model			
Load	Actual	Static	Hydraulic	Static	Hydraulic
TP, gP/m^2/year including:	8.51	0.07	0.37	0.13	0.75
Groundwater alimentation	4.57				
Overland flow	2.14				
Sewage	1.78				
Direct precipitation	0.02				
TN, gN/m^2/year including:	106	1.0	–	2.0	–
Groundwater alimentation	64.1				
Overland flow	17.9				
Sewage and other	23.1				
Direct precipitation	0.9				

and sewage. The calculated actual (external) load of the reservoir with nitrogen and phosphorus (Table 5) exceeded the acceptable and dangerous loads of the elements for water bodies functioning according to the so-called Vollenweider's static and hydraulic model [18, 19]. Therefore, biogenes supplied to the Zemborzyce Reservoir, and particularly mineral forms of N and P, contributed to the intensity of the process of eutrophication of waters and low ecological status of the water body.

Calculation results showed that the level of biogenes (N and P) in the waters of the Bystrzyca River was particularly determined by groundwater alimentation of the river (Tables 5 and 6). More than half of phosphorus and approximately 2/3 of nitrogen reaching the reservoir originates from this phase of water circulation. Their load introduced through overland flow and sewage was also of considerable importance for the supply of N and P (approximately 20% each). Therefore, non-point pollution sources (NPS) of agricultural character had the basic impact on the eutrophication of the waters of the reservoir. At the current low level of

Table 6 Balance of nitrogen, phosphorus, and carbon in the Zemborzyce Reservoir in the years 2008–2012

	TN	DIN	TP	DIP	TC	DIC	TOC
Index	t/year						
Inflow including:	296.8	229.2	23.8	16.2	5,983	5,678	304
Groundwater alimentation	179.5	160.7	12.8	11.1	5,641	5,555	43
Overland flow	49.8	26.7	6.0	3.3	158	90	68
Direct precipitation	2.6	1.9	0.05	0.04	5	2	3
Sewage and other	64.8	39.9	4.9	1.7	179	31	191
Outflow	182.6	32.8	11.2	1.9	4,393	3,915	478
Retention	114.1	–	12.5	–	–	1,763	–

fertilisation (approximately 80 kg/ha of nitrogen and 30 kg/ha of phosphorus), reduction of the supply of biogenes below values which confine the water eutrophication process in the reservoir is limited, even in the case of complete elimination of supply of biogenes discharged to the river with sewage.

6 Deposition of Biogenes in the Reservoir

The research showed that the reservoir was an important biogeochemical barrier in the case of migration of nitrogen and phosphorus (Table 6). In the study period, the average annual nitrogen retention in the reservoir amounted to approximately 114 tonnes, and that of phosphorus approximately 12 tonnes, and was higher than in the years 2005–2007 by approximately 15% and more than 80%, respectively [6]. In the currently deposited sediments of the reservoir, the content of nitrogen was at a level from several to a dozen g/kg N and phosphorus approximately 1 g/kg P [30, 31]. Further shallowing of the reservoir will presumably result in intensified resuspension of the deposited N and P and therefore an additional load of biogenes contributing to water eutrophication. Such unfavourable processes will particularly occur during intensive water mixing with sediments, e.g. during higher than average water stages or strong winds.

The Zemborzyce Reservoir was also an important biogeochemical barrier in the case of migration of dissolved inorganic carbon (DIC). Measurements performed in the years 2008–2012 showed that in the summer period, a reduction of carbonates in the reservoir reached 100 mg/L $CaCO_3$, and in the winter season, it was twice lower. On average, almost 14.7 thousand tonnes of $CaCO_3$ was deposited in the reservoir per year (1,760 t/year of DIC). Intensive deposition of carbonates in the reservoir is evidenced by results of research on the bottom sediments, where the content of $CaCO_3$ locally exceeded 50% of their dry mass [30]. Considerable oversaturation of waters with carbonates, to the point of their precipitation, occurred as a result of the biological breathing of the reservoir – use of carbon dioxide in the processes of photosynthesis. In the fluvial part of the reservoir (so-called sleeve), carbon dioxide was primarily used by macrophytes. In the basin of the reservoir, where limnic processes prevailed, deposition was primarily caused by phytoplankton. The potential possibility of deposition of carbonates was suggested by the index of water saturation with calcite SI_{cal} which on the inflow to the reservoir was at a level of approximately SI_{cal} +1 and in the reservoir showed values from SI_{cal} +1 to SI_{cal} +2. Deposition of carbonates resulted in a decrease in electrolytic conductivity of water in the reservoir by approximately 30%.

7 Concluding Remarks

The ecological status of the Zemborzyce dam reservoir was determined by the inflow of biogenes with the waters of the Bystrzyca River. The main load of biogenes supplied to the reservoir constituted forms of DIP and DIN originating from the groundwater alimentation of the Bystrzyca River. Considering the present agricultural land use in the catchment, and morphological parameters of the reservoir, reduction of their level to values which would not negatively affect water quality is limited.

For the purpose of reduction of the load of biogenes supplied to the reservoir, a change in the way of its functioning was proposed, from the dam to lateral type, as well as deepening (extraction of peats and currently deposited sediments). The altered reservoir should be constantly or periodically fed with waters from the small preliminary reservoir with reduced discharge. In the preliminary reservoir, hydrobiological and chemical manipulations should ensure a reduction of nutrients to a level limiting the processes of water eutrophication in the main reservoir. Areas suitable for the construction of the preliminary reservoir are located below the Cienista Street (Fig. 3). A new river channel should be formed on the side of the Zemborzyce village. Existing pump station for flood waters can be used for main reservoir supplying. Such a solution will eliminate direct supply of polluted waters from the built-up area and overland flow from cultivated fields to the reservoir.

The proposed changes in the functioning of the reservoir will limit the supply of pollutants from point and non-point sources directly reaching the reservoir and therefore will contribute to an improvement of water quality. They will reduce the inflow of waters from groundwater alimentation with high trophic potential. The negative effect of hydrotechnical infrastructure on the discontinuation of the river, and particularly its self-cleaning processes, will be eliminated. Such measures will be incorporated into the strategy of sustainable management of surface waters in the Bystrzyca River catchment.

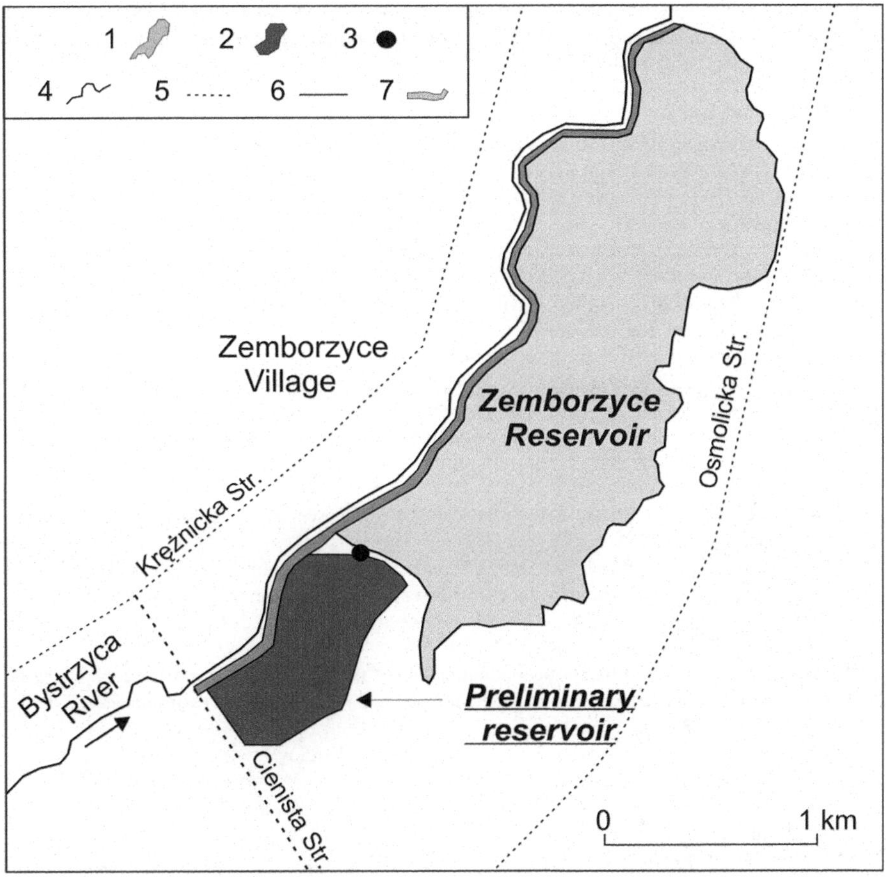

Fig. 3 The scheme of the functioning of the Zemborzyce Reservoir. 1, reservoir; 2, preliminary reservoir; 3, outflow control point (pump station); 4, river; 5, streets; 6, canal; 7, embankment

Acknowledgements This work was supported by the Polish Ministry of Science and Higher Education No. N N306181237.

References

1. Vannote RL et al (1980) The river continuum concept. Can J Fish Aquat Sci 37(1):130–137
2. Kajak Z (2001) Hydrobiology-limnology. Ecosystems of inland waters. PWN, Warszawa, p 360
3. Woyciechowska J, Dojlido J (1982) Changes in the quality of surface waters under the influence of hydrotechnical infrastructure. Gospodarka Wodna 5:47–51
4. Siuta W, Chróst JR (2015) Hydrobiotechnology – biological basis, the current state of knowledge and development prospects. Technologia Wody 5(45):31–41
5. Cooke GD et al (2005) Restoration and management of lakes and reservoirs. Taylor and Francis, Boca Raton

6. Chmiel S et al (2009) Some issues in the assessment of eutrophication of river waters as a consequence of the construction of a storage reservoir (on the example of the Bystrzyca River). Ecohydrol Hydrobiol 9(2–4):175–179
7. Zalewski M (2012) Ecohydrology – process oriented thinking for sustainability of river basins. Ecohydrol Hydrobiol 12(2):89–92
8. Pütz K, Benndorf J (1998) The importance of pre-reservoirs for the control of eutrophication of reservoirs. Water Sci Technol 37(2):317–324
9. Mazur A (2013) Performance evaluation of the pre-dam reservoir on the Por River. Infrastruct Ecol Rural Areas 1(4):299–310
10. Michalczyk Z (2012) Assessment of the conditions of water occurrence and development of surface runoff in Lublin. UMCS, Lublin, p 268
11. Skowron P, Igras J (2012) Anthropogenic sources of nitrogen in the Bystrzyca River catchment. Przemysł Chemiczny 91(5):970–977
12. Skowron P, Igras J (2013) Anthropogenic sources of phosphorus in the catchment of the Bystrzyca river. Preliminary analysis of the share of agriculture in the water pollution. Przemysł Chemiczny 92(5):787–790
13. Radwan S (2004) Zemborzyce reservoir. Ecological structure anthropogenic threats and protection. Polska Akademia Nauk – Komitet Inżynierii Środowiska, Akademia Rolnicza, Lublin
14. How to improve the ecological state of the Zemborzyce Reservoir in Lublin (2007) Conference proceedings, Lublin, 17 Apr 2007
15. Zemborzyce Reservoir in Lublin (2015) Chances and threats, Lublin City Office, Lublin
16. Carlson RE (1977) A trophic state index for lakes. Limnol Oceanogr 22:361–369
17. Kratzer CR, Brezonik PL (1981) A Carlson-type trophic state index for nitrogen in Florida lakes. Water Resour Bull 17:713–715
18. Vollenweider RA (1968) Scientific Fundamentals of the Eutrophication of lakes and flowing waters with particular reference to nitrogen and phosphorus as factors in eutrophication. OECD, Paris, DAS/SCI/68.27
19. Vollenweider RA (1976) Advances in defining critical loading levels for phosphorus in lake eutrophication. Mem Ist Ital Idrobiol 33:53–83
20. PN-EN ISO 10304-1 (2009) Water quality – determination of dissolved anions by means of ionic chromatography – part I
21. PN-EN ISO 14911 (2002) Water quality – determination of dissolved cations by means of ionic chromatography – method for waters and sewage
22. PN-ISO 6058 (1999) Water quality – determination of calcium content – titration methods with EDTA
23. PN-EN ISO 9963-1 (2001) Water quality – determination of total alkalinity
24. PN-EN 872 (2007) Water quality – determination of suspensions
25. Method 8048, Hach, Phosphorus reactive. Ascorbic acid
26. Method 10071, Hach, Nitrogen total. Alkaline persulfate digestion
27. Method 8190, Hach, Phosphorus total. Acid persulfate digestion
28. Pawlik-Skowrońska B, Toporowska M, Kalinowska R (2015) Dynamics research of blue-green algae and cyanotoxins production in Zemborzyce reservoir against the background of environmental conditions. In: Zemborzyce reservoir in Lublin. Chances and threats. Lublin City Office, Lublin, pp 63–67
29. Dobrowolski R et al (2015) Geological reconnaissance of bottom deposits and bedrock structure of Zemborzyce reservoir. In: Zemborzyce reservoir in Lublin. Chances and threats. Lublin City Office, Lublin, pp 49–62
30. Rodzik J, Dobrowolski R, Melke J (2009) Estimation of kind, amount and mechanism of sedimentation in the Zemborzyce reservoir near Lublin. Teka Kom Ochr Kszt Środ Przyr 6:261–276
31. Smal H et al (2013) Nitrogen and phosphorus in bottom sediments of two small dam reservoirs. Pol J Environ Stud 22(5):1479–1489

Heated Konin Lakes: Structure, Functioning, and Succession

Bogusław Zdanowski, Agnieszka Napiórkowska-Krzebietke, Konrad Stawecki, Aleksander Świątecki, Roman Babko, Elżbieta Bogacka-Kapusta, Bartosz Czarnecki, and Andrzej Kapusta

Contents

Abstract The Konin lakes are a system that has been artificially heated by power stations since 1958 and has continuously been the subject of ecological research. The coolest of the lakes remains Lake Ślesińskie, while the warmest is Lake Licheńskie. Both of the lakes are eutrophic, and their ecological potential is decidedly below good. Low water retention in the lakes facilitates the increase of easily absorbed organic matter flowing in from the catchment area, and the higher temperatures of the waters intensify processes of their decomposition. Summer phytoplankton blooms are noted in both lakes; they are more intense in Lake Licheńskie. A significant increase in primary production, higher cyanobacteria biomass, and the limited development of large cladocerans and copepods and the domination of small

B. Zdanowski · A. Napiórkowska-Krzebietke (✉) · K. Stawecki · E. Bogacka-Kapusta · B. Czarnecki · A. Kapusta
Department of Ichthyology, Hydrobiology and Aquatic Ecology, Inland Fisheries Institute in Olsztyn, Olsztyn, Poland
e-mail: a.napiorkowska-krzebietke@infish.com.pl

A. Świątecki
Faculty of Biology and Biotechnology, University of Warmia and Mazury in Olsztyn, Olsztyn, Poland

R. Babko
Institute of Zoology, National Academy of Sciences, Kyiv, Ukraine

E. Korzeniewska, M. Harnisz (eds.), *Polish River Basins and Lakes – Part I*, The Handbook of Environmental Chemistry 86,
https://doi.org/10.1007/978-3-030-12123-5_16

rotifers are confirmed in this lake. The lake system is inhabited by approximately 100 alien species. These frequently include invasive plant and animal species such as *Vallisneria spiralis*, *Dreissena polymorpha*, *Sinanodonta woodiana*, and *Pseudorasbora parva*, which contribute to the structure and functioning of the lakes. The best water quality is in the post-mining water basin, which was created in 2010 by inundating it with Early Cretaceous and Tertiary waters, the physicochemical and biological aspect of which indicate good or maximum ecological potential.

Keywords Alien species · Phytoplankton blooms · *Sinanodonta woodiana* · *Vallisneria spiralis*

1 Introduction

The Konin lakes (Gosławskie–Ślesińskie) are a compact system of five lakes connected by a system of canals located in Central Poland some 40 km from Konin and heated artificially by power stations; the system was created in an early glacial group of lakes in the Kujawski Lake District in the Warta River catchment (Fig. 1). The two small Biskupia and Kleczewska streams are the only natural inflows into the lakes.

The lakes are characterized by diverse morphometry (Table 1). Lakes Gosławskie and Pątnowskie are shallow and polymictic, and Lake Licheńskie is slightly deeper and stratified only in summer, while lakes Wąsosko-Mikorzyńskie and Ślesińskie are deep and remain stratified in both the summer and winter seasons.

The lakes are in basins located on the outwash plain above the level of the Warta River. Lakes Pątnowskie, Wąsosko-Mikorzyńskie, and Ślesińskie are the top segment of the Morzysławski Canal (Warta-Gopło) that links the Warta River with Lake Gopło (Fig. 1). In the lower segment of the canal, water from the Warta is usually pumped into the lakes to maintain the appropriate level and to facilitate power station water uptake. Excess water in the lakes can be returned to the Warta or directed to Lake Gopło through Lake Ślesińskie at the upper segment of the canal and then further through the Noteć River to the Warta or the Brda and Vistula. These hydrological conditions mean that the Morzysławski Canal can be recognized as an important transfer pathway for alien species migration.

The configuration of the lake catchment (surface area 418 km^2) was shaped during the last Baltic glaciation. It created moraine and kames, sandy outwash plains, and troughs occupied by the lakes that are located on Mesozoic deposits of the Upper Cretaceous and Tertiary sediments of the Miocene and Pliocene and cover lignite deposits. Coal exploitation began in 1942, and open-pit mining began in 1956. Consequently, the landscape of this region was altered over time by the mining and energy industries. New forms of exploiting the land were created including the system of canals for water uptake and post-cooling effluent discharge of a combined total length of 26 km, various types of fish ponds, excavated areas

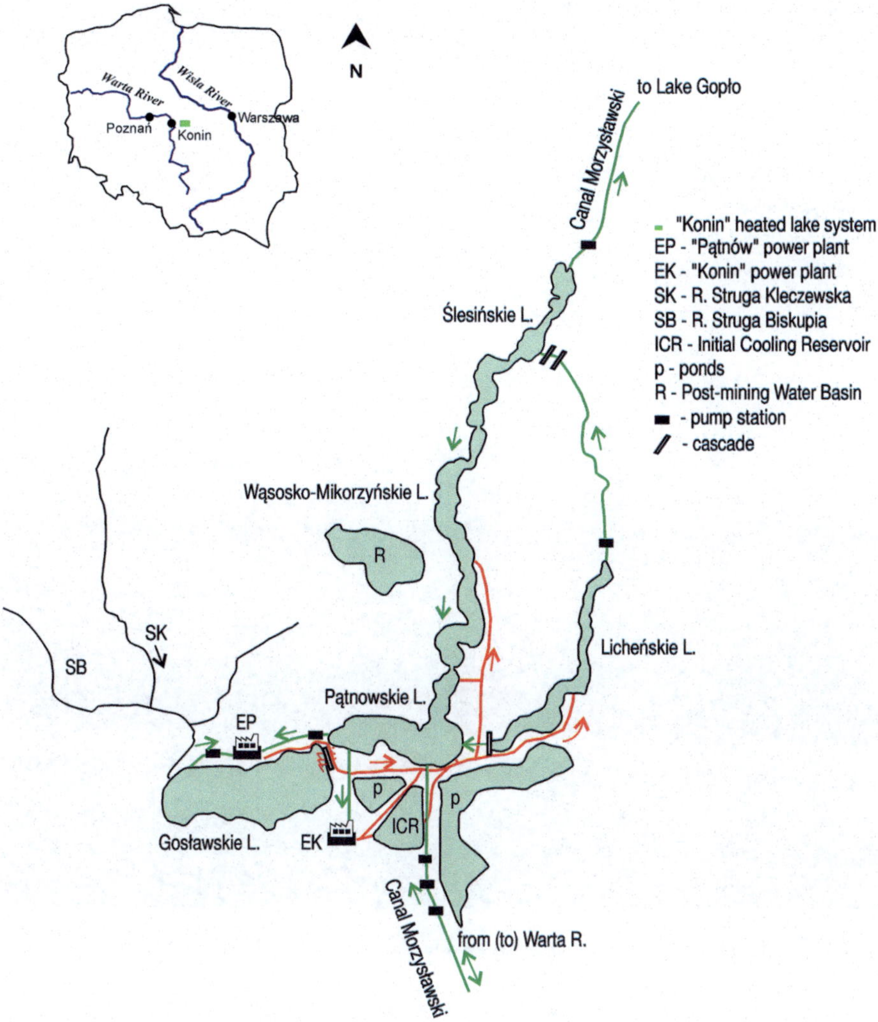

Fig. 1 Scheme of the Konin lake system

of exploited coal deposits, reclamation of land surpluses, isolated landfills, Konin power station retention reservoirs, and pit lakes. Lakes Licheńskie and Gosławskie are pile up in relation to the others.

Lakes Pątnowskie, Licheńskie, and Wąsosko-Mikorzyńskie were included in the cooling system of post-cooling effluents of the Konin power station with an output of 600 MW in 1958, while lakes Gosławskie and Ślesińskie were included in 1970 after the Pątnów power station with an output of 1,200 MW (Fig. 2) became operational. The initial cooling reservoir is an artificial reservoir that was built in 1976 to receive warm post-cooling effluents from the Konin power station. The post-mining water

Table 1 Konin heated lake system[a]

	Area, ha	Depth, m
Gosławskie lake	454.4	5.3
Licheńskie lake	147.6	12.6
Pątnowskie lake	282.6	5.5
Wąsosko-Mikorzyńskie lake	251.8	36.5
Ślesińskie lake	152.3	24.5
Initial cooling reservoir	75.0	4.2
Post-mining water basin	364.0	48.0
Ponds (p)	272.0	2.0

[a]See Fig. 1

Fig. 2 Pątnów power station by Lake Gosławskie

basin is an artificial reservoir with no outlet that was built in 2010 in a depleted pit mine, and it was inundated with Early Cretaceous and Tertiary waters (Fig. 3).

Lake Pątnowskie has always been the reservoir for the Konin power station, which can take up water in nominal quantities of 30 $m^3\ s^{-1}$. Post-cooling effluents from the power station, which are heated to a maximum temperature of 35°C in summer, are discharged into the canal that connects to the Pątnów power station canal and then further to lakes Licheńskie and Pątnowskie and Wąsosko-Mikorzyńskie (Fig. 1). During spring, these waters are directed to fish ponds. Because of the critical degree to which the water is heated in the system, during the summer the so-called extended water cooling circuit for post-cooling effluents is used that includes Lake Ślesińskie, which receives waters from Lake Licheńskie.

Lakes Gosławskie and Pątnowskie are water reservoirs for the Pątnów power station and have always been its basic water cooling circuit (Fig. 1). The power station can take up water in nominal quantities of up to 50 $m^3\ s^{-1}$ from the uptake

Fig. 3 Post-mining water basin (in the foreground: remains of the ramp to transport coal from the open-pit mine)

located on the western shore of this lake and then discharge warm effluents (up to 35°C) on the eastern shore of Lake Gosławskie, while 50% of these waters are redirected again to lakes Pątnowskie, Wąsosko-Mikorzyńskie, and Licheńskie.

Post-cooling effluents from the power stations are discharged into the lake system through a series of overflows, gravitational canals, and cascade canals. Close to 50% of post-cooling effluents from both of the power stations were received by the most intensely heated Lake Licheńskie until 2000. The retention time of the post-cooling effluents in the lake in the summer were reduced to that range of 3 (Lake Gosławskie) to 7 (Lake Wąsosko-Mikorzyńskie) days.

After 2010 and following the modernization and reconstruction of the Pątnów power station, which now has an output of 1,700 MW, the nominal load of post-cooling effluents discharged into Lake Gosławskie and the other lakes has been maintained. After the reconstruction of the Konin power station to a thermal power plant with an output of 180 MW, the quantity of post-cooling effluents discharged into the lakes has been reduced significantly to 7 $m^3 s^{-1}$. Consequently, the water retention time in the lakes during the summer season has increased in lakes Licheńskie and Pątnowskie to 9 days and in lakes Ślesińskie and Wąsosko-Mikorzyńskie to 36 days. The initial cooling basin, like Lake Ślesińskie, is only included in the post-cooling effluent cooling circuit in periods when temperatures are high. Lake Licheńskie, which formerly did not freeze in winter, now only lacks ice cover in the zone where post-cooling effluents are discharged.

Patalas [1] conducted SWOT analysis of the consequences of including the Konin lakes in the Konin power station water cooling circuit based on results of studies performed in the 1960s. This indicated that heating the lake waters had a negative impact and limited the occurrence of cold-water fishes (*Coregonus albula* L.) and those that prefer colder environments (*Esox lucius* L., *Perca fluviatilis* L., *Sander lucioperca* L.). Increasing the water temperature in the lakes (>28°C) also negatively impacted indigenous populations of cyprinid species [2, 3]. The vegetative season in the heated lakes commenced one and a half months earlier than in unheated lakes in this region of Poland, and the peak of the season was in late June and early July. Balanced high energy transfer at specific trophic levels was positively assessed, and it limited water eutrophication. Phytoplankton blooms in Lake Licheńskie dominated by cyanobacteria were noted in the late 1960s. These were the consequence of grass carp (*Ctenopharyngodon idella* Val.), which had been introduced into the lakes, destroying the hydrophytes that had formed highly productive submerged meadows in the littoral and sublittoral zones of the lakes. With time, these lake zones came to be inhabited by the green algae *Cladophora glomerata* (L.) Kütz., which supported developing concentrations of *Dreissena polymorpha* (Pall.), followed by *Najas marina* L. and in recent years the thermophilic alien aquarium species *Vallisneria spiralis* L.

The post-mining waters rich in minerals that were discharged into the lakes and emissions of pollutants from the power plants and the aluminum smelter occurred to be key in shaping the physicochemical and trophic relations of the waters in the lakes in the late 1970s and early 1980s [4, 5]. Atmospheric emissions of sulfur (0.130 mg m^{-3}) and deposition of alkaline dusts (200 t km^{-2}) from brown coal incineration caused increased water salinity and progressive siltation in the lakes. The increased alkalinity of the water provided conditions for binding the phosphorous load (up to 2.9 g m^{-2}), which the lakes received from the catchment area and was already significant in that period, into sedimenting forms of calcium carbonate that dangerously exceeded the annual lake load of phosphorus (up to 0.2 g m^{-2}).

The short water retention time and the forced intensive water exchange, which equalized the natural trophic differences among the lakes, limited the disadvantageous occurrence of troublesome cyanobacteria blooms. Lake bottoms covered with sedimented minerals proved to be an excellent substrate for the then vegetatively developing communities of *Vallisneria spiralis*.

Substances in surface runoffs from the catchment area and municipal pollution from recreation and aquaculture played decisive roles in altering the environmental and trophic statuses of the lakes that were noted in the decade from the late twentieth to early twenty-first centuries [6, 7]. The persistent, significant phosphorous loading of Lake Gosławskie from post-mining waters discharged by Biskupia stream (up to 3.0 g m^{-2} annually) and the pollutants flowing into lakes Licheńskie (1.3 g P m^{-2} annually) and Wąsosko-Mikorzyńskie (0.5 g m^{-2} annually) from fish culture in discharge canals were still considered to be dangerous.

Protasov and Zdanowski's [8] SWOT analysis indicated that the key factors contributing to the degradation of the ecological status of the lakes included their settlement by alien, invasive species of plants and animals and progressing pollution with organic compounds. Increasing concentrations of organic material in waters taken up by the power plants could provide excellent substrate for the development of microorganisms on the heat exchangers of power station energy blocks [9–12]. The formation of biofilms, 0.1 mm thick, can reduce heat exchange rates by as much as fourfold, and this is particularly dangerous for nuclear power plants located on rivers and large dam reservoirs. Until recently, the growth of biofilms on boiler limescale did not force the Konin power stations to renovate energy blocks more frequently than once annually.

Since 1965 the Konin lakes have been under continuous study. Based on its results, the directional, long-term changes of environmental and trophic in the lakes that result from discharging warm post-cooling effluents that affect the thermal regimes and hydrology of the waters have been defined. These results also permit identifying the consequences of the expansive settlement of the lakes by alien, invasive, species of plants and animals. The newest results, unpublished, from the last decade of research with only retrospective references to the rich publications of the previous period were presented. Further, key theses are presented that could be helpful in assessing changes shaped by the Earth's changing climate as a consequence of the greenhouse effect.

2 Environment

The average quantity of post-cooling effluents discharged into the lake system by the Pątnów power station remained unchanged in the 2008–2015 period at approximately 40.0 $m^3 s^{-1}$. The volume of heated water discharged into the system by the Konin power station decreased significantly to just 7.0 $m^3 s^{-1}$. Lake Ślesińskie was used less intensively as part of the extended water cooling circuit. The average amount of post-cooling effluents pumped from Lake Licheńskie to Lake Ślesińskie decreased to 3.4 $m^3 s^{-1}$ (Table 2).

Table 2 Changes in quantities of discharged post-cooling effluents – mean (range) in the cooling system of the Pątnów and Konin power stations and water retention times and mean (range) in lakes Licheńskie and Ślesińskie in 1995–2015

	Discharge of cooling waters, $m^3 s^{-1}$		Retention time, d	
Period	Pątnów power station	Konin power station	Lake Licheńskie	Lake Ślesińskie
1995–1999[a]	40.0 (34.1–46.0)	20.4 (19.2–21.9)	4 (3–5)	18 (3–26)
2000–2007	42.3 (36.3–45.7)	15.6 (13.1–17.1)	4 (4–5)	22 (5–32)
2008–2015	38.8 (32.0–42.5)	7.0 (3.5–8.3)	5 (1–9)	26 (7–36)

[a]According to Stawecki et al. [13]

The average level of water heating in the Konin power station (4.2°C) and in the Pątnów power station (6.3°C) was lower than in previous years by 2.9 and 1.4°C, respectively. The water retention time in Lake Licheńskie lengthened by 1 day, on average, while in the cooler seasons of the year, it was 4 days longer and in Lake Ślesińskie it was from 8 to 10 days longer (Table 2).

The hydrology of Lake Gosławskie, as the main reservoir for post-cooling waters from the Pątnów power station, remained stable. The coolest lake in the system (approximately 14°C on average) is still Lake Ślesińskie. The oxygen concentration and saturation in the entire mass of water in this lake decreased (Table 3). The most heated of the lakes, with a mean temperature of approximately 15°C at the surface and 11°C at the bottom, is Lake Licheńskie, which is heated year round. This lake still receives the largest load of post-cooling effluents at about 60% of the total water discharge from both power stations. The temperature of the surface layer water in this lake decreased only slightly. The oxygen content and saturation of the waster did not change significantly either. The occurrence of extremely high temperatures exceeding 30°C was limited to episodic instances that were noted only during heat waves. The near-bottom water layer was colder, on average, by more than 2°C. Oxygen concentration decreased by an average of 4.4 mg dm^{-3}, while saturation decreased even more to 37%. During summer stratification there were substantial deficits in oxygen content in the water layers beneath the thermocline in both lakes. In the post-mining water basin, which had better oxygen conditions, fluctuations in oxygen content were not large.

Once summer thermal stratification was established, the distribution of oxygen content in the vertical profile was described by a heterograde oxygen curve with the oxygen maximum in the metalimnion. The mean oxygen concentration and saturation in the near-bottom water layers did not decrease below <6.0 mg O_2 dm^{-3} or < 50%, respectively (Fig. 4).

The waters of lakes Licheńskie and Ślesińskie were characterized by calcium carbonate salinity. The content of carbonates and bicarbonates in the total concentration of ions dissolved in the water was approximately 86%, sulfates 9%, and chlorides 5%. The linearly increasing electrolytic conductivity of the water (up to 650 μS cm^{-1}) indicates the continued accumulation of clay minerals in lakes that are supplied to them in the runoff from the catchment area [14–17].

The pH of the surface water layer in the lakes was alkaline (pH >8.3) throughout the season. Higher pH values (up to pH 8.7) were noted during spring-summer phytoplankton blooms. The water of the post-mining basin was characterized by less mineralization (calcium carbonate) and lower electrolytic conductivity (approximately 550 μS cm^{-1}) and, as a rule, lower pH (7.7–8.0).

In terms of nutrient resources, lakes Licheńskie and Ślesińskie can be classified as highly eutrophic. The mean content of phosphorous in the surface water layer increased in the 2004–2015 period to approximately 0.140 mg dm^{-3}. This is determined by the substantial phosphorus load from aquaculture pollutants that is estimated to be approximately 23 g m^{-2} annually in Lake Licheńskie. The low nitrogen to phosphorus ratio (N:P ~7–8) indicates that phytoplankton development is already dependent on the availability of nitrogen.

Table 3 Changes (mean ± SD) of temperature and oxygen content and saturation in the surface and near-bottom layers in lakes Licheńskie and Ślesińskie and the post-mining water basin in 1995–2015

		Lake Licheńskie		Lake Ślesińskie		Post-mining water basin[a]	
Parameter	Period	Surface	Bottom	Surface	Bottom	Surface	Bottom
T (°C)	1995–1999	15.6 ± 8.0	12.6 ± 5.6	13.6 ± 7.9	4.9 ± 0.9	nd	nd
	2000–2007	16.6 ± 7.7	11.7 ± 3.4	13.9 ± 7.7	4.6 ± 0.6	nd	nd
	2008–2015	15.6 ± 8.0	10.3 ± 3.0	13.4 ± 7.7	4.9 ± 0.9	17.4 ± 5.4	5.6 ± 0.4
O_2 (mg dm^{-3})	1995–1999	8.8 ± 2.1	5.9 ± 4.6	10.0 ± 2.5	3.3 ± 2.7	nd	nd
	2000–2007	9.1 ± 2.1	11.7 ± 3.4	10.7 ± 2.8	4.6 ± 0.6	nd	nd
	2008–2015	9.1 ± 2.3	4.4 ± 2.3	9.4 ± 2.1	2.3 ± 1.6	9.1 ± 1.8	7.2 ± 1.5
Saturation (%O_2)	1995–1999	88.9 ± 12.4	51.7 ± 27.2	96.1 ± 18.9	25.5 ± 16.5	nd	nd
	2000–2007	91.8 ± 18.5	40.9 ± 38.9	102.1 ± 24.2	17.5 ± 14.7	nd	nd
	2008–2015	89.8 ± 24.5	36.8 ± 35.5	89.6 ± 23.8	17.8 ± 15.3	90.5 ± 8.0	57.4 ± 12.0

nd no data

[a]Data from April to October 2014–2015, data from 1995 to 1999 according to Stawecki et al. [13]

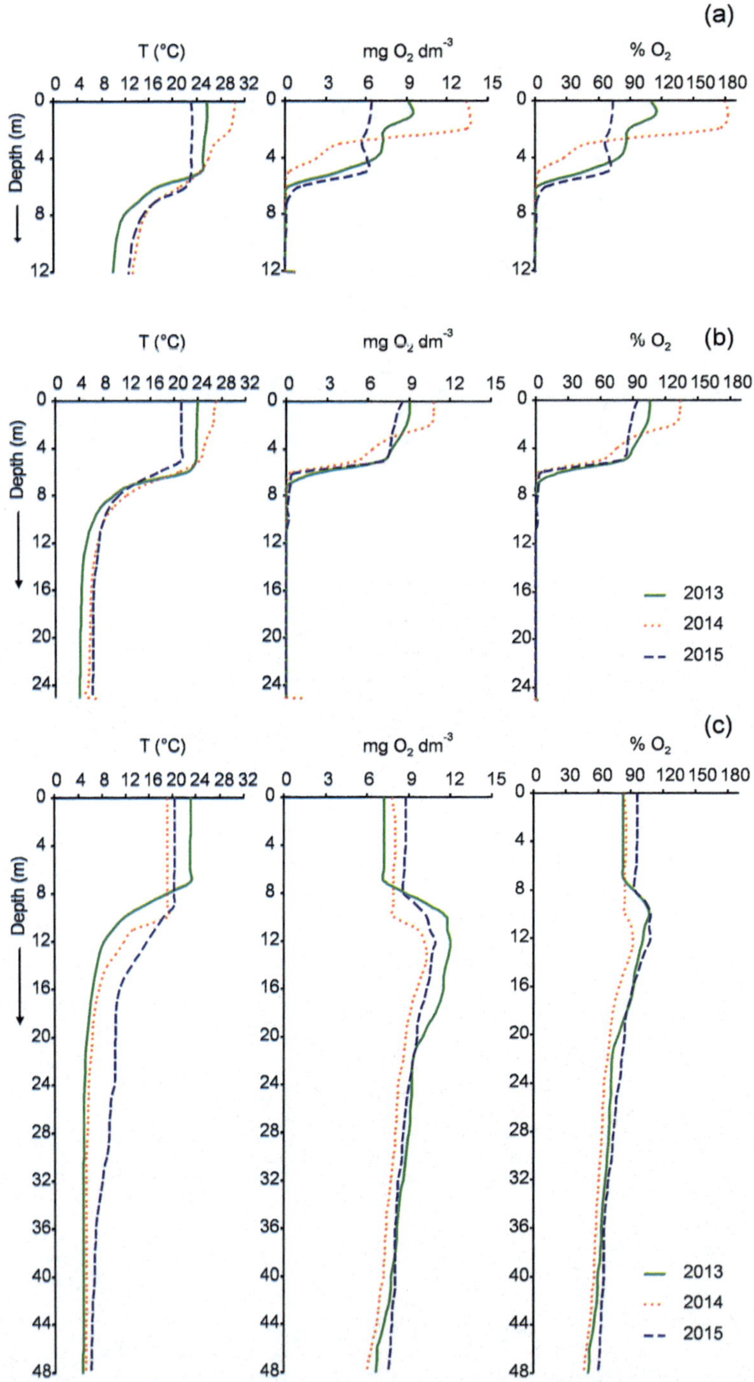

Fig. 4 Summer (August) thermal-oxygen stratification of the waters of Lake Licheńskie (**a**) and Lake Ślesińskie (**b**) and the post-mining water basin (**c**) in 2013–2015

3 Microorganisms

The microorganisms in the Konin lakes constitute a highly structurally and functionally diversified grouping. The low temperatures and relatively high water retention in the fall-winter period facilitate the domination of the microbiological decomposition of organic compounds and their mineral decomposition. These processes are conducted by specialized groups of heterotrophic and chemolithotrophic bacteria with relatively low reproduction rates. The intensity of these processes determines the lowering of concentrations of dissolved forms of organic matter and the increase of the mineral forms of nitrogen and phosphorus. The greater availability of nutrients impacts the intensity of primary production, as is reflected in more abundant blooms of phytoplankton in early spring.

Including the lakes in the cooling circuit for warm waters discharged by power stations radically changed the structure and functioning of microorganisms in the spring-summer period. Low water retention in lakes facilitated significantly increasing quantities of easily absorbed organic matter flowing in from the catchment area. Significant increases in water temperature affected the intensity of microbiological processes characterized by, inter alia, the development of bacterioplankton, high bacterial production, and the initiation of a cascade of phenomena within the microbiological loop [18–21]. The intense bioconversion of organic matter, in which organisms from various trophic levels participate, was essential in stabilizing the structure and functioning of the entire ecosystem of the lakes. The multidimensional reaction of the biocenosis to anthropogenic pressure combined with the simultaneous existence of strong, effective trophic connections among particular groups of organisms played fundamental roles in these processes.

4 Phytoplankton

From the perspective of almost 50 years of research, the greatest phytoplankton biomass formed in Lake Licheńskie at a maximum of approximately 82 mg dm^{-3} in the 1960s, and the summer blooms consisted primarily of cyanobacteria (Table 4). Power station discharges of waters with elevated temperatures (up to 35°C) were at the time one for the factors that modified the abundance and structure of the phytoplankton, and the strong pressure exerted by grass carp compounded this effect. The phytoplankton biomass in Lake Ślesińskie prior to its being included in the cooling circuit was 14 mg dm^{-3}. In the 1970s and 1980s, the maximum was 12 mg dm^{-3} and 7 mg dm^{-3}, respectively, for the two lakes. At this time the functioning of the cooling devices stabilized, and intense water flow limited the development of phytoplankton.

The maximum phytoplankton biomass of up to approximately 50 mg dm^{-3} was noted in these lakes in the 1990s. At this time, eutrophication processes intensified as were manifested in oxygen depletion in the hypolimnion of the lakes, shallowing

Table 4 Annual total phytoplankton biomass and cyanobacteria relative biomass in lakes Licheńskie and Ślesińskie in 1965–2015[a]

	Lake Licheńskie		Lake Ślesińskie	
Study period	Total biomass, mg dm^{-3}	Cyanobacteria, %	Total biomass, mg dm^{-3}	Cyanobacteria, %
1965–1969	16.1 (0.9–82.1)	40.9	4.0 (0.1–14.0)	28.2
1977–1980	5.9 (2.4–12.0)	30.6[b]	3.6 (0.6–7.2)	36.1[b]
1983–1984	1.5[b] (0.6–1.9)[b]	16.4	1.2[b] (0.6–4.0)[b]	11.6
1989–1991	4.8 (0.4–17.9)	10.9	5.7 (0.2–24.0)	8.5
1992–1994	9.4 (0.4–49.3)	7.0	10.0 (0.6–35.1)	4.2
1995–1999	4.2[b] (0.3–18.6)	4.3	5.6 (0.2–46.7)	5.1
2000–2003	5.0 (0.3–24.8)	3.6	5.4 (0.2–30.0)	3.4
2004–2005	3.0 (0.2–8.6)	4.7	3.6 (0.4–13.2)	5.0
2011–2015	19.6[b] (5.2–35.1)[b]	14.4[b]	8.9[b] (4.3–13.7)[b]	8.9[b]

[a]Data from 1965 to 2005 according to Spodniewska [22], Sosnowska [23], Simm [24], Socha [25–27], Socha, Hutorowicz [28] and Napiórkowska-Krzebietke [29]
[b]Summer period

and siltation, increased salinity, and deteriorating sanitary status of the waters [6, 30, 31]. In the early 2000s, large biomass was still noted (at a maximum of 25 mg dm^{-3} and 30 mg dm^{-3}, respectively, in lakes Licheńskie and Ślesińskie); however, by 2004–2005, this had decreased by approximately two- to threefold.

In recent years, the phytoplankton biomass in summer increased to 35 mg dm^{-3} in Lake Licheńskie, which was the effect of increased phosphorus loading and the extended water retention times. At this time, the significance of cyanobacteria in the assemblage also increased, which was similar to observations made in 1965–1966 [23]. The maximum biomass exceeded significantly the biomass limit value (8 mg dm^{-3}) that is characteristic for phytoplankton blooms and typical lakes in the temperate zone in advanced stages of eutrophication [32, 33]. The domination or co-domination of diatoms throughout the vegetative season [22–29], and especially in summer, was undoubtedly the consequence of artificially halting water flow through both lakes. The higher share of cyanobacteria in the phytoplankton of Lake Licheńskie has recently been linked with increased trophy and increased water retention times and also with decreased zooplankton pressure on the phytoplankton.

In the most heated of the lakes, Licheńskie, which has been loaded with phosphorous from aquaculture in recent years, the total maximum biomass of the summer phytoplankton assemblage ranged from 6.1 to 35.1 mg dm^{-3} in 2011–2015 (Fig. 5). Similarly, significant changes were noted in primary production, the value of which, in such a short period of time, increased approximately threefold, i.e., from 2,700 kcal m^{-2} 150 d^{-1} in 2011 to over 7,000 kcal m^{-2} 150 d^{-1} in 2013–2015 (Fig. 6). The comparison of nearly 50 years of research results indicated that such high primary production values as those recorded in 2014 had not been noted previously in this lake. In Lake Ślesińskie, the coolest of the system, the summer phytoplankton biomass maximum ranged from 5.0 to 12.8 mg dm^{-3} (Fig. 5).

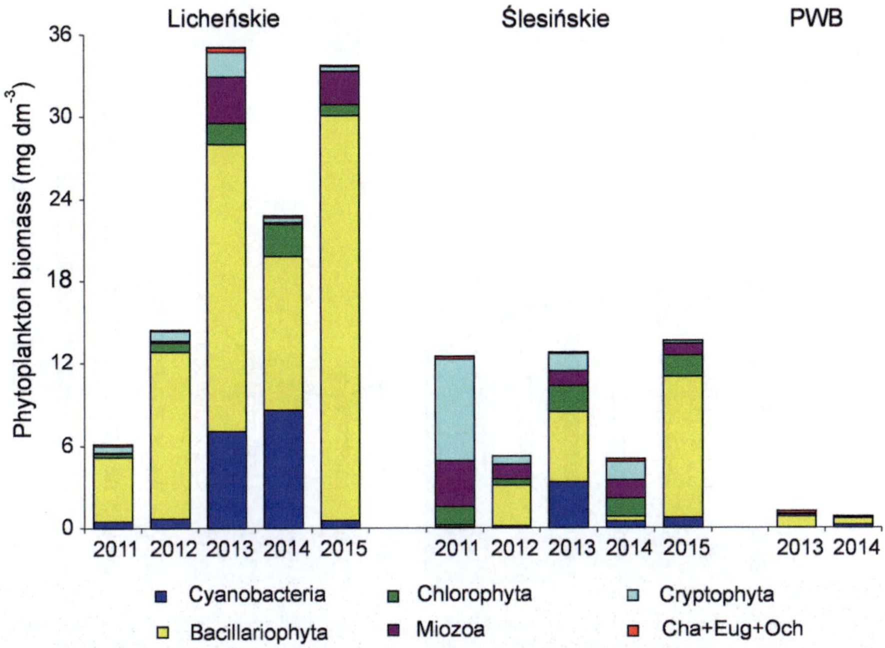

Fig. 5 Summer phytoplankton biomass and structure in lakes Licheńskie and Ślesińskie and in the post-mining water basin (PWB) in 2011–2015; Cha, planktonic charophyta; Eug, Euglenophyta; Ochr, Ochrophyta

Differences in the structure of the phytoplankton assemblages of these two lakes were noted (Fig. 5). The most intense diatom development consisting of approximately 50 to 88% of the overall biomass was noted in the flow-through of Lake Licheńskie. The diatoms were dominated by *Aulacoseira granulata* (Ehrenberg) Simonsen, and the associate species were mainly *Ulnaria ulna* (Nitzsch) Compère, *Fragilaria crotonensis* Kitton, and species of the genus *Cyclotella*. Cyanobacteria was present in summer contributing from 5 to 38% of the total biomass; the dominants were *Anabaena oscillarioides* Bory ex Bornet & Flahault and picoplankton colonies of species from the genera *Aphanothece* and *Aphanocapsa*. On the other hand, another phytoplankton group that contributed a maximum of 10% was formed by dinophytes, mainly *Naiadinium polonicum* (Woloszynska) S. Carty and the green algae *Binuclearia lauterbornii* (Schmidle) Proschkina-Lavrenko and *Eudorina elegans* Ehrenberg. The cryptophytes *Cryptomonas erosa* Ehrenberg and *Cryptomonas curvata* Ehrenberg had a relative biomass of up to approximately 8%.

The phytoplankton assemblage of Lake Ślesińskie was dominated by diatoms (2–75%), cryptophytes (10–59%), green algae (10–27%), and dinophytes (9–27%). Cyanobacteria constituted a large share of the overall biomass that ranged from

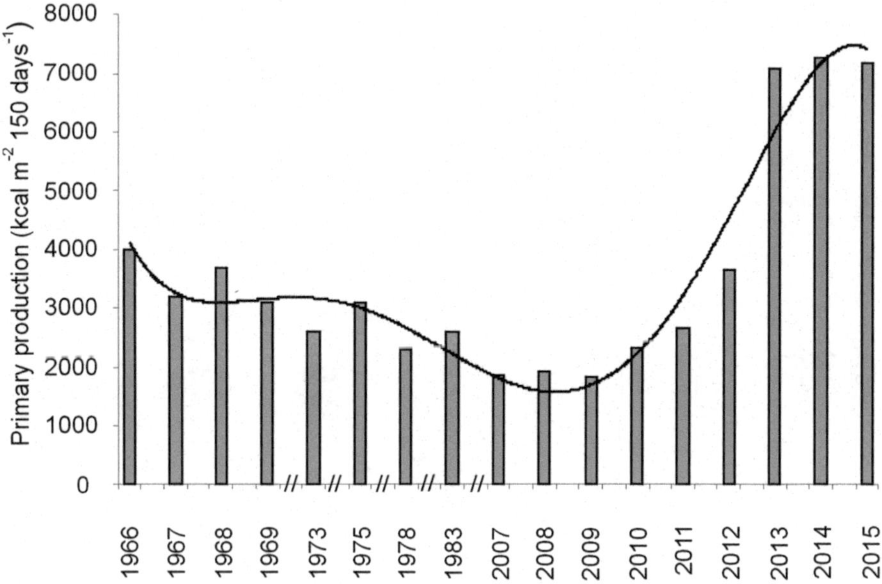

Fig. 6 Trends of changes in primary production (gross) in Lake Licheńskie in the vegetative period (May–September) in 1966–2015 (data from 1966 to 1983 according to Patalas [34], Hillbricht-Ilkowska et al. [35], Zdanowski [36, 37])

less than 1 to 27% (maximum in 2013). The same species noted in Lake Licheńskie dominated, e.g., *Aulacoseira granulata*, *Cryptomonas erosa*, *C. curvata*, *Anabaena oscillarioides*, and *Binuclearia lauterbornii*, along with the dinophytes *Parvodinium umbonatum* (Stein) S. Carty, *Peridinium willei* Huitfeldt-Kaas, and *Kolkwitziella* sp.

The summer phytoplankton biomass that formed in the post-mining water basin was substantially smaller at less than 1 mg dm^{-3} (Fig. 5). The assemblage was dominated at this time by the medium-sized diatom *Ulnaria acus* (Kützing) Aboal and the larger *U. ulna* as well as small nanoplanktonic species of the genus *Cyclotella*. The picoplanktonic colony-forming chroococcalean cyanobacteria of the genera *Aphanocapsa* and *Aphanothece* and the chrysophytes of the genera *Dinobryon* contributed a large share. The very low summer phytoplankton biomass dominated by diatoms in the post-mining basin is comparable to the biomass noted in oligo-mesotrophic lakes [38].

5 Ciliates

Among the protozoans, the ciliates (Ciliata) are distinguished by the greatest species richness. They occur in both oxygenated and anoxic waters. The greatest number of ciliates was confirmed in the bottom sediments and on the surfaces of aquatic plants (Fig. 7), and 150 species of ciliate occur on the macrophytes of the Konin lakes [39].

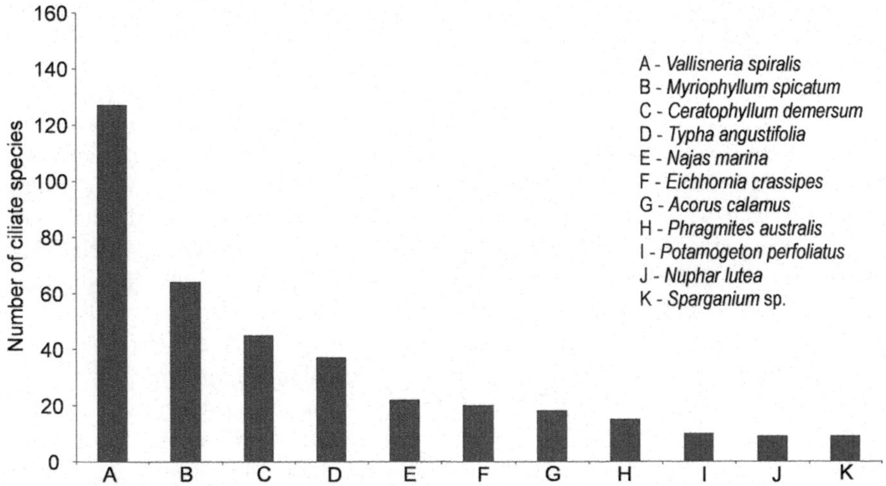

Fig. 7 Number of ciliated protozoa species found on different macrophytes in three Konin lakes (data according to Babko et al. [39])

It was noted most frequently on *Vorticella campanula* Ehrenberg, *Vorticella convallaria* complex, *Coleps hirtus* (Müller), *Cinetochilum margaritaceum* (Ehrenberg), and *Stylonychia mytilus* (Müller).

Twenty-nine species were confirmed in the bottom sediments and ten in the water column. Just three species – *Pelagovorticella mayeri* (Faure-Fremiet), *Tintinnopsis cylindrata* Kofoid & Campbell, and *Urotricha pelagica* Kahl – were noted exclusively in the plankton. Of the 29 species confirmed in the bottom sediments, only ten do not occur in other habitats. *Coleps elongatus* Ehrenberg, *Colpoda steinii* Maupas, *Loxodes magnus* Stokes, and *Spirostomum minus* Roux were indicator species of poorly oxygenated conditions, while *Brachonella spiralis* (Smith), *Metopus contortus* Levander, 1894, and *Metopus es* Müller were indicative of anoxic conditions.

Ciliates of the genus *Stentor* are among the largest. Of the eight common species [39], six were confirmed in the Konin lakes, e.g., *Stentor coeruleus* (Pallas), *S. igneus* Ehrenberg, *S. muelleri* Ehrenberg, *S. multiforme* (Müller), *S. polymorphus* (Müller), and *S. roeseli* Ehrenberg. All of these species occurred in the coolest of the lakes, Ślesińskie, while in the heated Lake Licheńskie, only four species of this genus were confirmed.

Overall, the occurrence of over 163 species of ciliates from 10 classes was confirmed in the Konin lakes [40, 41], of which 113 in Lake Licheńskie, 85 in Lake Ślesińskie, and 100 in Lake Wąsosko-Mikorzyńskie. The faunal similarity of ciliates in the lakes, which was calculated with the Jaccard similarity index, fluctuated from 42 to 45%. The greatest number of species was confirmed in Lake Licheńskie, the warmest of the lakes. The elevated water temperature increased ciliate diversity. The protozoan fauna was highly diversified and changed across the thermal gradient.

6 Zooplankton

The zooplankton species density, biomass, and domination structure differed among the lakes. Zooplankton most similar to that in biocenoses of natural lakes with similar trophic conditions was noted in the least and sporadically heated Lake Ślesińskie [42, 43]. The highest zooplankton biomass was noted in the northern part of this lake, which is beyond the impact of the main stream of heated waters, and also in spring when this lake is not yet included in the water cooling circuit. The higher share of Cladocera (*Daphnia cucullata* Sars, *Daphnia hyalina* Leydig) and juvenile stages of Copepoda [44], similarly to that in the late 1970s and early 1980s [44–46], are evidences of the maintenance of the transfer of matter from the primary link which limits the occurrence of summer phytoplankton blooms.

The natural trophic system of the plankton was transformed the most in lakes Licheńskie and Gosławskie, where higher temperatures and water flow limited plankton development, increased its mortality, and altered the species domination structure. In addition to heterotrophic bacteria and protozoans, detritivorous planktonic rotifers participated in the transformation of organic matter [42, 47, 48]. In early spring, cryophilic species typical of this season were displaced [49]. Declines in zooplankton abundance were noted in summer when the water in the lakes was heated by 5–6°C and the discharged water temperature exceeded 30°C. At this time, large cladocerans and copepods were eliminated, and the dominants were small rotifers [47, 50].

The dominant rotifer species in Lake Licheńskie in the last decade were *Keratella cochlearis* f. *tecta* (Gosse) and *Polyarthra longiremis* Carlin. Among crustaceans, the dominants were juvenile copepod forms (nauplii), the small cladocerans of *Bosmina coregoni* Baird and *Bosmina longirostris* var. *typica* (Müller), and the thermophilic cladoceran predators of *Ceriodaphnia quadrangula* (Müller) and *Diaphanosoma brachyurum* (Lieven).

The zooplankton in the warm discharge canals was poorer because of high thermal and hydraulic selection occurring in this habitat. The mortality of organisms passing through the cooling system is as high as 80%. The highest mortality is noted in cladocerans, which, in comparison to rotifers, are larger and they have delicate swimming and filtration appendages [42, 44].

The extensive littoral area overgrown with *V. spiralis* currently provides refuge for zooplankton [48, 51]. Species not formerly noted in Poland were identified here including [52], inter alia, *Asplanchnopus hyalinus* Harring, *Beauchampia crucigera* (Dutrochet), *Lecane inopinata* Harring & Myers, *Lecane shieli* Segers & Sanoamuang, *Lecane undulata* Hauer, and *Lepadella apsida* Harring. The abundantly occurring predatory species of *Cupelopagis vorax* (Leydig), which is known in Europe, Asia, North America, and Australia but which occurs rarely in Poland, was also noted.

7 Invertebrates

The diversity of Chironomidae has been lower that was reported by Leszczyński [53] for the 1960s. The number of snail taxa (Gastropoda) noted in the lakes was also lower in comparison to basins with a natural thermal regime. More species of snail were only noted [54] during the initial period of lake heating.

As early as the mid-1990s, progressive declines in the species abundance of Gastropoda and Bivalvia were apparent [55, 56]. The primary Bivalvia representative in the zone with moderate water heating (<28°C) was *Dreissena polymorpha*, while in the zone with intensely heated waters, it was *Sinanodonta woodiana* Lea [57]. Indigenous bivalve species inhabited the cooler parts of the system. Declining species diversity was also noted within other taxonomic groups, including the Hirudinea (leeches), Isopoda, and Hydrachnidia (water mites). Thirteen Hirudinea species still occurred in the 1960s, while only seven did by 1970 [58], and in 2000–2001, there were just three [57]. The diversity of Hydrachnidia during this period was smaller than that of Biesiadka confirmed in the 1970s [59, 60].

The bottom fauna of the flow zones that were heated intensively were the poorest. Mainly Oligochaeta occurred here as they are more tolerant of high temperatures. The development of bottom macrofauna was more intense in environments that were more oxygenated. The highest concentration of macrophytes was noted in the littoral and sublittoral zones of Lake Licheńskie [56].

The primary epibiontic fauna on *S. woodiana* was colonies of *D. polymorpha*. More intense zoobenthos development was confirmed in the zone dominated by these two species [57]. Phytophilous fauna, represented mainly by Chironomidae, developed more decisively in the *Ceratophyllum demersum* L. and *Myriophyllum spicatum* L. assemblage than in the *V. spiralis* assemblage. A newly discovered cryptogenic species is the predatory *Chaetogaster zdanowski* that occurs on *C. demersum* only in Lake Ślesińskie, while the alien species *Urnatella gracilis* Leidy, originating from Brazil, was found to have settled on the shells of dead bivalves; this is the only representative in Poland of the class of goblet-shaped Kamptozoa, that inhabit fresh waters and reproduce in reservoirs with heated waters [55].

8 Alien Species

The heterogeneous habitats the Konin Lake system offers provide conditions that facilitate the settlement of alien, often invasive, species. According to the assessment by Najberek and Solarz [61], as many as 41 alien species are noted in the system, and there are at least 58 cryptogam species, the origin of which has not been determined. These were dominated by chordates (*Chordata*), mollusks (Mollusca), and flat worms (Platyhelminthes). The ichthyofauna of the lakes has been enriched with 12 alien fish species, three of which have formed stable populations, inter alia,

Pseudorasbora parva (Temminck and Schlegel), *Carassius gibelio* (Bloch), and *Carassius auratus* (L.) [62].

The circumstances under which the lake system was settled were only established for 58% of the species. Most of the species originate from the Pontocaspian and Asian regions. Species originating from North and South America were also noted, as well as a few from Australia, Oceania, and Central America. A significant number of species have settled the lakes after having arrived as "stowaways." Some were introduced to the lakes purposefully, e.g., the Asian herbivorous fish species of grass carp, silver carp [*Hypophthalmichthys molitrix* (Val.)], and bighead carp [*Aristichthys nobilis* (Rich.)], or were the "escapees" from culture facilities and fishkeeping, for example, *Acipenser* spp., *Oncorhynchus mykiss, Oreochromis niloticus*, and *Carassius auratus* [63–65], while others were introduced along with so-called plantings (Fig. 8).

Most of these species are unable to reproduce in the heated lakes. The effect they have on the environment and the already weakened indigenous ichthyofauna of the lakes is unknown. Commercial fish catches in the lakes, primarily of common bream and common roach, excluding introduced alien species, decreased, on average, from 26 kg ha^{-1} noted in 1958–1992 [66] to barely 3 kg ha^{-1} in 2005–2016. The fish fauna of the lakes is currently under strong angling pressure. Heated lakes may also be accessed by piscivorous birds and mammals in winter when many natural water bodies and rivers are covered with ice and they are unable to access their usual source of prey.

Fig. 8 Alien water lily *Nymphaea x hybrida*, with plantings in Lake Ślesińskie (2016)

The key species that shape the structure and functioning of the lakes are tape grass, the zebra mussel, the Chinese pond mussel, and topmouth gudgeon.

Tape Grass (*Vallisneria spiralis* L., Hydrocharitaceae) naturally occurs in the tropic and sub-tropic zones [67, 68]. In Europe, it reaches the southern edge of the Alps [63]. It occurs in Konin lakes from 1993 [7, 55, 69, 70]. *V. spiralis* covers the entire bottom of lakes Gosławskie and Pątnowskie. It inhabits the littoral zones of lakes Licheńskie and Wąsosko-Mikorzyńskie up to depths of 4 m. It does not occur in Lake Ślesińskie. It is the main structural element of aquatic ecosystems (Fig. 9). It dominates among submerged vegetation. Its wet weight in Lake Licheńskie is estimated to be 30 kg m^{-2} [7]. It accumulated nutrients and heavy metals and stabilizes the bottom sediments [71]. It accelerates sedimentation and limits water flow. The dead parts of the plant pollute the shores of the lakes and decrease the throughput of the canals [7]. Significant diel changes in water physicochemical parameters are noted, specifically increased water pH and oxygen content, in habitats with dense *V. spiralis* stands. Water oxygen saturation in summer can increase to 300%, while at night it can fall to as low as 20%.

V. spiralis stands in the littoral zones of lakes offer habitats for diverse groups of animals [57, 69], including new species of aquatic fauna [52, 62]. It determined the occurrence and diversity of juvenile fish assemblages [69]. Habitats that were totally

Fig. 9 *Vallisneria spiralis* L. assemblages in Lake Licheńskie (in the background: the Sanctuary of Our Lady of Licheń in Licheń Stary, which is visited by approximately 1.5 million tourists annually)

overgrown by *V. spiralis* were characterized by the highest fish biomass. Diel variations in their species structure and abundance were also noted [72]. Fish reacted to the high water temperature and the significant oxygen oversaturation by moving to areas with less dense macrophytes during the day.

Zebra Mussel (*Dreissena polymorpha* (Pallas, 1771), Bivalvia) is an invasive Pontocaspian species that began settling in Polish waters in the second half of the nineteenth century (Fig. 10). It is an excellent filter feeder that maintains water cleanliness in better condition.

The first information about the occurrence of zebra mussel in the Konin lakes was published by Leszczyński [53] and Berger and Dzięczkowski [54]. Stańczykowska [73], Kornobis [74], and Stańczykowska et al. [75] studied the occurrence of larvae in the lake pelagic zone, the growth rates of settled bivalves, and the densities of various age groups in the late 1980s and early 1990s. These researchers noted significantly decreased growth rates in bivalve communities in the bottom sublittoral zone in comparison to those in lake littoral zone and also decreased life spans of just 3 to 4 years in waters that were intensely heated and the domination in these environments of juvenile specimens (0–1+).

The peak of zebra mussel population development in the Konin lakes recorded in 1997–2001 (Fig. 11 [76, 77]) occurred during the period when there was constant, intense water exchange among the lakes. The highest bivalve biomass was noted in the initial cooling reservoir (up to 18 kg m^{-2}). The mussel population, which was genetically homogeneous [78], formed subpopulations that differed in phenotype depending on the environmental conditions [55, 79]. In environments that were intensively heated, the populations were unstable, the share of females was greater, life span shortened, and individual maximum mussel mass decreased [76, 77].

Fig. 10 *D. polymorpha* Pall. aggregations on Chinese pond mussel (*Sinanodonta woodiana* Lea, 1863)

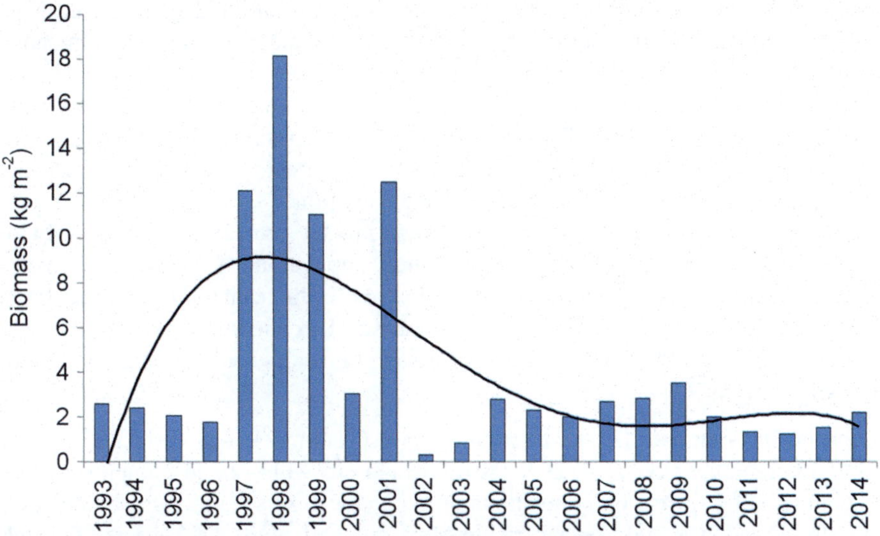

Fig. 11 *Dreissena polymorpha* Pall. biomass in the Konin lake system in 1993–2014 (data from 1993 to 2006 according to Sinicyna and Zdanowski [76, 77])

The survival of populations in these environments was possible only through increased reproductive potential, including the participation of females, and increased number of reproduction cycles. Under conditions of stress, inter alia, long-term increased water temperatures (>28°C), massive bivalve mortality was noted.

Zebra mussel settlement in the lake system has been less intense in recent years. The maximum mussel biomass in different years ranged from 0.3 to 3.5 kg m^{-2} (Fig. 11). Stable populations occurred only in the cooler parts of the system. Bivalves of greater size (>2.5 cm) and of heavier weights (>1.5 g) were rarely noted.

Chinese Pond Mussel (*Sinanodonta woodiana* Lea 1864, Bivalvia) (Fig. 10), which occurs in the Amur and Yangtze rivers, was introduced to Poland from Hungary along with herbivorous fishes in the early 1980s. It was found by accident in the summer of 1993 during underwater exploration of the bottom discharge canal of the Pątnów power station. The aim was to determine the degree of silting from the intensive cage fish culture that was being conducted there. Its biomass at the time was estimated to be 50 kg m^{-2}. Because of the excellent feeding conditions at this location, individuals occurred here that weighed approximately 1 kg with a shell length of 25 cm. Upstream from the cages, the occurrence of the African snail *Melanoides tuberculatus* (O. F. Müller, 1774) was also noted at this time [64].

The Chinese pond mussel is an expansive species. It already occurs in the lower Oder River near power station heated water discharges and in carp ponds in various regions of Poland. It is gonochoristic. Its main reproduction site in the Konin lakes was the initial cooling reservoir [65]. The mussels avoided stagnant water zones, those with strong turbulence, and coarse-grained substrates. They preferred sites with higher water temperatures (up to 35°C). They did not displace indigenous *Unionidae* species. They are very effective filter feeders [80]. In 1993–2002, this species settled the warmer parts of the Konin system, forming cohorts that biomass of which ranged from 30 to 50 kg m^{-2} (Fig. 12). This species was more dispersed in the littoral of the warmer lakes [55, 56, 65, 81], while in the coolest lake, Ślesińskie, they occurred in small numbers and in worse condition.

The maximum biomass of Chinese pond mussel decreased progressively in 2007–2014 to 15 kg m^{-2} (Fig. 12). Stabile populations occurred only in the discharge canals of the Pątnów power station and in the canal that connects lakes Licheńskie and Pątnowskie. The dominant mussels here measured 150 cm in length and weighed 200–300 g. The bivalve biomass in the initial cooling reservoir did not exceed 10 kg m^{-2}.

Topmouth Gudgeon [*Pseudorasbora parva* (Temminck and Schlegel, 1846)] (Fig. 13) inhabits over 50 sites, most of which are within pond complexes [82]. The impact of *P. parva* on populations of indigenous fish species included decreasing the abundance and species richness of autochthonous fish species through predation, food competition, disrupted spawning, and the destruction of spawning sites [62, 83, 84].

P. parva is one of the 12 species of alien fish that has been very successful in colonizing the Konin lakes. With its growing abundance, changes have been noted in the structure of small-sized fish assemblages that inhabit the shallow littoral zones [62]. Particularly drastic declines in abundance have been noted

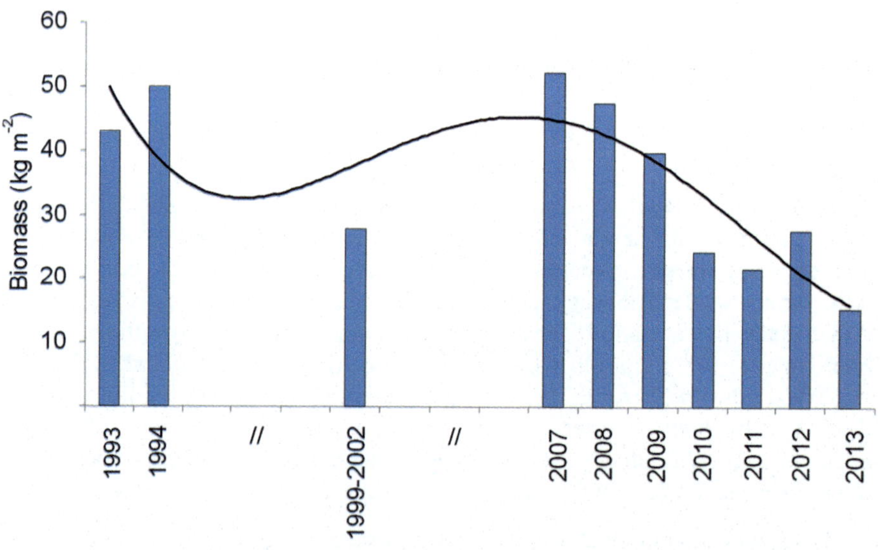

Fig. 12 *Sinanodonta woodiana* (Lea, 1863) biomass in the Konin lake system in 2003–2014 (data from 1993 to 1994 according to Protasov et al. [55, 56], data from 1999 to 2002 according to Kraszewski [81], Kraszewski and Zdanowski [65])

Fig. 13 Topmouth gudgeon [*Pseudorasbora parva* (Temminck and Schlegel, 1846)]

in *Alburnus alburnus* (L.) and *Tinca tinca* (L.). Adult *P. parva* specimens were noted for the first time in Lake Licheńskie in 2002 [85, 86].

P. parva prefers more shallow lake littoral zones (<1.0 m). It occurred most abundantly in the zone covered with *V. spiralis* stands. Reproduction occurred from April to September at water temperatures of 15–30°C. *P. parva* spawns eggs in several portions most frequently onto aquatic vegetation. The smallest sexually mature females measured 18.9 mm. The mean body length of mature females was 33.9 mm [87, 88].

9 Conclusions

The Konin lakes can be classified according to Water Framework Directive (WFD) [89, 90] as "heavily modified bodies of water," or, in other words, "bod[ies] of surface water which as a result of physical alterations by human activity [are] substantially changed in character" (Table 5). Classification based on physicochemical elements indicated that the ecological potential of the waters of lakes Licheńskie and Ślesińskie was markedly deteriorated. This was determined primarily by the mean total phosphorus content and the mean oxygen saturation of the hypolimnion waters during the vegetative season.

Table 5 Ecological potential assessment of the water of the Konin lakes according to selected physicochemical and biological parameters

Lake/ Basin	Study year	Hypolimnion oxygenation %	Conductivity in 20°C µS cm^{-1}	Total nitrogen mg dm^{-3}	Total phosphorus mg dm^{-3}	Secchi disk visibility m	Physicochemical assessment	Chlorophyll Index[a]	Biological assessment
Lake Licheńskie	2004	1.6	536	0.83	0.121	1.8	BGP	2.8	III
	2005	0	526	1.02	0.129	2.1	BGP	3.3	IV
	2006	6.8	551	1.03	0.138	1.8	BGP	3.1	IV
	2007	8.1	640	0.94	0.192	1.9	BGP	2.4	III
	2008	6.7	645	1.17	0.196	2.0	BGP	1.7	II
	2009	5.7	663	1.18	0.132	2.1	BGP	1.8	II
	2010	24.2	666	0.85	0.130	1.7	BGP	2.2	III
	2011	0	638	0.99	0.129	1.8	BGP	2.3	III
	2012	0	657	1.16	0.114	2.0	BGP	2.4	III
	2013	0	634	1.40	0.132	1.6	BGP	4.2	V
	2014	0	633	1.26	0.141	2.2	BGP	3.6	IV
	2015	0	563	1.03	0.135	2.3	BGP	2.7	III
Lake Ślesińskie	2004	1.8	513	0.65	0.117	2.4	BGP	2.6	III
	2005	8.2	515	1.09	0.140	2.3	BGP	2.7	III
	2006	10.0	532	1.08	0.140	1.9	BGP	2.996	III
	2007	4.9	596	1.09	0.218	2.2	BGP	2.2	III
	2008	3.2	621	1.24	0.192	2.7	BGP	1.8	II
	2009	3.1	629	1.16	0.126	2.7	BGP	1.4	II
	2010	21.2	634	0.96	0.148	2.6	BGP	2.4	III
	2011	0	624	1.39	0.134	2.1	BGP	2.5	III
	2012	0	646	1.07	0.125	2.6	BGP	2.1	III
	2013	0	604	1.18	0.109	2.2	BGP	2.1	III
	2014	0	612	1.13	0.147	2.8	BGP	2.4	III
	2015	0	524	0.99	0.155	3.0	BGP	2.5	III
Post-mining Water Basin	2012	53.2	549	0.63	0.059	5.8	II	0	I
	2013	51.5	554	0.54	0.069	5.8	II	0.3	I
	2014	62.7	541	0.47	0.069	6.4	II	1.1	II
	2015	78.8	488	0.49	0.104	6.2	BGP	1.1	II

Legend:

Biological assessment

class I | class II | class III | class IV | class V | potential

Physicochemical assessment

class I — for TP and SDV

classes I-II — class II for TP and SDV

BGP — below good potential

[a]partial metric of phytoplankton multimetric PMPL

A simplified biological assessment based on chlorophyll indicated that Lake Licheńskie is classified as class II through classes III and IV to class V, which correspond to waters that range from good to bad ecological potential. However, the waters of Lake Ślesińskie were classified as classes II and III, which correspond to waters that range from good to moderate ecological potential.

The best water quality was in the post-mining water basin, which, based on physicochemical elements, was classified as class II, except in 2015 because of elevated phosphorous levels, but water transparency was very high. The value of the chlorophyll index indicated classes I or II, which correspond to waters of high to good ecological potential.

The reduced heating of the waters and the increased water residence times and phosphorous loading in the past decade have facilitated the occurrence of phytoplankton blooms in Lake Licheńskie. Currently, the impact of point source and dispersed pollution, mainly from fish culture facilities, and the expansive development of *Vallisneria spiralis* shape the functioning of the lake system. Among the system's lakes, only in Lake Gosławskie is the ecological potential stable.

Acknowledgments The material was collected and analyzed primarily as a part of statutory research (S-009) conducted at the Stanisław Sakowicz Inland Fisheries Institute in Olsztyn.

References

1. Patalas K (1976) Commentary on keynote paper by Dr Ulf Grimas: Same aspects on cooling water discharge and environmental enhancement. Proceedings workshop held at the Chalk River Nuclear Labs October 6–10, 1975, vol 1, pp 161–166
2. Horoszewicz L (1969) The effect of higher temperature on fish. Ekol Pol B 15:299–321
3. Horoszewicz L, Backiel T (1979) Biology of fish as test for heated effluents. Pol Ecol Stud 5:5–120
4. Hillbricht-Ilkowska A, Zdanowski B (1988) Main changes in the Konin lake system (Poland) under the effect of heated-water discharge pollution and fishery. Ekol Pol 36(1-2):5–21
5. Hillbricht-Ilkowska A, Zdanowski B (1988) Changes in lake ekosystems connected with the power-generating industry (the outline of problem); the Konin lakes (Poland) as the study sites. Ekol Pol 36(1–2):23–45
6. Zdanowski B (1994) Characteristics of. heated Konin lakes, pollution sources, main results and conclusions. Arch Pol Fish 2(2):139–160
7. Socha D, Zdanowski B (2001) The aquatic ecosystem near Konin. Bibl Mon Środ, Poznań, pp 1–75 (in Polish)
8. Protasov AA, Zdanowski B (2003) Determination of impacts of nuclear and thermal power plants on hydroecosystems using expert assessment. Hydrobiol J 39(1):48–58
9. McCoy WF (1991) Microbial fouling control in heat exchangers. Trans Am Nucl Soc 64:63
10. Afanasiev SA, Sidersky AV, Shatokhina AV (1990) Fouling in the intake conduit of the atomic power plant as a factor of biological interferences. Gidrobiol Zh 26(2):25–29. in Russian
11. Afanasiev SA, Romanova EG, Slepnev AE (1991) Bacterial fouling of heat exchange equipment. Gidrobiol Zh 27(5):56–61. in Russian
12. Afanasiev SA, Świątecki A, Zdanowski B (1998) Biodisturbances in heated systems – technical and ecological consequence. In: Zdanowski B (ed) Konin lakes – 40 years of study. Current state and conclusions for protect. Bibl Mon Środ, Konin, pp 1–124 (in Polish)
13. Stawecki K, Zdanowski B, Pyka JP (2013) Long-term changes in post-cooling water loads from power plants and thermal and oxygen conditions in stratified lakes. Arch Pol Fish 21:331–342
14. Korycka A, Zdanowski B (1976) The influence of heated effluent waters on the water chemism of Konin lakes. Rocz Nauk Roln Ser H 97(3):89–107. in Polish
15. Zdanowski B, Korycka A, Gębicka A (1988) Long-term variation in habitat and trophic factors in the Konin lakes (Poland) under the influence of heated-water discharge and pollution. Ekol Pol 36(1–2):47–77
16. Zdanowski B, Prusik S (1994) Temperature-oxygen relations and chemical composition of water in heated Konin lakes. Arch Pol Fish 2:161–178
17. Pyka JP, Zdanowski B, Stawecki K (2013) Long-term trends in changes of the chemical composition of waters in lakes heated by electric power plants. Arch Pol Fish 21:343–355
18. Dunalska J, Pyka JP, Zdanowski B, Świątecki A (2007) Carbon dynamics in a heated water system (Wielkopolsko-Kujawskie Lake District, Poland). Arch Pol Fish 15(4):295–307
19. Marchlik WD, Świątecki A, Stawecki K (2005) Microbiological destruction dynamics in heated Konin lake. Acta Universitatis Nicolai Copernici Limnological Papers XXV, vol 113, pp 75–83
20. Świątecki A, Zdanowski B (2007) The seasonal dynamics of organic matter remineralization by bacterial consortia in the heated Konin lakes. Arch Pol Fish 15(4):309–320

21. Świątecki A, Zdanowski B, Dunalska J, Górniak D (2007) The dynamics of bacterioplankton versus resources of organic carbon in the heated Konin lakes. Arch Pol Fish 15(4):321–334
22. Spodniewska I (1984) Biomass and structure of phytoplankton in Konin lakes in the years 1976-1980. Ekol Pol 32(2):289–305
23. Sosnowska J (1987) Effect of heated effluents on the phytoplankton in some lakes near Konin. Rocz Nauk Roln Ser H 101(3):1–130. in Polish
24. Simm AT (1988) Changes in the composition and quantitative relations of the phytoplankton in heated lakes near Konin (Poland). Ekol Pol 36(1–2):97–113
25. Socha D (1994) Quantitative and qualitative changes of the phytoplankton in heated Konin lakes. Arch Pol Fish 2(2):219–234
26. Socha D (1994) Changes in the phytoplankton of warmed-up lakes near Konin (Central Poland) in 1987-1990. Idee Ekol 5 Ser Zeszyty 2:1–88. in Polish
27. Socha D (1997) Spatial and seasonal phytoplankton diversity in Licheńskie and Ślesińskie lakes, Konińskie District, in 1991-1993. Arch Pol Fish 5(1):117–136
28. Socha D, Hutorowicz A (2009) Changes in the quantitative relations of the phytoplankton in heated lakes. Arch Pol Fish 17(4):239–251
29. Napiórkowska-Krzebietke A (2009) Diversity and dynamics of phytoplankton in lakes Licheńskie and Ślesińskie. Arch Pol Fish 17(4):253–265
30. Zdanowski B (1994) Long term changes of phosphorus and nitrogen content and of trophic status in heated Konin lakes. Arch Pol Fish 2:179–191
31. Świątecki A (1994) Sanitary and bacteriological status of heated Konin lakes. Arch Pol Fish 2(2):207–218
32. Nebaeus M (1984) Algal water-bloom under ice-cover. Verh Internat Verein Limnol 22:719–724
33. Napiórkowska-Krzebietke A, Dunalska J (2015) Phytoplankton-based recovery requirement for urban lakes in the implementation of the Water Framework Directive's ecological targets. Oceanol Hydrobiol St 44(1):109–119. https://doi.org/10.1515/ohst-2015-0000
34. Patalas K (1970) Primary and secondary production in a lake heated by thermal power plant. Proceeding of the 1970 annual technical meeting of the institute of environmental sciences, Boston, pp 267–271
35. Hillbricht-Ilkowska A, Zdanowski B, Ejsmont-Karabin J, Karabin A, Węgleńska T (1976) Primary and secondary production of lakes with heated waters. Rocz Nauk Roln Ser H 97:69–88. in Polish
36. Zdanowski B (1976) Long-term changes of the pelagic primary production in heated lakes. Rocz Nauk Roln Ser H 97(3):123–139. in Polish
37. Zdanowski B (1988) Long-term and seasonal changes in the primary production and destruction in heated lakes near Konin (Poland). Ekol Pol 36:79–95
38. Heinonen P (1980) Quantity and composition of phytoplankton in Finnish island waters. Publ Wat Res 37:1–91
39. Babko R, Fyda J, Kuzmina T, Hutorowicz A (2010) Ciliates on the macrophytes in industrially heated lakes (Kujawy Lakeland, Poland). Vestnik Zoologii 44:483–493
40. Foissner W, Berger H, Kohmann F (1992) Taxonomische und ökologische Revision der Ciliaten des Saprobiensystems – Band II: Peritrichida, Heterotrichida, Odontostomatida – Informationsberichte des Bayer. Landesamt für Wasserwirtschaft 5/92:1–502
41. Lynn DH (2008) The ciliated protozoa. Characterization, classification, and guide to the literature. Springer Science+Business Media B.V., Berlin, p 605
42. Tunowski J (1994) The effect of heating and water exchange on the zooplankton composition in heated Konin lakes. Arch Pol Fish 2:235–255
43. Feniova IY, Razlutskij VI, Palash AL, Tunowsky J, Sysova EA, Dzialowski AR (2014) Cladoceran community structure in three meso-eutrophic Polish lakes with varying thermal regimes. Limnetica 33:13–30
44. Tunowski J (1988) Zooplankton losses during the passing through the cooling system of a power station. Ekol Pol 36:231–243

45. Hillbricht-Ilkowska A, Zdanowski B (1978) Effect of thermal effluents and retention time on lake functioning and ecological efficiencies in plankton communities. Int Rev Ges Hydrobiol 63(5):609–617
46. Hillbricht-Ilkowska A, Ejsmont-Karabin J, Węgleńska T (1988) Long-term changes in the composition, productivity and trophic efficiency in the zooplankton community of heated lakes near Konin (Poland). Ekol Pol 36:5–21
47. Hillbricht-Ilkowska A, Simm AT (1988) Spatial pattern of temperature, oxygen and nutrient concentration in two lakes of different heated-water discharge systems. Ekol Pol 36(1-2):165–182
48. Bogacka-Kapusta E, Kapusta A (2013) Spatial and diurnal distribution of cladocera in beds of invasive *Vallisneria spiralis* and open water in heated lake. Acta Zool Bulg 65(2):225–231
49. Ejsmont-Karabin J, Węgleńska T (1988) Disturbances in zooplankton seasonality in lake Gosławskie (Poland) affected by permanent heating and heavy fish stocking. Ekol Pol 36(1-2):245–260
50. Bogacka E, Paturej E (2006) Responses of zooplankton communities to elevated water temperature in lakes affected by water discharge from power stations. Acta Sci Pol Formatio Circumiectus 5(2):111–120
51. Ejsmont-Karabin J, Hutorowicz A (2011) Rotifera communities associated with invasive *Vallisneria spiralis* (Hydrocharitaceae) versus native macrophytes in the lakes heated by power stations (Konin lakes, W. Poland). Pol J Ecol 59:569–576
52. Ejsmont-Karabin J (2011) Does invasion of *Vallisneria spiralis* L. promote appearance of rare and new rotifer (Rotifera) species in littoral of the lakes heated by power station (Konin Lakes, W. Poland)? Pol J Ecol 59:201–207
53. Leszczyński L (1976) The influence of effluent heated waters on the fauna of lakes in the vicinity of Konin. I. Quantitative relations and qualitative composition of the bottom fauna of the Konin lakes complex. Rocz Nauk Roln Ser H 97(3):7–27. in Polish
54. Berger L, Dzięczkowski A (1977) Mollusca. In: Wróblewski A (ed) Bottom fauna of heated Konin lakes, vol 7. Monografie Fauna Polski, PWN, Warszawa, pp 151–179
55. Protasov AA, Afanasjev SA, Sinicyna OO, Zdanowski B (1994) Composition and functioning of benthic communities. In: Zdanowski B (ed) Ecological disturbances in heated Konin lakes, vol 2(2), Arch Ryb Pol, pp 257–284
56. Protasov AA, Zdanowski B, Sinicyna OO, Afanasjev SA, Tunowski J (1997) Structure and functioning of zooperiphyton and benthos communities of the channels of heated lakes of konińskie district. Arch Pol Fish 5:77–99
57. Ciemiński J, Zdanowski B (2009) Changes in the zoobenthos structure in a system of heated lakes in central Poland. Arch Pol Fish 17:221–238
58. Kasprzak K (1977) Hirudinea. In: Wróblewski A (ed) Bottom fauna of the heated Konin Lakes, vol 7, Monogr Fauny Pol, pp 147–150 (in Polish)
59. Biesiadka E (1977) Coleoptera. In: Wróblewski A (ed) Bottom fauna of the heated Konin lakes, vol 7, Monogr Fauny Pol, pp 259–280 (in Polish)
60. Biesiadka E (1977) Hydracarina. In: Wróblewski A (ed) Bottom fauna of the heated Konin lakes, vol 7, Mogr Fauny Pol, pp 281–350 (in Polish)
61. Najberek K, Solarz W (2008) Konin Lakes as hot spot for biological invasions in Poland. In: Głowaciński WZ, Okarma H, Pawłowski J, Solarz W (eds) Book of invasive alien species in Polish fauna. Wyd Intern Instytutu Ochrony Przyrody PAN, Kraków, pp 615–620 (in Polish). www.iop.krakow.pl/gatunkiobce
62. Kapusta A, Bogacka-Kapusta E (2015) Non-native fish species in heated lakes: origins and present status. Arch Pol Fish 23:121–129
63. Gąbka M (2002) *Vallisneria spiralis* (Hydrocharitaceae) – a new species to the Polish flora. Fragm Flor Geobot Polonica 9:76–73. in Polish
64. Piechocki A, Wawrzyniak-Wydrowska B, Zdanowski B (2003) *Melanoides tuberculatus* (O. F. Müller, 1774) (Orthogastropoda: Thiaridae), a gastropod speciec new for the fauna of Poland. Folia Malacol 11(1/2):39–41

65. Kraszewski A, Zdanowski B (2007) *Sinanodonta woodiana* (Lea, 1834) (Mollusca) – a new mussel species in Poland: occurrence and habitat preferences in a heated lake system. Pol J Ecol 55(2):337–356
66. Ciepielewski W (1994) Long-term changes in commercial fish catches and bream and roach biomass in heated Konin lakes. Arch Pol Fish 2(2):345–358
67. Lowden RM (1982) An approach to the taxonomy of *Vallisneria* L. (Hydrocharitaceae). Aquat Bot 13:269–298
68. Les DH, Jacobs SW, Tippery NP, Chen L, Moody ML, Wilstermann-Hildebrand M (2008) Systematics of *Vallisneria* (Hydrocharitaceae). Syst Bot 33:49–65
69. Kapusta A (2004) Relationship between the abundance of larvae and juvenile stage assemblages and the occurrence of macrophytes in the shallow littoral of Lake Licheńskie. Arch Pol Fish 12:163–176
70. Hutorowicz A, Dziedzic J, Kapusta A (2006) Nowe stanowisko *Vallisneria spiralis* (Hydrocharitaceae) w jeziorach konińskich (Pojezierze Kujawskie). Fragm Flor Geob Pol 13:89–94
71. Hutorowicz A, Hutorowicz J, Brzozowska R (2007) Chemical composition of the top-layer bottom deposits from underneath assemblages of the subtropical species *Vallisneria spiralis* L. in the phytolittoral zone of two heated lakes. Arch Pol Fish 15:457–464
72. Kapusta A, Bogacka E (2006) Diel variability in the occurrence of juvenile fish in the shallow littoral zone: do macrophytes impact the structure of fish assemblages? Arch Pol Fish 14:301–308
73. Stańczykowska A (1976) Occurrence and individual growth of *Dreissena polymorpha* (Pall.) in lakes included in a cooling system. Rocz Nauk Roln Ser H 97(3):109–123. in Polish
74. Kornobis S (1977) Ecology of *Dreissena polymorpha* (Pall.) (Dreissenidae, Bivalvia) in lakes receiving heated water discharges. Pol Arch Hydrobiol 24:531–545
75. Stańczykowska A, Lewandowski K, Ejsmont-Karabin J (1988) The abundance and distribution of the mussel *Dreissena polymorpha* (Pall.) in heated lakes near Konin (Poland). Ekol Pol 36(1-2):261–276
76. Sinicyna OO, Zdanowski B (2007) Development of the zebra mussel, *Dreissena polymorpha* (Pall.), population in a heated lakes ecosystem. I. Changes in population structure. Arch Pol Fish 15(4):369–385
77. Sinicyna OO, Zdanowski B (2007) Development of the zebra mussel, *Dreissena polymorpha* (Pall.), population in a heated lakes ecosystem. II. Life strategy. Arch Pol Fish 15(4):387–400
78. Soroka M, Zdanowski B (2001) Morphological and genetic variability of the population of *Anodonta woodiana* (Lea, 1834) occurring in the heated Konin lakes system. Arch Pol Fish 9(2):239–252
79. Sinicyna O, Protasov A, Zdanowski B, Kraszewski A (2001) Ecological niche of *Dreissena polymorpha* (Pall.) aggregations in the heated Konin lakes system. Arch Pol Fish 9(1):133–142
80. Kiss A (1995) The propagation, growth and biomass of the Chinese huge mussels (*Anodonta woodiana woodiana* 1984) in Hungary. Privat, Second edn. University of Agricultural Sciences of Godollo, Hungary, pp 1–33
81. Kraszewski A (2007) The continuing expansion of *Sinanodonta woodiana* (Lea, 1834) (Bivalvia, Unionidae) in Poland and Europe. Folia Malacol 15(2):65–69
82. Witkowski A (2009) On the expansion and occurrence of an invasive species – *Pseudorasbora parva* (Temminck et Schlegel, 1846) (Teleostei: Cyprinidae: Gobioninae) in Poland. Fragmenta Faunistica 52:25–32
83. Gozlan RE, Andreou D, Asaeda T, Beyer K, Bouhadad R, Burnard D, Caiola N, Cakic P, Djikanovic V, Esmaeili HR, Falka I, Golicher D, Harka A, Jeney G, Kovác V, Musil J, Nocita A, Povz M, Poulet N, Virbickas T, Wolter C, Tarkan SA, Tricarico E, Trichkova T, Verreycken H, Witkowski A, Zhang CG, Zweimueller I, Britton RJ (2010) Pan-continental invasion of *Pseudorasbora parva*: towards a better understanding of freshwater fish invasions. Fish Fish 11:315–340

84. Jackson MC, Britton JR (2013) Variation in the trophic overlap of invasive *Pseudorasbora parva* and sympatric cyprinid fishes. Ecol Freshw Fish 22:654–657
85. Kapusta A, Bogacka-Kapusta E, Czarnecki B (2008) The significance of stone moroko, *Pseudorasbora parva* (Temmick and Schlegel), in the small-sized fish assemblages in the littoral zone of the heated Lake Licheńskie. Arch Pol Fish 16:49–62
86. Kapusta A, Kutsokon Y, Bogacka-Kapusta E (2014) Comparisons of morphometrics recently established population of topmouth gudgeon (*Pseudorasbora parva*) from a heated lakes in Poland. Acta Universitatis Prešoviensis Folia Oecol 6:4–8
87. Záhorská E, Balážová M, Šúrová M (2013) Morphology, sexual dimorphism and size at maturation in topmouth gudgeon (*Pseudorasbora parva*) from the heated Lake Licheńskie (Poland). Knowl Manag Aquat Ec 411:07. https://doi.org/10.1051/kmae/2013074
88. Záhorská E, Kováč V, Švolíková K, Kapusta A (2014) Reproductive parameters of topmouth gudgeon (*Pseudorasbora parva*) from a heated Lake Licheńskie (Poland). Cent Eur J Biol 9:212–219
89. European Commission (2000) Directive of the European Parliament and of the Council 2000/60/EC establishing a framework for community action in the field of water policy. Off J 2000 L 327/1, European Commission, Brussels
90. Regulation of the Minister of the Environment of 21 July 2016 on status classification of surface water bodies and environmental quality standards for priority substances. Off J Laws, Item 1187 (in Polish)

Index

E. Korzeniewska, M. Harnisz (eds.), *Polish River Basins and Lakes – Part I*,
The Handbook of Environmental Chemistry 86,
https://doi.org/10.1007/978-3-030-12123-5